Lightning Effects on Carbon Fiber-Reinforced Polymer Composites

Jinru Sun · Xueling Yao · Shu Li

Lightning Effects on Carbon Fiber-Reinforced Polymer Composites

Damage Behavior, Mechanisms, and Simulation

西安交通大學出版社
XI'AN JIAOTONG UNIVERSITY PRESS

Springer

Jinru Sun
School of Electrical Engineering
Xi'an Jiaotong University
Xi'an, Shaanxi, China

Xueling Yao
School of Electrical Engineering
Xi'an Jiaotong University
Xi'an, Shaanxi, China

Shu Li
School of Electrical Engineering
Xi'an Jiaotong University
Xi'an, Shaanxi, China

ISBN 978-981-95-3195-0 ISBN 978-981-95-3196-7 (eBook)
https://doi.org/10.1007/978-981-95-3196-7

Jointly published with Xi'an Jiaotong University Press
The print edition is not for sale in China (Mainland). Customers from China (Mainland) please order the print book from: Xi'an Jiaotong University Press.

This Springer imprint is published by the registered company Springer Nature Singapore Pte Ltd.
The registered company address is: 152 Beach Road, #21-01/04 Gateway East, Singapore 189721, Singapore

Preface

On February 9, 2006, the State Council of China released the National Medium- and Long-Term Science and Technology Development Plan (2006–2020), designating the large commercial aircraft program as one of 16 major national science initiatives. On December 9, 2022, China's first domestically developed large airliner, the C919, was delivered to China Eastern Airlines, marking a milestone in the practical deployment of homegrown aviation technology. The 14th Five-Year Plan for the Raw Materials Industry has since emphasized advancing high-performance fiber and composite materials. In the next-generation wide-body aircraft, the C929, composites are expected to comprise over 50% of the airframe by weight, setting new benchmarks for lightning and electromagnetic protection.

Lightning strikes represent an intricate transient phenomenon, combining high-magnitude current impulses, intense electromagnetic forces, and severe thermal effects—that can induce fiber fracture, resin decomposition, and deep delamination within carbon-fiber laminate structures. Unlike metals such as aluminum, steel, and titanium alloys, carbon-fiber composites exhibit poor electrical and thermal conductivity, posing critical challenges when they serve as load-bearing elements. Therefore, elucidating the damage mechanisms and developing effective protection strategies for carbon-fiber composites is essential for their widespread adoption in large aircraft.

This monograph consolidates over a decade of research by our team in high-voltage and high-current testing, carbon-fiber-composite lightning-direct-effect mechanisms, and protective technologies. We begin by characterizing the severe electromagnetic environment experienced by aircraft and introduce our groundbreaking test technique to apply multiple, continuous, time-sequenced lightning current components. Validated by the Ministry of Education, our system "meets and surpasses international standards for continuous multi-component lightning testing, placing its measurement and control capabilities at the forefront of global practice."

Building upon this foundation, we analyze the anisotropic electrical properties of carbon-fiber composites and their impact on multi-physics simulation accuracy, detail the design and operation of multi-component lightning test equipment, and reveal the damage modes and progression laws under various current combinations. A comprehensive multifactor assessment then compares traditional metallic meshes, metal–insulator hybrids, metal–semiconductor systems, and innovative non-metallic composite structures. We propose a 3D porous conductive network based on carbon-fiber/graphene hybrids as a lightweight, efficient protective architecture for next-generation composite airframes.

Chapters 1, 4, and 6 were authored by Jinru Sun; Chaps. 2 and 3 by Xueling Yao; and Chap. 5 by Shu Li. All three authors jointly verified all figures and tables. We gratefully acknowledge the support of the High-Voltage and High-Current Testing Innovation Team at Xi'an Jiaotong University and the many students and engineers who contributed, as well as funding from the National Natural Science Foundation of China and the Civil Aviation Administration.

Xi'an, China
May 2025

Jinru Sun
Xueling Yao
Shu Li

Contents

Chapter 1
Introduction

1.1 Formation of Lightning

Lightning is a transient high-voltage, high-current, long-distance flash discharge phenomenon that occurs in the atmosphere, most lightning discharges are generated in thunderstorms with characteristic extents of 5–10 km and, in extreme cases, can exceed 100 km. Lightning flashes usually originate from charge centers in cumulonimbus clouds, although they can occur under other atmospheric conditions. The charges in clouds are produced by complex processes of freezing and melting, and by collision and fragmentation resulting from movements of raindrops and ice crystals. Typically, positive charges accumulate at the top of cumulonimbus clouds, leaving the lower regions negatively charged, although a small positive region may exist near the cloud base. The electrical charge distribution in a typical cumulonimbus cloud is shown in Fig. 1.1 [1, 2].

Accumulated charges in clouds generate an electric field. When the electric field strength reaches a critical threshold, air ionization and electrical breakdown occur, initiating the lightning discharge, as shown in Fig. 1.2.

Lightning discharges are classified as follows:

1. Cloud flashes, comprising:
   (1) Intracloud discharges: flashes between regions of opposite polarity within in a cloud.
   (2) Intercloud discharges: flashes between regions of opposite polarity in different clouds.
2. Ground flashes, including:
   (1) Cloud-to-air discharges: flashes between charged regions within a cloud and the surrounding air.

J. Sun et al., *Lightning Effects on Carbon Fiber-Reinforced Polymer Composites*,
https://doi.org/10.1007/978-981-95-3196-7_1

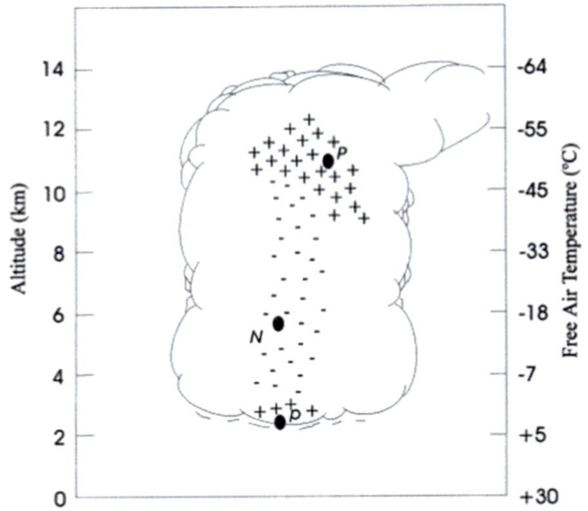

Fig. 1.1 Generalized diagram showing the distribution of electrical charge in a typical cumulonimbus cloud

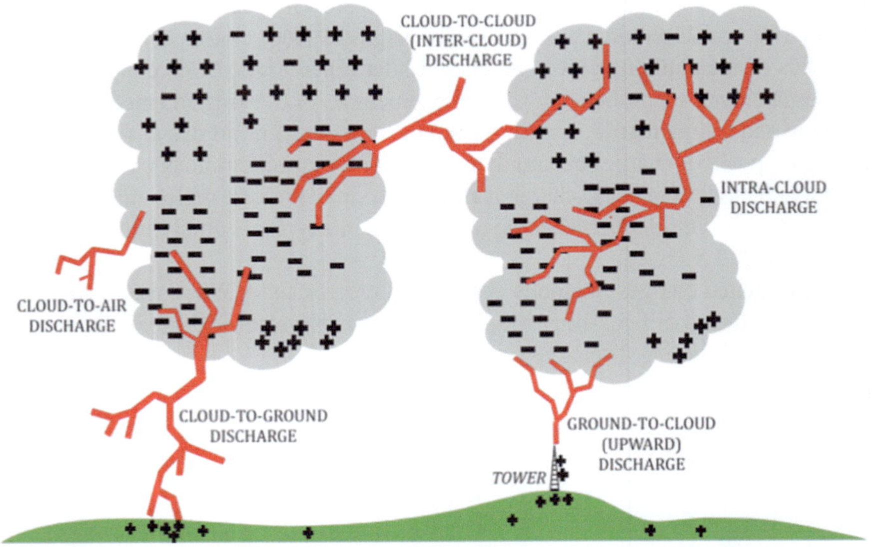

Fig. 1.2 Charge distribution of two isolated thunderclouds and potential lightning discharge locations [2]

(2) Cloud-to-ground and ground-to-cloud discharges: flashes between clouds and the ground of either polarity. Ground-to-cloud flashes, however, occur only for tall structures (e.g., towers or mountains).

More than 50% of lightning discharges are internal cloud discharges, while the lightning flashes commonly observed are generally cloud-to-ground discharges. The process that culminates in a lightning flash begins with the formation of an ionized column called a leader, which originates from a region where the electric field is sufficiently high to initiate progressive ionization. This occurs when the field is about 500 kV/m. The leader advances in zigzag steps, each approximately 50 m long and separated by pauses of 40–100 ms (hence the name stepped leader).

The diameter of the stepped leader is between 1 and 10 m, although the current, which is relatively low (about 100 A), is probably concentrated in a small, highly ionized core of about 1 cm diameter. The average velocity of propagation is 1.5×10^5 m/s. The leader may form branches on its downward path to the ground. When a branch approaches the ground, it enhances electrical fields at protrusions such as trees and buildings. These fields may ionize the air and initiate upward connecting leaders, one or more of which will attach to the tip of the downward-propagating leader. This connection closes a switch and the location where the leaders meet is known as the attachment point. At this moment, charge in the leader flows into the earth and the condition propagates rapidly up the leader. This is known as a return stroke, and this process discharges the leader at a velocity of about 5×10^7 m/s. This initial return stroke is characterized by a high-amplitude current pulse accompanied by a bright flash. After the first return stroke, subsequent strokes may occur as higher portions of the negative charge regions are discharged; the dart leaders for these strokes usually traverse the same path as the first but in one continuous sweep at a velocity of 2×10^6 m/s [1, 2].

1.1.1 The Negative Flash to Ground

The return-stroke current in a severe negative flash is illustrated in Fig. 1.3.

The number of strokes in a negative flash is usually between 1 and 11, the mean value being 3; the aximum number is up to 24. The total duration is between about 20 ms and 1 s, with a mean value of 0.2 s. The time interval between strokes is typically about 60 ms. There is some correlation among these parameters, with flashes having the greatest number of strokes tending also to have the longest duration. The rise time of the first stroke is about 2 μs, with a decay time (to half the peak amplitude) of 40 μs. Subsequent strokes in the flash tend to have a higher rate of rise although lower peak amplitudes than the initial stroke, and they can, therefore, be important for inducing voltages in wiring, as the induced voltages is proportional to the rate of change of the lightning current.

Near the end of some strokes in a negative flash, there is often a lower-level current of a few kA persisting for several milliseconds, known as an "intermediate current

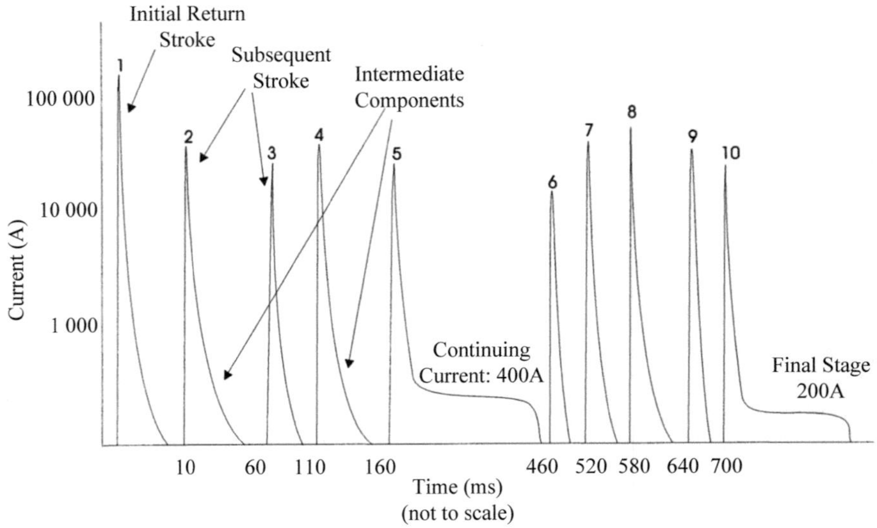

Fig. 1.3 A negative lightning flash current waveform [1, 2]

component," as shown in Fig. 1.3. After some strokes, a "continuing current" of 100–400 A flows for 100–800 ms, resulting in substantial charge transfer during this phase. It is particularly common for a continuing current to follow the final stroke.

It is generally thought that before a restrike can occur, the continuing current must cease, as illustrated after stroke 5 in Fig. 1.3.

1.1.2 The Positive Flash to Ground

Positive flashes to ground generally occur less frequently than negative flashes, however, in certain geographic locations there may be a higher proportion of positive ground flashes. Current standards assume that approximately 10% of ground flashes are positive. Positive flashes are usually initiated by upward-moving leaders and occur more frequently over mountainous terrain than over flat terrain. They typically consist of a single stroke. They have slower rise times than negative flashes, but with higher peak currents and greater charge transfer. The duration of a positive stroke is longer than a single negative stroke but shorter than the total duration of a negative flash. A continuing current may follow the stroke.

The lightning current in a positive flash is shown in Fig. 1.4. This is a moderately severe example, not the occasional "super flash." Typically, the rise time of a positive flash is 20 μs, and the total duration is 0.1 s. Although positive flashes are much less frequent globally than negative flashes, they must be considered when selecting design and test parameters.

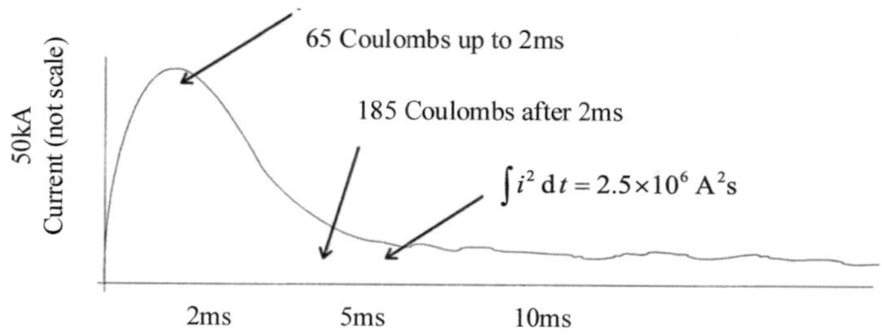

Fig. 1.4 A positive lightning flash current waveform [1, 2]

1.1.3 Flash Parameters

Most available statistical data pertain to cloud-to-ground and ground-to-cloud lightning flashes. The relevant data are presented in Tables 1.1 and 1.2, categorizing negative and positive flashes. The tables include statistics for the lightning currents and related parameters of interest for defining the external environment. For a given flash or stroke parameter, the tables show that as the magnitude increases, the frequency of occurrence decreases. The extreme parameters do not occur together in a single flash [1].

In addition to lightning currents, electric fields exist before and during a lightning strike. Initially, these fields cause air breakdown to form the attachment and may also cause dielectric breakdown of materials on an aircraft. The magnitudes of these fields depend on air breakdown thresholds and range between 400 and 3000 kV/m, with rates of rise up to 1000 kV/m/μs.

1.2 Lightning Interactions with Aircraft

1.2.1 Aircraft Lightning Attachment and Swept Flash Phenomena

Lightning initially attaches to an airplane, entering at one point and exiting at another point. Generally, lightning strike attachment points occur at the nose or wing tips of an aircraft, and these attachment points are called entry or exit points. The attachment point at the front or upper part of the aircraft is usually called the entry point, and the point at the tail or lower part of the aircraft is called the exit point. Since the aircraft travels a distance greater than its length during a lightning strikes, the entry point progressively shifts towards the tail, and the exit point likewise moves rearward. Therefore, there will be many attachment points, as shown in Fig. 1.5.

The terms in Fig. 1.5 are as follow:

Table 1.1 Parameters for negative lightning flashes measured at ground

Parameters	Unit	Lightning parameters 95%	Lightning parameters 50%	Lightning parameters 5%
Negative flashes				
Number of strokes	/	1–2	3–4	12
Time intervals between strokes	ms	8	35	140
Flash duration	s	0.03–0.04	0.2	1
Charge in flash	C	1.3	7.5	40
Negative first stroke				
Peak current	kA	14	30	80
Peak rate-of-rise	A/s	5.5×10^9	1.2×10^{10}	3.2×10^{10}
Time to peak	μs	1.8	5.5	18
Time to half value	μs	30	75	200
Impulse charge	C	1.1	5.2	24
Action integral	A^2s	6×10^3	5.5×10^4	5.5×10^5
Negative subsequent strokes				
Peak current	kA	4.6	12	30
Peak rate-of-rise	A/s	1.2×10^{10}	4×10^{10}	1.2×10^{11}
Time to peak	μs	0.22	1.1	4.5
Time to half value	μs	6.5	32	140
Impulse charge	C	0.2	1.4	11
Action integral	A^2s	5.5×10^2	6×10^3	5.2×10^4
Continuing current		98%	50%	2%
Amplitude	A	33	140	520
Duration	s	0.058	0.16	0.40
Charge	C	7	26	110

Note 1 The above lightning parameters may not occur together in a single flash
Note 2 The percentage figures represent percentiles, i.e., the percentage of events with amplitude exceeding those shown

(i) Lightning attachment point: the place where the lightning flash touches (attaches to) the aircraft.
(ii) Initial entry point: the place where the lightning flash channel first “enters” the aircraft (usually an extremity).
(iii) Final entry point: the place where the lightning flash channel last “enters” the aircraft (typically a trailing edge).
(iv) Initial exit point: the place where the lightning flash channel first “exits” from the aircraft (usually an extremity).
(v) Final exit point: the last place where the lightning flash “exits” from the aircraft (usually a trailing edge).

Table 1.2 Parameters for positive lightning flashes measured at ground

Parameters	Unit	Lightning parameters 95%	Lightning parameters 50%	Lightning parameters 5%
Positive flashes				
Flash duration	ms	14	85	500
Total charge	C	20	80	350
Positive stroke				
Peak current	kA	4.6	35	250
Peak rate-of-rise	A/s	2×10^8	2.4×10^9	3.2×10^{10}
Time to peak	μs	3.5	22	200
Time to half value	μs	25	230	2000
Impulse charge	C	2	16	150
Action integral	A^2s	2.5×10^4	6.5×10^5	1.5×10^7

Note 1 The above lightning parameters listed above do not necessarily occur together in one flash
Note 2 The percentage figures represent percentiles, that is, the percentage of events having a greater amplitude than those given

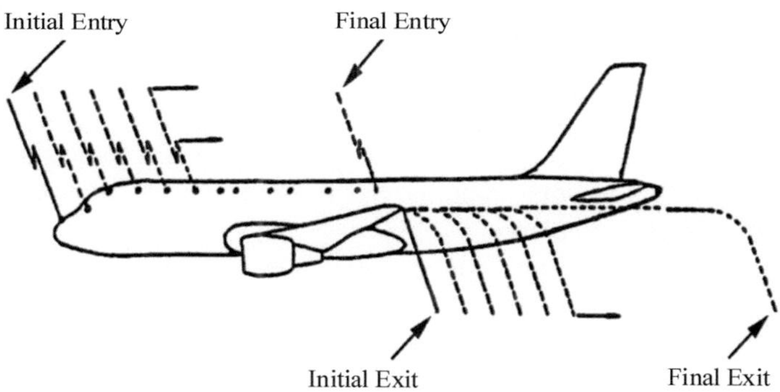

Fig. 1.5 Typical path of swept-channel attachment points [3]

(vi) Swept "flash" (or "stroke") points: spots where the flash channel reattaches between the initial and final points, usually associated with the entry part of the flash channel.

1.2.2 Aircraft Lightning Experience

The lightning probability of an aircraft is related to various factors such as flight altitude, climate, flight phase (such as climb, level flight, or descent), and so on. Statistical data shows that airplanes are particularly vulnerable to lightning strikes at flight altitudes of 3–5 km, as shown in Fig. 1.6 [4]. This is within the main cloud

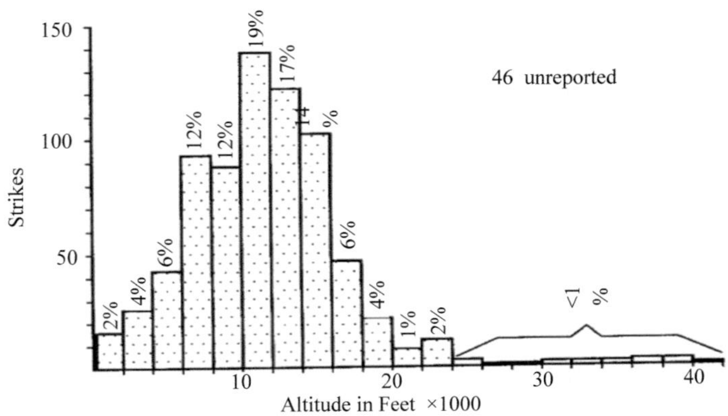

Fig. 1.6 Aircraft lightning strikes versus altitude [4]

layer, and most of the lightning accidents occur during the takeoff and landing phases when aircraft pass through clouds.

This data indicates that there are more lightning strikes experienced at intermediate altitudes than at cruise altitudes for transport airplanes. This fact indicates (1) that there are more lightning flashes to be intercepted below about 20,000 ft than above this altitude, and (2) that jet aircraft are being struck below cruise altitudes: that is, during climb, descent, or hold operations. Figures 1.7, 1.8, and 1.9 show the immediate environment of the aircraft at the times of the 881 strikes reported. In over 80% of the strikes reported, aircrafts were within a cloud and were experiencing precipitation and some turbulence.

The incident reports above also indicate that most aircraft strikes occur when an aircraft is near the freezing level. Figure 1.10 shows the distribution of lightning strikes to aircraft as a function of outside air temperature. Freezing temperatures (and below) are required for the electrical charge separation process to function. Of course, strikes to aircraft at temperatures higher than 10°C have occurred when the

Fig. 1.7 Aircraft location with respect to clouds when lightning strikes have occurred [4]

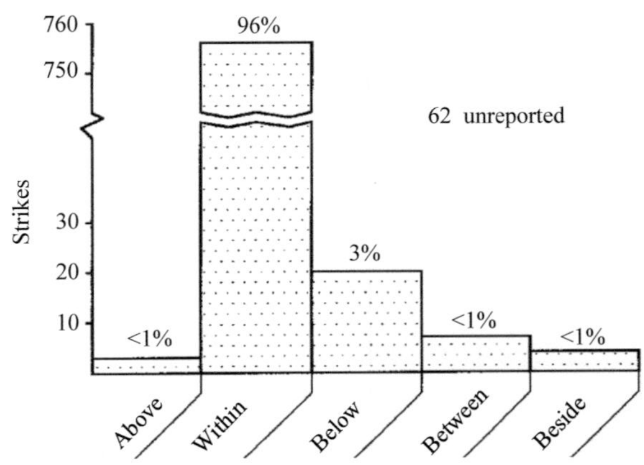

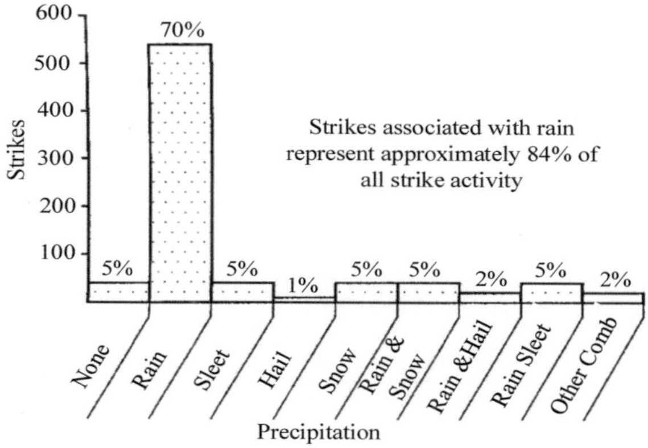

Fig. 1.8 Precipitation at time of aircraft lightning strikes [4]

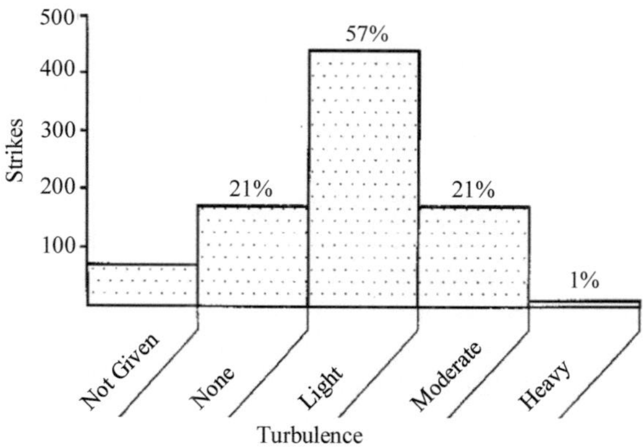

Fig. 1.9 Turbulence experienced when lightning strikes have occurred [4]

aircraft was close to (or on) the ground, where the ambient air temperature may be as high as about 25°C [4].

1.2.3 Aircraft Lightning Strike Mechanisms

At the beginning of lightning flash formation, when a stepped-leader propagates outward from a cloud charge center, the ultimate destination of the leader, at an opposite charge center in the cloud or on the ground, has not yet been determined. The potential difference between the stepped leader and the opposite charge(s) establishes an electrostatic field, represented by imaginary equipotential surfaces shown as lines in the two dimensional drawing of Fig. 1.11. The field intensity, commonly expressed

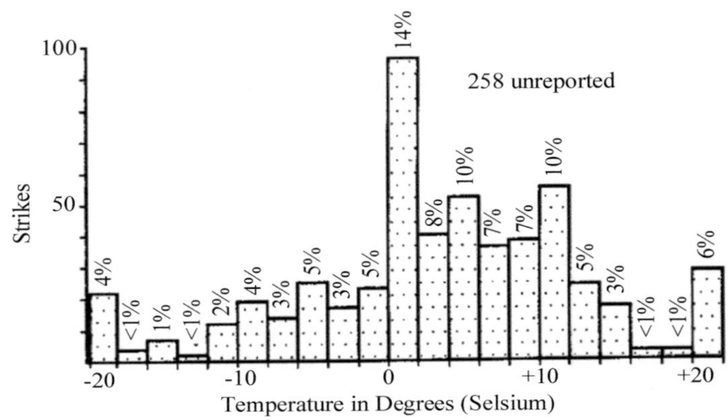

Fig. 1.10 Outside air temperature during lightning strikes [4]

in kilovolts per meter, is greatest where equipotential surfaces are closest together. It is this field that ionizes the air and forms the conductive spark which is the leader. Because the electrostatic force is normal to the equipotentials and strongest where they are closest together, the leader is most likely to progress toward the regions of highest field intensity.

Since the aircraft is typically a large conductor whose surfaces are all at this same potential, it will deflect and compress adjacent equipotentials, thus increasing the electric field intensity at its extremities, and especially between it and other charge sources, such as the advancing leader. If the aircraft is far from the leader, its effect on the field near the leader is negligible; however, if the aircraft is within tens or hundreds of meters of the leader, the increased field intensity between them may be sufficient to direst subsequent leader propagation toward the aircraft. As this happens, the local field will become even more intense, and the leader will advance more directly toward the aircraft.

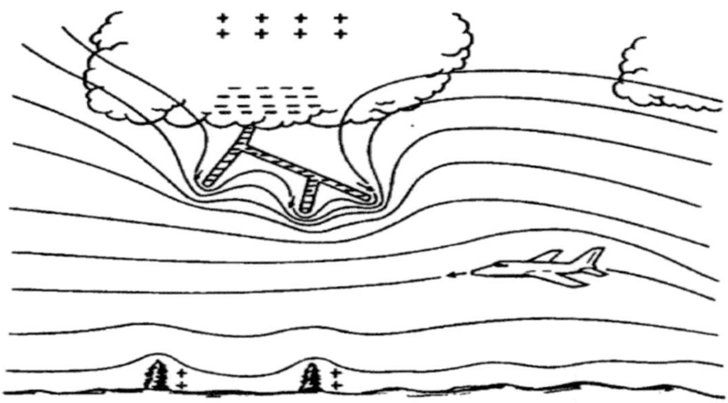

Fig. 1.11 Stepped leader approaching an aircraft [4]

The highest electric fields at the aircraft will occur around extremities, where the equipotential lines are compressed most closely together, as shown in Fig. 1.12. Typically, these are the nose, wingtip, and empennage tips, and also smaller protrusions, such as antennas or air-data probes. When the leader advances to the point where the field adjacent to an aircraft extremity is increased to approximately 30 kV/cm (at sea level pressure), the air will ionize and electrical sparks will form at the aircraft extremities, extending in the direction of the oncoming leader. Several of these sparks, called streamers, usually occur nearly simultaneously from several extremities of the aircraft. These streamers will continue to propagate outward as long as the field remains above about 5–7 kV/cm. One of these streamers, called the junction leader, will meet the nearest branch of the advancing leader and form a continuous spark from the cloud charge center to the aircraft. Thus, when the aircraft is close enough to influence the direction of the leader propagation, it will very likely become attached to a branch of the leader system.

Streamers may simultaneously propagate from two or more extremities of the aircraft. If so, the oncoming leader has split, and the two (or more) branches will continue from the aircraft independently until one or both reach their destination. This process of attachment and continued propagation from an aircraft is shown in Fig. 1.13.

When the leader reaches its destination and a continuous ionized channel between charge centers is established, recombination of electrons and positive ions occurs along the leader channel, forming the high-amplitude return-stroke current. This stroke current, and any subsequent strokes or continuing-current components, must flow through the aircraft, which has now become part of the conductive path between charge centers.

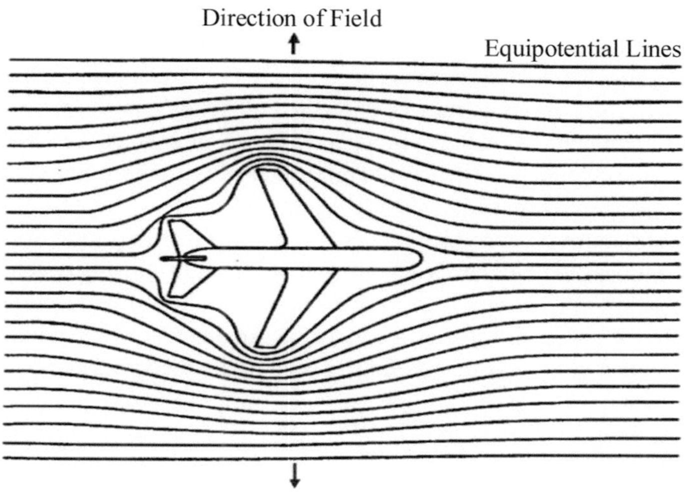

Fig. 1.12 Compression of electric field around an aircraft [4]

(a) Aircraft as part of the ionized channel (b) Ionized channel bypassing the aircraft

Fig. 1.13 Return stroke paths [4]

If another branch of the original leader reaches the ground before the branch that involves the aircraft, the return stroke will follow that branch, and all other branches will terminate. In such cases, no substantial currents flow through the aircraft, and any damage to the aircraft is minimal.

1.2.4 Direct Effect on Skin Structure

The effects of lightning on aircraft skins, both metallic and composite include: (i) Melting or burning at lightning attachment points; (ii) Resistive heating; (iii) Magnetic force effects; (vi) Acoustic shock effects; (v) Arcing and sparking at bonds, hinges and joints; (vi) Ignition of vapors within fuel tanks. Not all materials experience the same damage under equal lightning-current intensity. Metallic skins are most susceptible to melting at lightning attachment points, while composite materials are particularly vulnerable to acoustic shock waves. CFRP composites, however, are resistive conductors and thus subject to the same phenomena as metal structures, albeit to different degrees, including magnetic forces, arcing, and resistive heating. Non-conductive composites, such as fiberglass and aramid-fiber-reinforced plastics, are subject to dielectric breakdown and puncture.

Aircraft structures comprise the outer skins and the internal framework, such as spars, ribs, frames, and bulkheads. Lightning currents must flow between entry and exit points on an aircraft and tend to disperse as they travel between attachment points, using the entire airframe as a conductor. Any conductive material, whether metal or a conductive composite, forming these structures becomes part of the current path for lightning.

1.3 The Lightning Test Environment

In order to simulate the real-world conditions of natural lightning strikes, standards such as SAE-APR 5412 [1], MIL-STD-464A [5], and EUROCAE ED-14E/RTCA-DO-160 [6] have developed detailed test specifications and requirements for aircraft, and have defined different lightning zones.

1.3.1 Lightning Strike Zones

Because of the lightning attachment process, not all locations on an aircraft are subjected to the same lightning environment components. To optimize protection, the aircraft is divided into different lightning strike zones, each of which is protected against the relevant components of the lightning environment. Based on the probability of lightning attachment, the location of attachment points, and the swept channel paths, the aircraft is divided into lightning strike Zones 1, 2, and 3. Furthermore, Zone 1 is subdivided into Zones 1A, 1B, and 1C, and Zone 2 into 2A and 2B, as shown in Fig. 1.14.

In general, an aircraft can be divided into the following zones [2],

(i) Zone 1A: First Return-Stroke Zone, all surface areas where a first return stroke is likely during lightning channel attachment, with a low expectation of flash hang-on.
(ii) Zone 1B: First Return-Stroke Zone with Long Hang-On; areas where a first return stroke is likely, with a high expectation of flash hang-on.
(iii) Zone 1C: Transition Zone for First Return Stroke; surfaces where a first return stroke of reduced amplitude (i.e., at higher altitudes) is likely, with a low expectation of flash hang-on.
(iv) Zone 2A: Swept Stroke Zone; surfaces where a subsequent return stroke is likely to be swept, with a low expectation of flash hang-on.
(v) Zone 2B: Swept-Stroke Zone with Long Hang-On; i.e., surfaces into which a lightning channel carrying a subsequent return stroke is likely to be swept, with a high expectation of flash hang-on.
(vi) Zone 3: Current-Conduction Zone; surfaces not in Zones 1A, 1B, 1C, 2A, or 2B, where attachment of the lightning channel is unlikely, and which conduct substantial current between direct or swept-stroke attachment points.

1.3.2 Idealized Lightning Current Components

Lightning Current Environment

The environment waveforms represent idealized conditions applied to the aircraft for analysis and testing. The waveforms are not intended to duplicate a specific

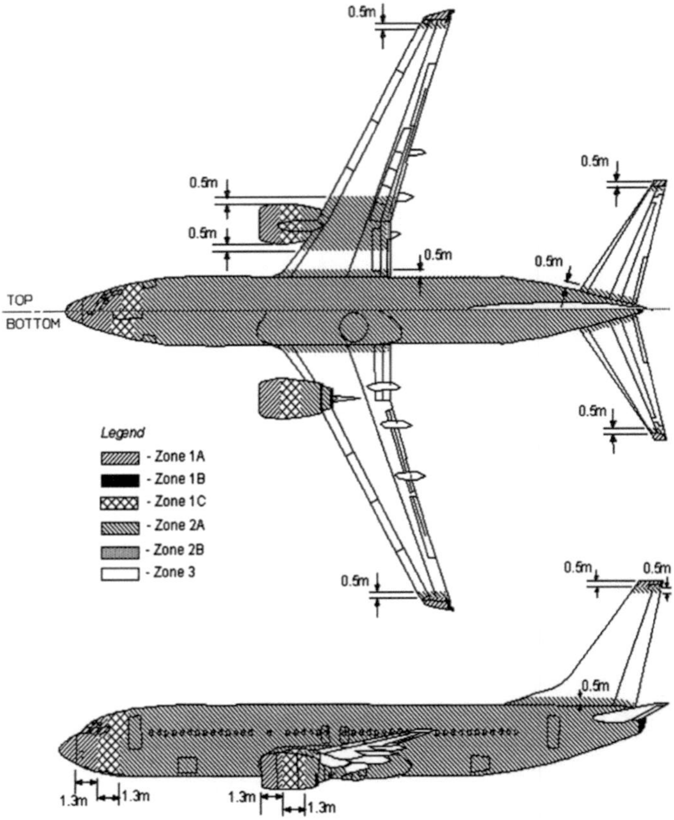

Fig. 1.14 Lightning strike zone details for transport aircraft [3]

lightning event, but are composite waveforms whose interactions with the aircraft mimic those expected from natural lightning. The standard lightning environment comprises distinct voltage and current waveform components that capture the key characteristics of natural lightning flashes. In the following waveform descriptions, parameters critical to the direct or indirect effects under consideration are included, while less relevant parameters are omitted. For example, in direct-effects evaluations, peak current, action integral, and duration are of primary importance, whereas in indirect-effects evaluations, rates of current rise and decay and peak amplitude are crucial.

Not all aircraft surfaces need to withstand the same lightning threat. The assignment of lightning environment waveforms and components to specific zones, as described in ARP5414, is shown in Table 1.3 [3].

The voltage waveforms presented in this ARP5412 [2] are intended to evaluate possible lightning attachment locations and/or dielectric breakdown paths through nonconducting surfaces or structures. In contrast, the current waveform represents the direct effect of lightning on aircraft. The external lightning environment consist

Table 1.3 Lightning environment to aircraft zones

Aircraft zone	Voltage waveforms	Current component
1A	A, B, D	A, B, C*, H
1B	A, B, D	A, B, C, D, H
1C	A	Ah, B, C*, D, H
2A	A	D, B, C*, H
2B	A	D, B, C, H
3 (Conducted)	–	A, B, C, D, H
3 (Direct attachment)	C	

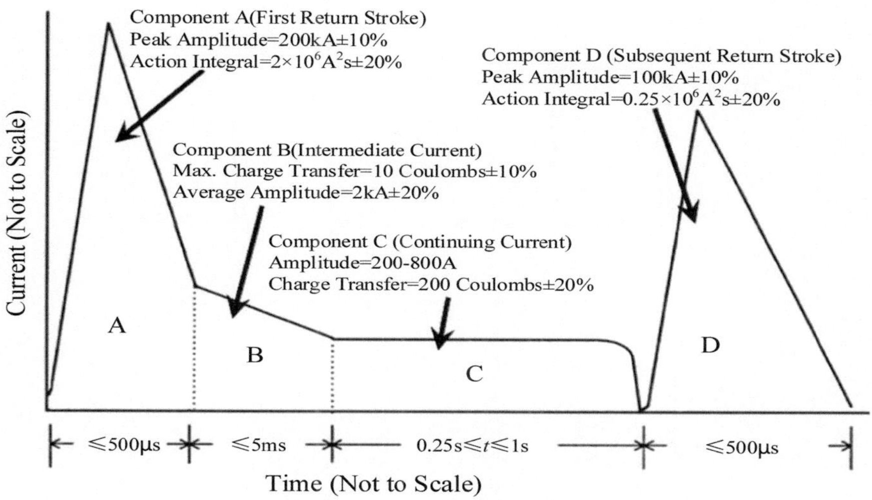

Fig. 1.15 Current components A through D for direct-effects testing

of current components A, Ah, B, C, D, and H, as well as the Multiple Stroke (MS) and Multiple Burst (MB) Waveform sets. The MS includes components D and D/2, and the MB set is comprises sequences of H pulse.

(i) Components A, B, C, and D from the lightning flash current waveform for direct-effects evaluation, as shown in Fig. 1.15.

(ii) Components A and D, plus the MS and MB sets, are used for indirect-effects evaluation. The latter two are shown in Figs. 1.16 and 1.17.

Lightning Current Components

(1) Current Component a (First Return Stroke)

This waveform combines the extreme parameters of both the negative and positive first return strokes. It most often occurs on aircraft flying at lower altitudes. For

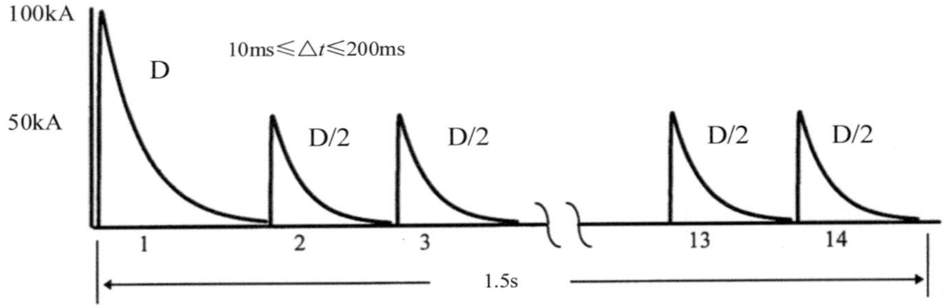

Fig. 1.16 Multiple stroke waveform set for direct-effects testing

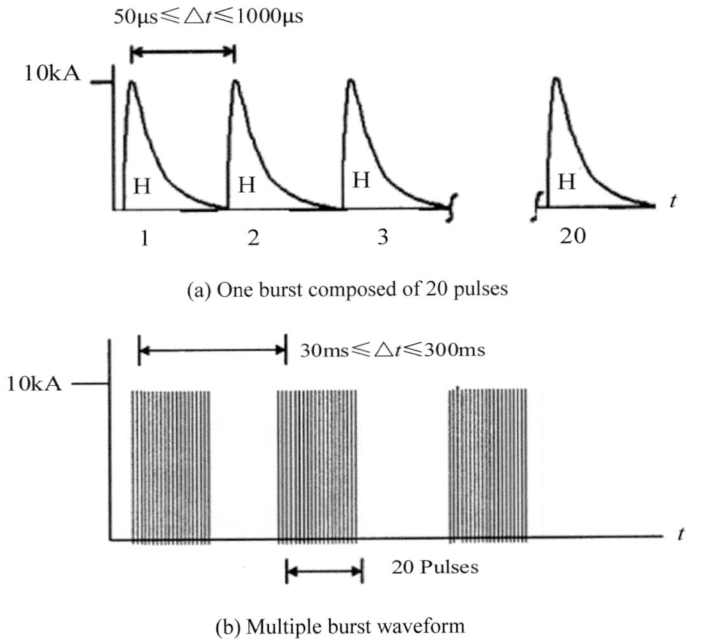

Fig. 1.17 Multiple burst waveform set for indirect-effects testing

indirect-effect analysis, the double-exponential waveform shown in Fig. 1.18 shall be applied.

$$i(t) = I_0(e^{-\alpha t} - e^{-\beta t}) \tag{1.1}$$

where, $I_0 = 218{,}810$ A, double exponential function waveform factor; $\alpha = 11{,}354$ s^{-1}, double exponential function frequency parameter; $\beta = 647{,}265$ s^{-1}, double exponential function frequency parameter; t: time.

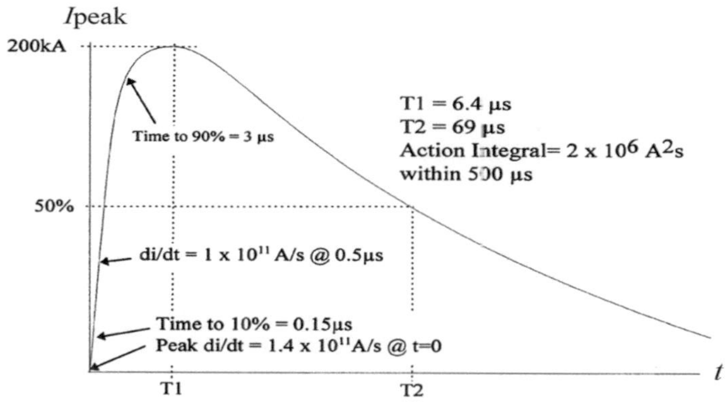

Fig. 1.18 Current component A for analysis and indirect-effects test purposes [1]

For direct-effects testing, component A can be simulated using either an oscillatory waveform (Fig. 1.19a) or a unipolar waveform (Fig. 1.19b). The waveform must have an amplitude of 200 kA (±10%) with a rise time up to 50 µs (measured between 10 and 90% of peak). The action integral must be 2×10^6 A^2 s (±20%), and the total duration to 1% of peak must not exceed 500 µs.

(2) Current Component Ah (Transition Zone First Return Stroke)

This waveform applied to Zone 1C and represents the estimated shape of the first return stroke (Component A) at higher altitudes. For analysis, the double-exponential waveform in Fig. 1.20 is used; it is defined mathematically by the following double-exponential function:

$$i(t) = I_0(e^{-\alpha t} - e^{-\beta t}) \tag{1.2}$$

where, $I_0 = 164{,}903$ A, double exponential function waveform factor; $\alpha = 16{,}065$ s^{-1}, double exponential function frequency parameter; $\beta = 858{,}888$ s^{-1}, double exponential function frequency parameter; t: time.

For direct effects testing, component Ah can be simulated by an oscillatory or unidirectional waveform as shown in Fig. 1.21a, b.

(3) Current Component B (Intermediate Current)

This component captures the intermediate currents that follow some negative initial return strokes and/or restrikes. For analysis purposes, a double-exponential current waveform can be used, as presented in Fig. 1.22. This waveform is described mathematically by the following expression:

$$i(t) = I_0(e^{-\alpha t} - e^{-\beta t}) \tag{1.3}$$

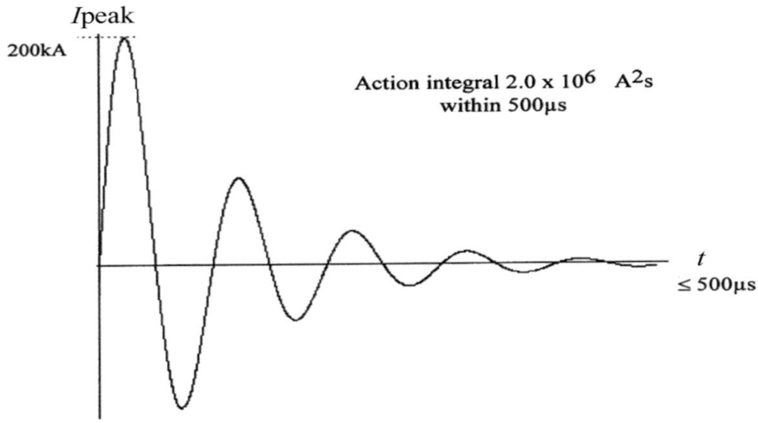

(a) Damped Sinusoidal Current

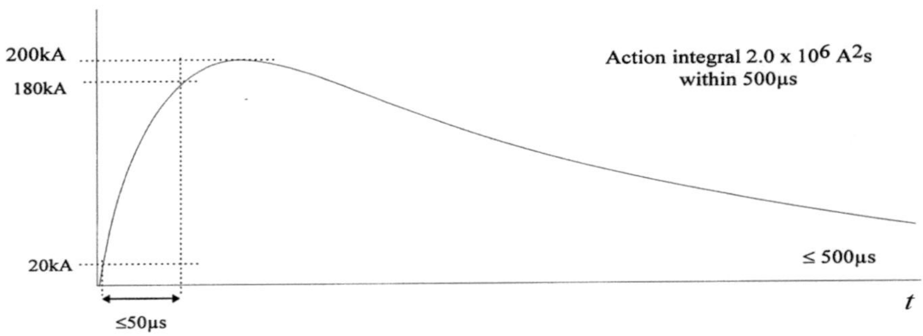

(b) Unipolar Current

Fig. 1.19 Current component A for direct-effects test purposes [1]

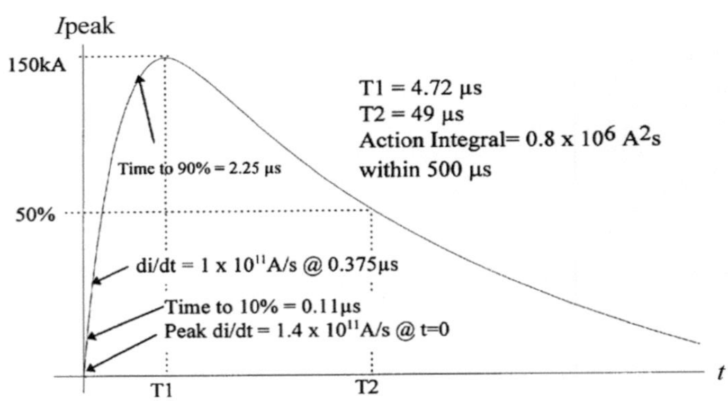

Fig. 1.20 Current component Ah for analysis purposes [1]

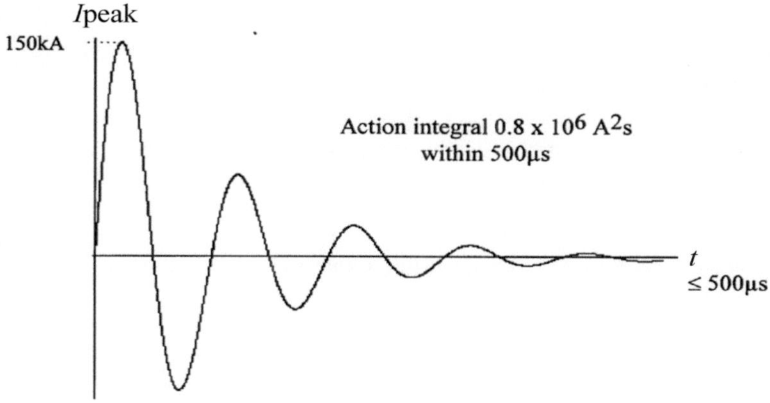

(a) Damped Sinusoidal Current

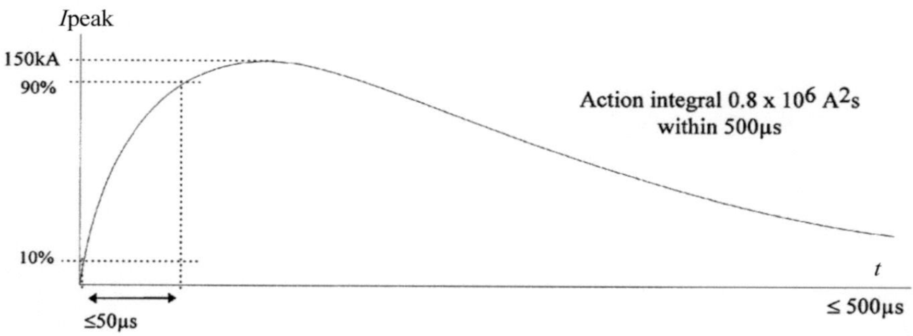

(b) Unipolar Current

Fig. 1.21 Current component Ah for direct effects test purposes [1]

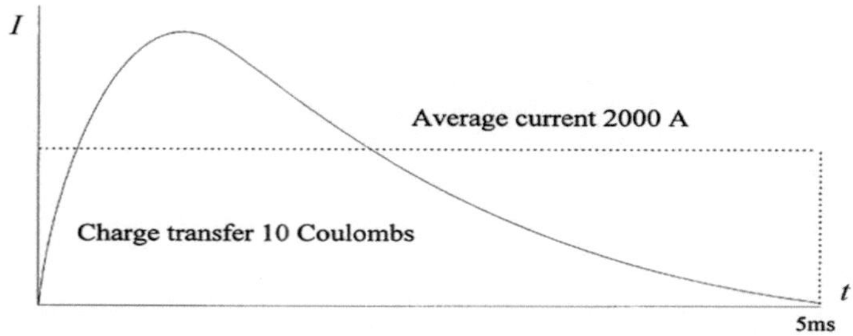

Fig. 1.22 Current component B for analysis purposes [1]

where, $I_0 = 11{,}300$ A, double exponential function waveform factor; $\alpha = 700\ \mathrm{s}^{-1}$, double exponential function frequency parameter; $\beta = 2{,}000\ \mathrm{s}^{-1}$, double exponential function frequency parameter; t: time.

For direct-effects testing, component B should be unipolar, for example rectangular, exponential, or linearly decaying, as shown in Fig. 1.23a, b. The mean amplitude must be 2 kA (±20%), flowing for a duration of 5 ms (±10%), with a charge transfer of 10 C (±10%).

(4) **Current Component C (Continuing Current)**

This current component represents the long-duration currents that may follow some restrikes of negative cloud-to-ground flashes and the return strokes of positive cloud-to-ground flashes. For analysis purposes, a square waveform of 400 A for 0.5 s should be utilized, as shown in Fig. 1.24.

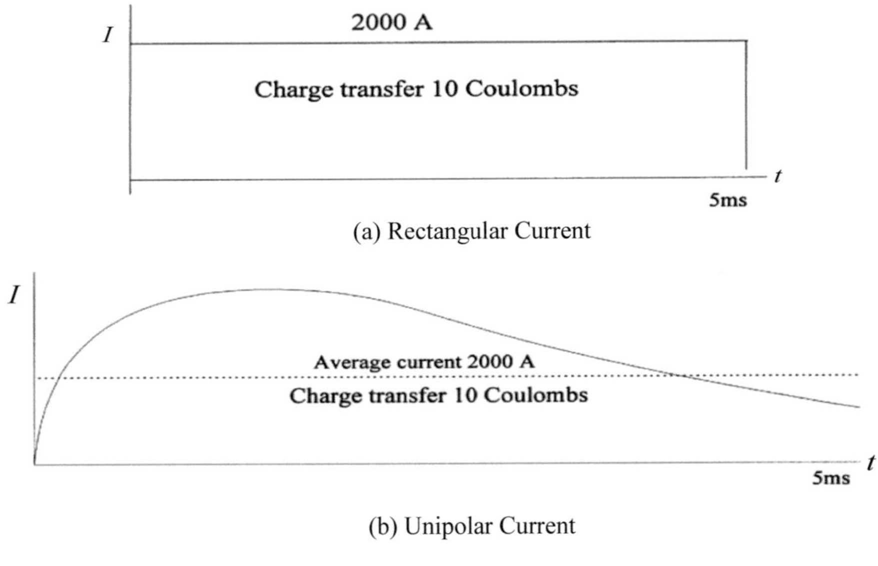

Fig. 1.23 Current component B for direct-effects test purposes [1]

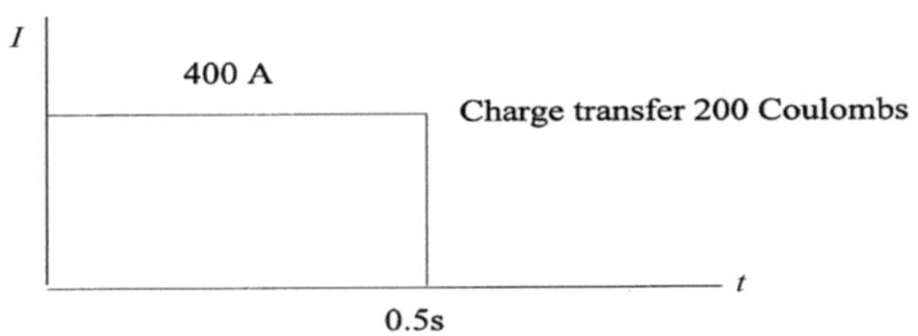

Fig. 1.24 Current component C for analysis purposes [1]

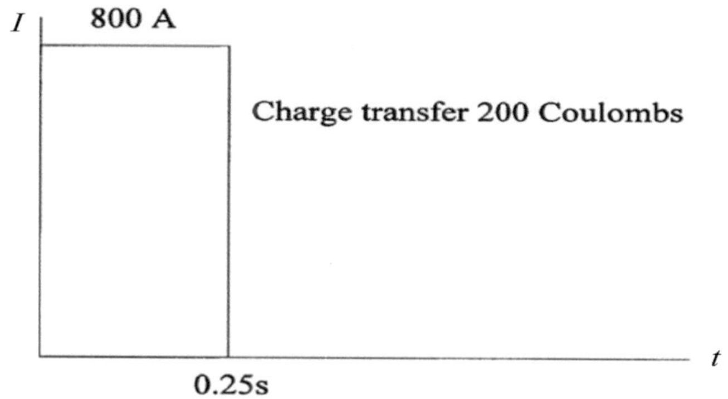

(a) Example of Current Component C (800 A, 0.25 s)

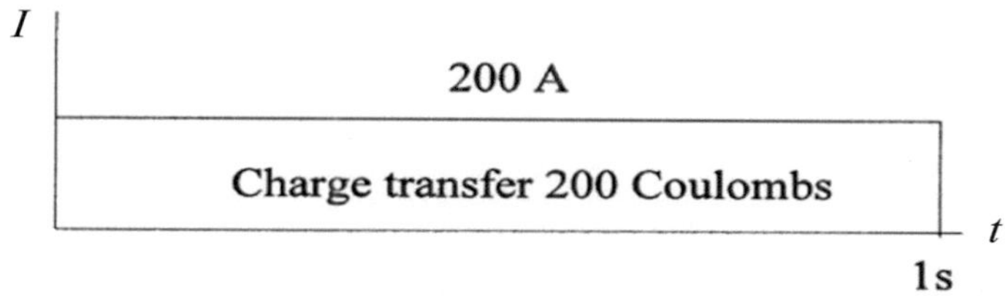

(b) Example of Current Component C (200 A, 1 s)

Fig. 1.25 Current component C for direct-effects test purposes [1]

For direct-effects testing, component C should be unipolar, for example rectangular, exponential, or linearly decaying, as shown in Fig. 1.25a, b. The mean amplitude must be 400 A (±20%) flowing for a duration of 0.5 s (±10%), with a charge transfer of 200 C (±10%).

(5) **Current Component C* (Modified Component C)**

This component represents the portion of component C that flows into an attachment point in Zone 1A or 2A if the dwell time at that point exceeds 5 ms. Component C* is primarily used for evaluating melt-through of metal skins. Component C* is a current of at least 400 A, averaged over a period equal to the dwell time minus the 5 ms duration of component B. The combination of components A or D, B, and C* therefore defines the dwell time, which may range from 1 to 50 ms. An example of component C* for test applications is shown in Fig. 1.26.

(6) **Current Component D (Subsequent Return Stroke)**

For analysis purposes, for indirect-effect considerations, the double-exponential waveform shown in Fig. 1.27 shall be applied.

The waveform is defined mathematically by the double-exponential expression below:

$$i(t) = I_0(e^{-\alpha t} - e^{-\beta t}) \tag{1.4}$$

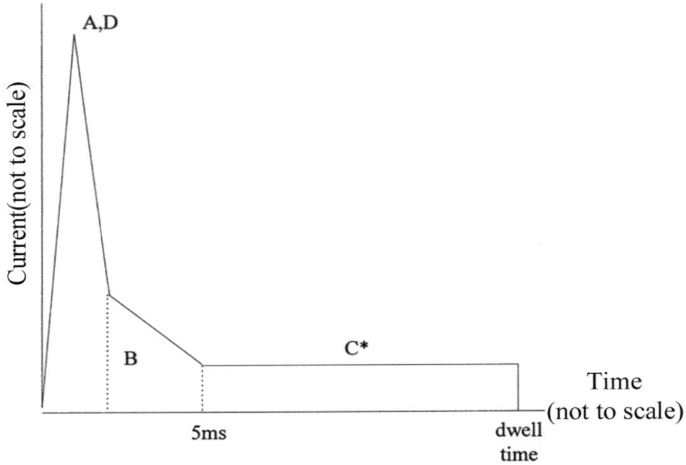

Fig. 1.26 Application of current component C* [1]

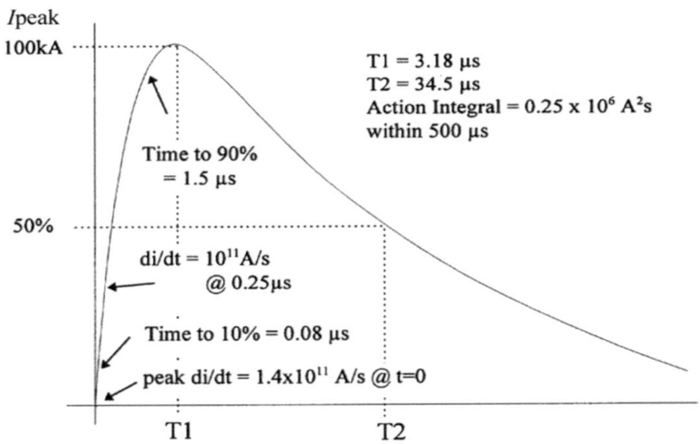

Fig. 1.27 Current component D for analysis and indirect-effects test purposes [1]

where, $I_0 = 109{,}405$ A, Double exponential function waveform factor; $\alpha = 22{,}708$ s^{-1}, Double exponential function frequency parameter; $\beta = 1{,}294{,}530$ s^{-1}, Double exponential function frequency parameter; t: time.

For direct-effects assessments current Component D represents a subsequent stroke, as shown in Fig. 1.28.

For direct-effects testing, Component D can be simulated by either oscillatory or unidirectional waveforms (Fig. 1.28a, b) with a total duration to 1% of peak not exceeding 500 μs. The amplitude shall be 100 kA (±10%), the rise time shall not exceed 25 μs (time between 10 and 90% of the amplitude), and the action integral is 0.25 × 106 A^2s(±20%).

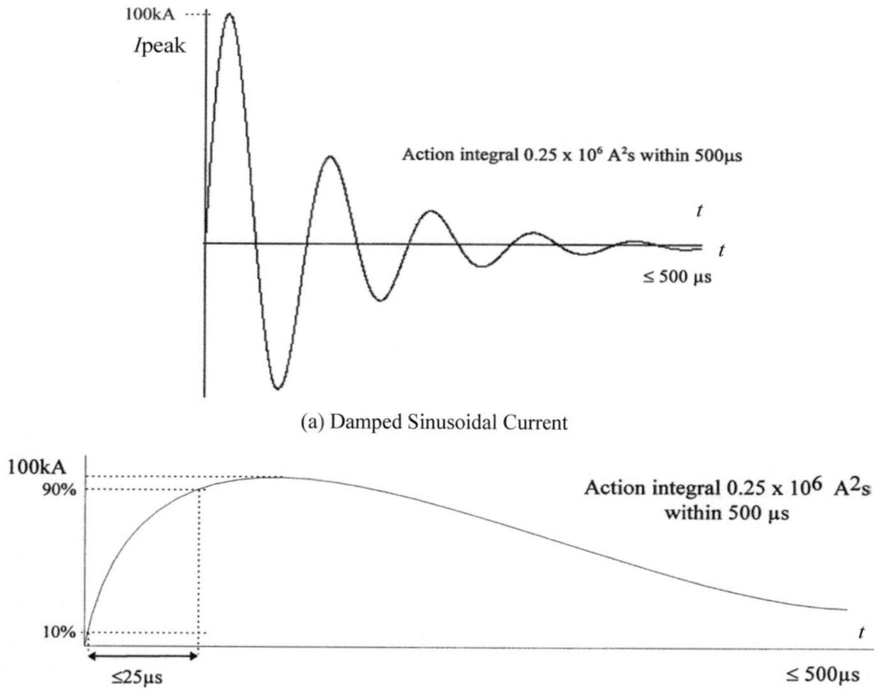

(a) Damped Sinusoidal Current

(b) Unipolar Current

Fig. 1.28 Current component D for direct-effects test purposes [1]

(7) **Multiple Stroke Waveform Set**

In many cases, up to 14 randomly spaced strokes have been observed in negative cloud-to-ground flashes. Additionally, several pulses of approximately 30 kA can occur in a random sequence in an intra-cloud event, as illustrated in Fig. 1.29.

The Multiple Stroke Waveform Set is used only for indirect-effects evaluation. The synthesized set is defined as a current component D followed by 13 components D/2, as shown in Fig. 1.29. The components D/2 are distributed randomly over a period of up to 1.5 s according to the following constraints:

(i) The minimum time between components is 10 ms;
(ii) The maximum time between components is 200 ms.

(8) **Multiple Burst Waveform Set**

The Multiple Burst Waveform set consists of Component H waveforms. Component H represents a high-rate-of-rise current pulse whose amplitude and duration are much smaller than those of a return stroke. Such pulses occur in groups at the initiation of a lightning strike to an aircraft and randomly throughout the lightning flash, along with other current components (Fig. 1.29). While unlikely to cause structural damage, the

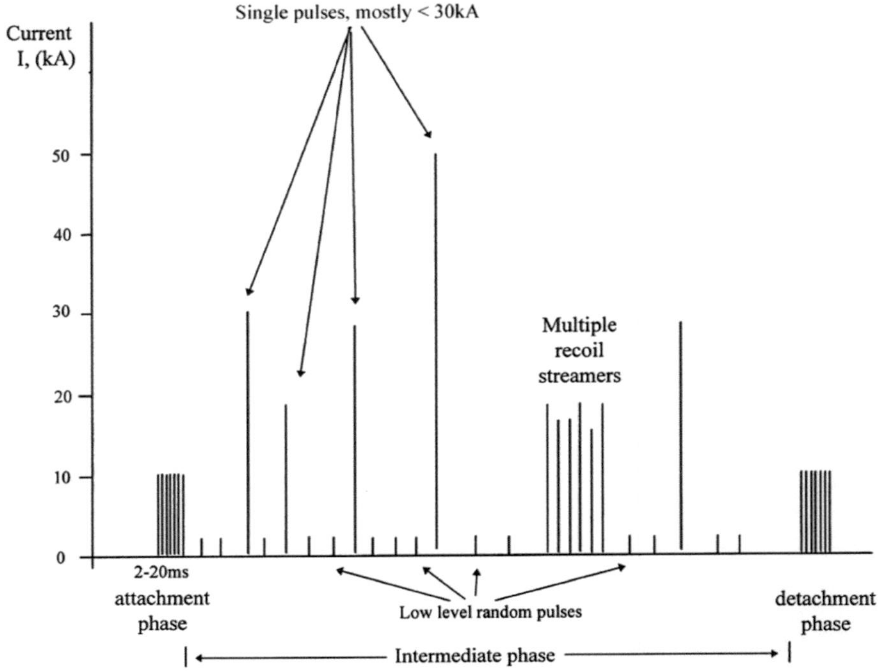

Fig. 1.29 Multiple Stroke Waveform Set for indirect-effects test purposes [1]

repetitive, random nature of these pulses may interfere with sensitive systems. The recommended set comprises three bursts of 20 Component H pulses each, as shown in Fig. 1.17. The interval between pulses within a burst is 50–1000 μs. The three bursts are spaced according to:

(i) Minimum time between bursts is 30 ms;
(ii) Maximum time between bursts is 300 ms.

Waveform H is defined mathematically by the following formula and is presented in Fig. 1.30:

$$i(t) = I_0(e^{-\alpha t} - e^{-\beta t}) \tag{1.5}$$

where, $I_0 = 10{,}572$ A, double exponential function waveform factor; $\alpha = 187{,}191$ s^{-1}, double exponential function frequency parameter; $\beta = 19{,}105{,}100$ s^{-1}, double exponential function frequency parameter; t: time.

Component H is presented in Fig. 1.30.

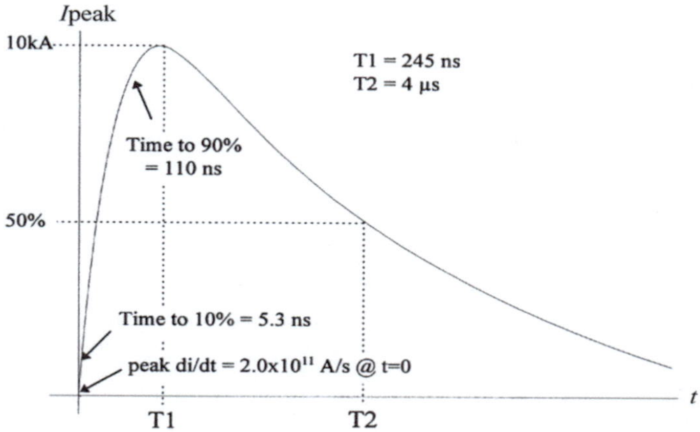

Fig. 1.30 Current component H for analysis and indirect-effects test purposes [1]

1.4 Carbon Fiber Reinforced Polymers in Aerospace

CFRP (Carbon Fiber Reinforced Polymers) composites are increasingly used in aerospace, military and civil application due to their excellent mechanical properties, low density, high strength, high modulus, temperature and chemical resistance, and fabrication versatility. With advances in aircraft design and composite manufacturing, the CFRP content in commercial, military, unmanned, and stealth aircraft continues to grow. CFRP is employed in structures such as wings, tail assemblies, fuselage, and floors. For example, CFRP comprises 50% of the Boeing B787's structural mass and 53% of the Airbus A350XWA's structure. The application areas of CFRP in aircraft are illustrated in Fig. 1.31.

The extensive application of CFRP composite materials promotes the miniaturisation, lightweighting, and performance enhancement of aircraft, and the CFRP content has become an important benchmark for aircraft structural advancement. Developed countries such as the US and Europe prioritize research on CFRP composites. To meet diverse design requirements and leverage advances in CFRP development, research

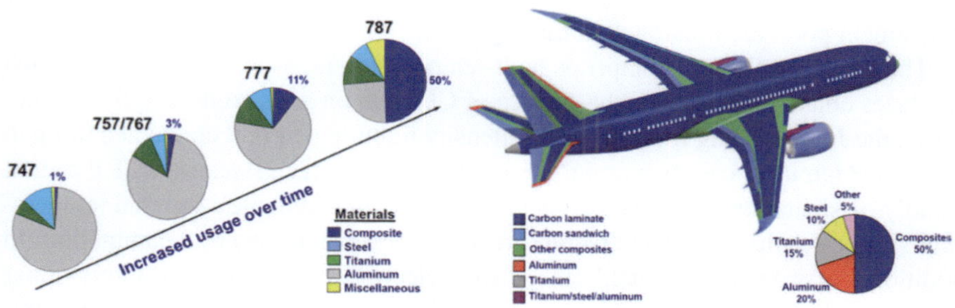

Fig. 1.31 The application of CFRP materials in aircraft

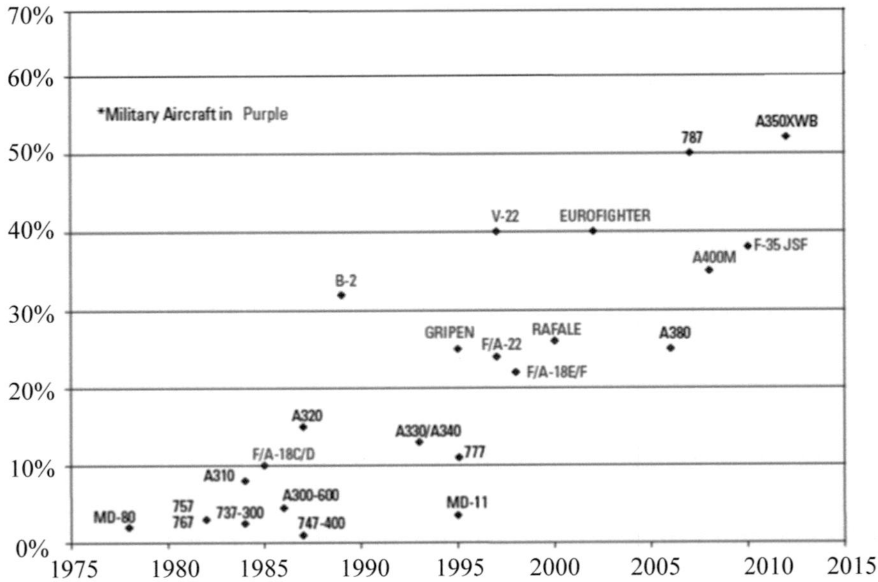

Fig. 1.32 The proportion of CFRP on the structure of the main types of aircraft

and development plans have been formulated to promote their structural application. The CFRP content in major Europe and US aircraft is shown in Fig. 1.32 [7–9].

Data indicate that the CFRP content in military aircraft is about 40%, whereas in civil aircraft such as in the Boeing B787 and Airbus A350XWB, it exceeds 50%. Compared to traditional metals, such as aluminium, steel and titanium alloys, CFRP composites exhibit poor electrical conductivity: carbon fibers have a resistivity on the order of 10^{-5} Ω m, while polymer matrices are electrical insulators with resistivities of 10^{11}–10^{14} Ω m. Typically, CFRP resistivity is 10^{-4} Ω m along the fiber alignment, 10^{-2} Ω m transverse to the fiber plane, and up to 10^{4} Ω m through the thickness.

Lightning strikes pose a significant safety risk to large aircraft and challenge the extension use of CFRP in aerospace. Lightning strikes are complex, transient events coupling multiple fields: mechanical impact from high currents, electromagnetic forces, and thermal effects. The multi-physics damage mechanisms at the lightning attachment point are illustrated in the Fig. 1.33.

The lightning channel comprises high-temperature, high-pressure plasma, which produces ablative damage on the surface of CFRP upon arc attachment. Simultaneously, the Joule heating from the high-intensity lightning current conducted through the CFRP laminate causes a rapid internal temperature rise. Because CFRP cannot rapidly dissipate the accumulated heat and charge, the laminate's internal temperature spikes, resulting in fiber fracture, resin pyrolysis, and severe delamination. In addition, the strong electric field can induce dielectric breakdown of the interlaminar medium, while the intense electromagnetic field exerts Lorentz force on current-carrying fiber bundles, forces perpendicular to the fiber direction, that, beyond a

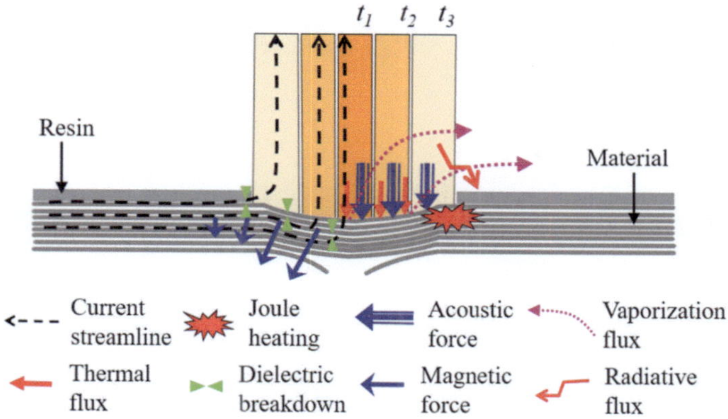

Fig. 1.33 The multi-physics field damage effect at the lightning attachment point

certain magnitude, cause mechanical damage. Thirdly, the lightning current pulse compresses the surrounding air, generating a violent shock wave that subjects the CFRP surface to extreme pressure, resulting in further mechanical damage.

Research on lightning damage and protection technology for carbon-fiber composites spans materials, electrical, and mechanics, making it highly challenging. Consequently, investigation into damage mechanisms and mitigation strategies holds significant scientific importance and engineering value, enhancing composite and protective material research and advancing the application of next-generation aerospace composites.

References

1. SAE ARP 5412. Aircraft Lightning Environment and Related Test Waveforms[S]. Warrendale, PA, United States: SAE International Group, 2005.
2. Reinaldo J. Perez, Handbook of Aerospace Electromagnetic Compatibility, 2019, IEEE Press, Wiley.
3. SAE ARP 5414. Aircraft lightning zoning[S]. Warrendale, PA, United States: SAE International Group, 2005.
4. Rupke E. Lightning direct effects handbook[R]. Pittsfield, MA, United States: Lightning Technologies Inc., 2002.
5. MIL-STD-464A. Electromagnetic environmental effects requirements for systems[S]. Washington DC, United States: Department of Defense Interface Standard, 2002.
6. EUROCAE ED-14E/RTCA DO-160F. Environmental conditions and test procedures for airborne equipment, section 23: lightning direct effects[S]. European Organization for Civil Aviation Equipment, Radio Technical Commission for Aeronautics, 2007.
7. Roeseler B, Sarh B, Kismarton M. Composite structures-the first 100 years[C]. 16th International Conference on Composite Materials, Kyoto, Japan, July, 2007.
8. Marsch G. Airbus A350 XWB update[J]. Reinforced plastics, 2010, 45(6): 20–24.
9. Taylor RP. Fiber composite aircraft capability and safety. Canberra, Australia: Australian Capital Territory, Australian Transport Safety Bureau, 2008.

Chapter 2
Dynamic Conductive Characteristics of CFRP Composites Under Lightning Strike

The lightning interaction with CFRP composites involves multiple complex processes, such as arc attachment, high-pressure shock waves, electromagnetic forces, and Joule heating. These processes progress through distinct phases: lightning leader development, arc attachment, surface streamers, internal ablation, and heat dissipation. During the arc attachment process, the surface electrical conductivity of CFRP plays a crucial role in dissipating lightning currents across the surface. Subsequently, the internal conduction paths and resulting Joule heating depend on the composite's dynamic behavior. The anisotropic dynamic electrical conductivity of CFRP composites directly influences the patterns and severity of lightning-induced damage. Importantly, the accuracy of numerical simulations of lightning damage on CFRP composites depends on accurate characterization of these dynamic electrical conductivity properties.

2.1 DC Conductivity Characteristics of CFRP Composites

CFRP composites consist of carbon fiber plies embedded in an epoxy matrix and exhibit semiconducting behavior. The direct-current (DC) conductivity of CFRP composites differs markedly from that of both pure carbon fibers and the insulating epoxy resin. Their DC resistivity depends strongly on laminate architecture, surface coatings, and manufacturing parameters. Todoroki et al. [1] investigated static DC conductivity of CFRP with varying fiber volume fractions measured both parallel and transverse to the fiber orientation, highlighting hybrid-material conduction effects. Cheng et al. [2] used eddy-current testing to quantify anisotropic volume conductivity, showing that cross-ply laminates conduct better than unidirectional laminates. They concluded that bulk conductivity is governed by the number of conductive pathways at fiber-overlap interfaces. Haider et al. [3] characterized anisotropic and nonlinear conductive behaviors under varying current densities for different laminate

J. Sun et al., *Lightning Effects on Carbon Fiber-Reinforced Polymer Composites*,
https://doi.org/10.1007/978-981-95-3196-7_2

layups. Park et al. [4] applied percolation theory and Kirchhoff-law–based Monte Carlo simulations to model electrical anisotropy in CFRP. Furthermore, embedded carbon fibers have been exploited as structural sensors: surface electrodes detect resistance changes, enabling integrity monitoring, damage localization, and quantitative assessment of delamination via resistivity variations [4, 5].

Research indicates that the electrical behavior of CFRP is strongly influenced by laminate architectures, which governs conduction pathways. The DC resistance of CFRP composites is typically measured using two-probe or four-probe methods. Two-electrode measurements fail to account for surface conductivity effects on bulk conductive behavior, whereas four-probe methods, designed to eliminate contact impedance in homogeneous materials, may not fully address the pronounced anisotropic conductivity of CFRP composites and warrant further validation.

Notably, Chap. 4 of this book discusses numerical simulation of lightning damage to CFRP composites, demonstrating that accurate DC resistivity/conductivity is essential for reliable damage predictions. Therefore, this section presents experimental studies of the CFRP DC resistivity, examines how measurement technique influences static resistivity results, investigates resistivity as a function of applied voltage, and elucidates the intrinsic conduction mechanism via experimental data and equivalent-circuit modeling.

2.1.1 DC Volume Resistivity Measurement Method for CFRPs

Test Specimens and Measurement Methodology

To assess the effect of laminate structures on CFRP's DC volume conductivity, specimens with varied surface conductivity patterns and layup configurations were designed, as detailed in Table 2.1.

To characterize the resistivity of CFRP composites, the three-electrode measurement system specified in the IEC 60093 standard [6] shown in Fig. 2.1 was used, and the DC resistance measurement circuit is shown in Fig. 2.2.

As shown in Fig. 2.1, the high-voltage electrode (A) is a copper disk of diameter $\Phi_A = 74$ mm, the guard (measurement) electrode (B) is a copper cylinder of diameter $\Phi_B = 50$ mm, and the protective electrode (C) is a copper ring with inner diameter $\Phi_{CI} = 54$ mm and outer diameter $\Phi_{CO} = 74$ mm, which captures surface leakage currents. All electrodes have a height of 20 mm. The high-voltage supply provides up to 11 kV at ≤ 10 mA, and a 1 MΩ series resistor R limits the current.

A layer of silicone grease was applied between the specimen and electrodes. Before testing, the electrodes were pressed firmly to minimize the grease layer thickness, ensuring reliable electrical contact via the electrode weight and the adhesive properties of the grease. A shunt resistor R_1 was connected between the guard electrode B and ground. By simultaneously measuring the voltage V_1 across R_1 and the

Table 2.1 Parameters of CFRP composite specimens

Number	Dimensions	Structure and process characteristics
A1	100 × 100 × 3 mm	$[0/90]_{4S}$, surface coated with high-conductivity polymer foam
A2	100 × 100 × 3 mm	$[0/90]_{4S}$, surface coated with high-conductivity nonwoven fabric
A3	100 × 100 × 3 mm	$[0/90]_{4S}$, surface coated with nano-modified nonwoven fabric
A4	100 × 100 × 3 mm	$[0/90]_{3S}$, surface coated with high-conductivity short fibers
A5	100 × 100 × 3 mm	$[0/90]_{3S}$, surface coated with low-conductivity short fibers
A6	100 × 150 × 5 mm	$[45/0/–45/90]_{4S}$, surface coated with micro carbon nanotubes
A7	100 × 150 × 5 mm	$[45/0/–45/90]_{4S}$, micro carbon nanotubes applied between 1/2 plies and surface
A8	100 × 150 × 5 mm	$[45/0/–45/90]_{4S}$, micro carbon nanotubes applied between 1/2, 2/3 plies and surface
A9	100 × 150 × 5 mm	$[45/0/–45/90]_{4S}$, micro carbon nanotubes applied between 1/2, 2/3, 3/4 plies and surface

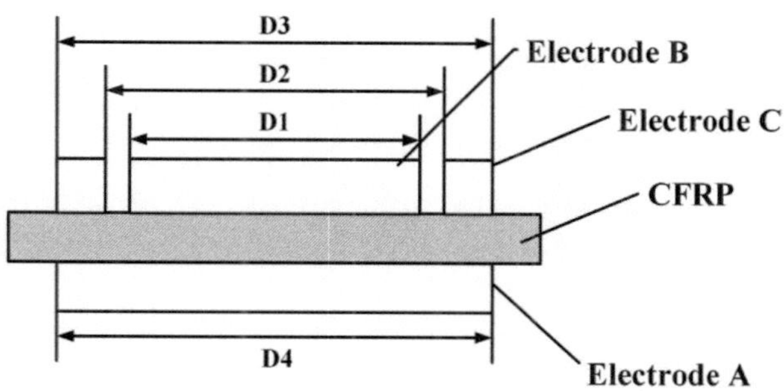

Fig. 2.1 Three-electrode system for DC resistance measurement of CFRP composites

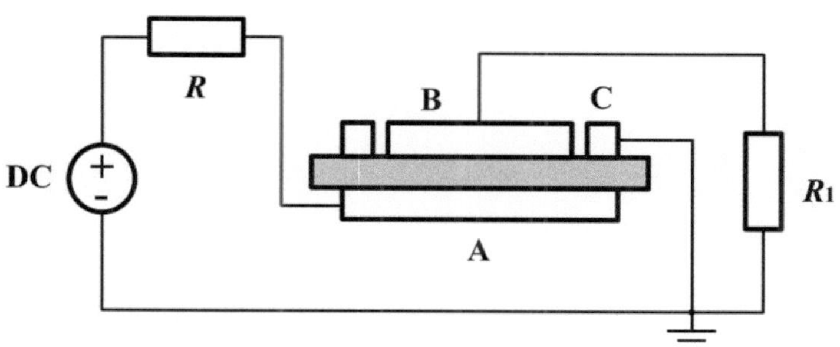

Fig. 2.2 DC volume resistance measurement circuit using the three-electrode method

voltage V between the high-voltage electrode A and ground, the volume resistance R_V of the CFRP composite is calculated using Eq. (2.1):

$$R_V = R_1(V - V_1)/V_1 \tag{2.1}$$

By varying the applied DC voltage V, the corresponding values of V and V_1 were recorded, establishing the relationship between R_V and V.

Conventional Three-Electrode Volume Resistivity Measurement

With $R_1 = 1\ \text{k}\Omega$, the DC volume resistivity of specimens A1 and A2 was measured using the circuit in Fig. 2.2. The effective conduction area was defined by the guard electrode (B) when calculating resistivity. The results are presented in Fig. 2.3. With $R_1 = 1\ \text{k}\Omega$, the DC volume resistivity of specimens A1 and A2 was measured using the circuit in Fig. 2.2. The effective conduction area was defined by the guard electrode base (B) when calculating resistivity. The results are shown in Fig. 2.3.

In Fig. 2.3, the following key observations are noted:

(1) Specimen A1's volume resistivity increases with applied DC voltage, reaching a maximum on the order of $10^5\ \Omega$. The bulk resistivity, calculated from the specimen's geometrical, is on the order of $10^5\ \Omega$ m, which deviates substantially from typical CFRP bulk conductivity values ($10^1\ \Omega$–$10^3\ \Omega$ m).

(2) Specimen A2 exhibits a volume resistivity on the order of $10^4\ \Omega$ m, with no consistent voltage dependence.

The significant discrepancy in DC volume resistivity between the two specimens, when measured by the conventional three-electrode method, demonstrates its unsuitability for CFRP composites. This limitation arises because:

The three-electrode method in IEC 60093 Standard [6] is intended for homogeneous insulating materials. Its guard electrode C and measurement electrodes A and

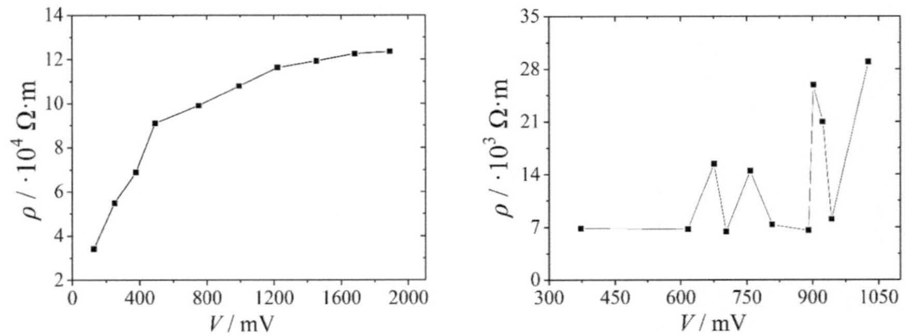

(a) Volume resistivity of specimen A1 (b) Volume resistivity of specimen A2

Fig. 2.3 DC Volume resistivity of CFRP composites measured using the traditional three-electrode method

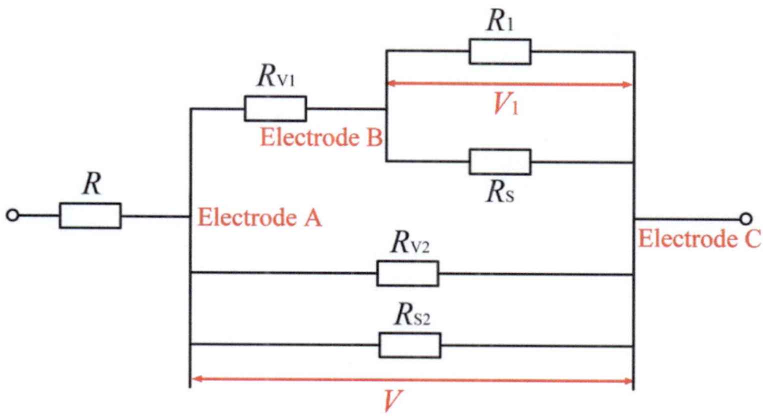

Fig. 2.4 Equivalent circuit diagram of traditional three-electrode method for measuring CFRP volume resistance

B form a three-terminal network that diverts parasitic surface currents away from the measurement circuit, thereby reducing interference and improving accuracy.

The measurement circuit of Fig. 2.2 can be represented by the equivalent network in Fig. 2.4. Here, R_{V1} is the volume resistance between the high-voltage electrode A and the guarded electrode B, the path from electrode A to guard electrode C is modeled as the parallel combination of volume resistance R_{V2} and surface resistance R_{S2}. R_S denotes the surface resistance between guard electrode B and guard electrode C, and R_1 is the sampling resistor.

The guard system effectively reduces measurement error under the following two conditions:

(i) The equivalent resistance of the parallel path containing the measured volume resistance must be sufficiently high to minimize current shunting;

(ii) The presence of the sampling resistor should not significantly disturb the original circuit condition, which requires that the equivalent resistance of the circuit branches in series or parallel with the sampling resistor be sufficiently enough.

For insulators, where current primarily arises from dielectric polarization and ionic displacement, the bulk resistivity typically exceeds 10^{10} Ω m. In such scenarios, R_1 is much smaller than the measured volume resistance R_{V1}, and the surface resistances R_S and R_{S2} are sufficiently high. Therefore, the circuit shown in Fig. 2.2 is well-suited to accurately measure the volume resistivity of insulating materials while minimizing the measurement errors. However, CFRP composites are semiconductive materials consisting of carbon fiber laminates with various fiber orientations bonded by interlayer resin. Their electrical conductivity is highly anisotropic, and the surface conductivity can vary significantly depending on the conductive modification coating used. For instance, when CFRP composite surfaces are coated with highly conductive materials such as high-conductivity polymer foams or high-conductivity nonwoven fabrics, the surface resistance (R_S) may become substantially lower than the volume resistances (R_{V1} and R_{V2}) and even lower than the sampling resistor (R_1). Under such conditions, the branches containing R_S and R_{S2} introduce a significant shunting effect, causing a substantial reduction in the current flowing through the measured resistances (R_{V1}, R_{V2}) and sampling resistor (R_1). When the current through R_1 drops to a level comparable to stray currents induced by system instability or external interference, the measurements becomes highly susceptible to noise. This current shunting effect also results in substantial deviations between the measured volume resistivity values of CFRP specimens and their actual values, as demonstrated by the experimental results shown in Fig. 2.3.

Improved Three-Electrode Volume Resistivity Measurement Method

Based on a comparative evaluation of the three-electrode and four-electrode methods for measuring the DC low-current volt-ampere characteristics of CFRP composites, an enhanced three-electrode method for precisely measuring CFRP volume resistivity is proposed. This method adds a second sampling resistor R_2 in series between the guard electrode C and the measurement system's low-voltage terminal [7–9]. The measurement circuit and its equivalent circuit are shown in Figs. 2.5 and 2.6, respectively.

In Fig. 2.6, R is current-limiting resistor; R_{V1} and R_{V2} are equivalent volume resistances in the measurement electrode (B) and guard-electrode (C) branches, respectively; R_S is equivalent surface resistance between measurement electrode (B) and guard electrode (C); R_1, R_2 are sampling resistors in the two branches; V_1 and V_2 are voltages across sampling resistors R_1 and R_2, respectively; V is voltage between the high-voltage electrode A and the measurement system's low-voltage termina. The volume resistances R_{V1} and R_{V2} of the CFRP composite are calculated by Eq. (2.2) and (2.3), respectively.

$$R_{V1} = R_1(V - V_1)/V_1 \tag{2.2}$$

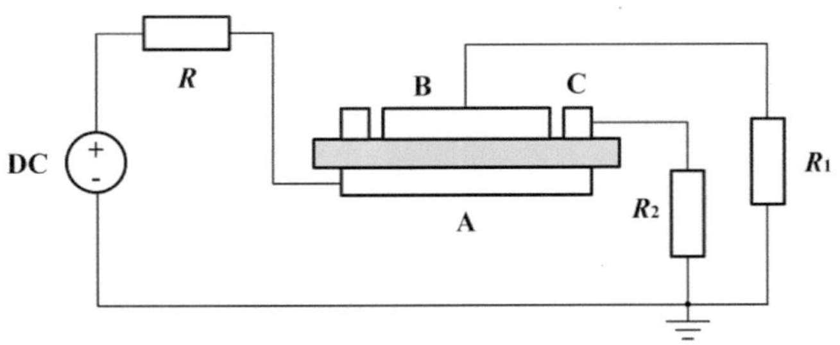

Fig. 2.5 Improved three-electrode measurement circuit for CFRP volume resistivity

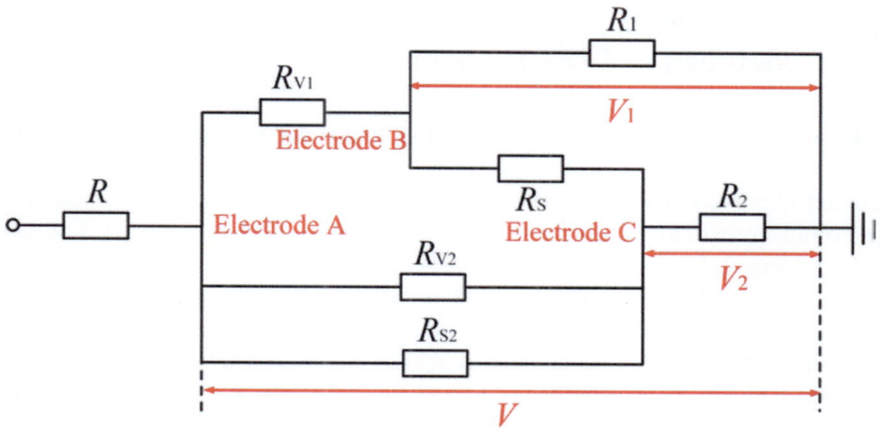

Fig. 2.6 Equivalent circuit diagram of the improved three-electrode method for CFRP volume resistivity

$$R_{V2} = R_2(V - V_2)/V_2 \tag{2.3}$$

Subsequently, using the specimen thickness l and the electrode areas S_1, (measurement electrode B) and S_2 (guard electrode C), the bulk resistivities ρ_{V1} and ρ_{V2} are determined via Eqs. (2.4) and (2.5).

$$\rho_{V1} = R_{V1} \cdot S_1/l \tag{2.4}$$

$$\rho_{V2} = R_{V2} \cdot S_2/l \tag{2.5}$$

By adding the sampling resistor R_2, even if the surface resistances R_S and R_{S2} are low, the high-voltage electrode A and the measurement electrode B cannot form a direct low-resistance path to the system ground through the surface. Thus, R_2 substantially reduces the shunting effect in the circuit. When the currents I_1 and I_2 through R_1 and R_2 greatly exceed system stray currents, measurement accuracy is preserved. To ensure precision and suppress stray-current interference, the sampling resistors should be as small as practical to increase measurable currents, but not so small as to reintroduce significant shunting via surface paths. Therefore, R_1 and R_2 must be carefully matched.

2.1.2 Measurement Results and Analysis of DC Bulk Resistivity in CFRP Composites

Influence of Sampling Resistors on Volume Resistivity Measurement

Combinations of sampling resistors with different resistance values were tested for DC volume resistivity measurements of CFRP composites. Experimental results are shown in Fig. 2.7 [10, 11].

As shown in Fig. 2.7a, the measured volume resistivity of specimen A3 along the thickness direction varies by more than a factor of two when using 1 kΩ versus 50 Ω sampling resistors. In contrast, specimen A4's bulk resistivities remains nearly constant across different sampling resistors (Fig. 2.7b). In Fig. 2.7c, d, both R_{V1} and R_{V2} for specimen A5 differ by approximately 10^2 Ω, consistently across all resistor pairs. When R_1 and R_2 are chosen as 1 kΩ and 50 Ω, values matched to the magnitudes of R_{V1} and R_{V2}, the stability of the measured resistances improves compared to using two 50 Ω resistors.

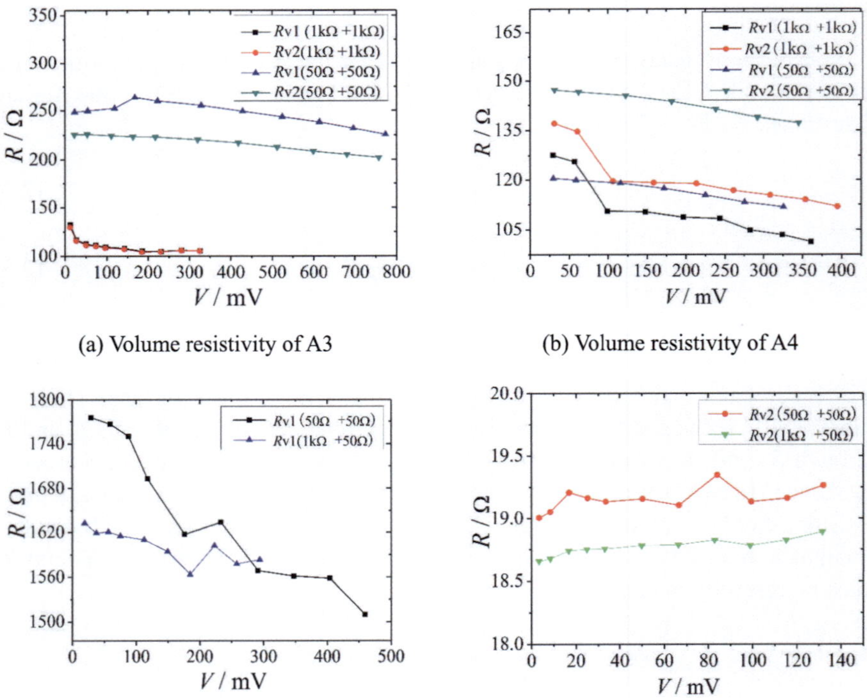

(a) Volume resistivity of A3

(b) Volume resistivity of A4

(c) Volume resistivity of A5

(d) Volume resistivity of A5

Fig. 2.7 Influence of sampling resistor on CFRP composites volume resistivity measurement

Excessively large sampling resistors cause an imbalanced voltage division between the CFRP volume resistance and the sampling resistor, while too-small sampling resistors are more susceptible to noise, amplifying calculation errors and producing unstable resistance readings. Therefore, R_1 and R_2 should be selected to match the order of magnitude of the expected R_{V1} and R_{V2}, thus minimizing measurement errors.

Measurement Results of Volume Resistivity Using the Improved Three-Electrode Method

The enhanced three-electrode method with optimized sampling resistors was applied to measure the volume resistivity of CFRP composites (three specimens per configuration, each tested at least twice). The trends are plotted in Fig. 2.8 [10, 11].

From Fig. 2.8, the volume resistivity of CFRP composites displays distinct behaviors depending on manufacturing process, layup architecture, and surface treatment:

(i) The through-thickness volume resistivity of specimens ranges from 10^2–$10^3 \Omega$ m, indicating semiconductive behavior arising from the combination of conductive fibers and the insulating resin.
(ii) Nonlinear DC resistivity is observed: as the applied DC voltage increases, resistivity decreases, with variations up to 10^3 Ω m under the experimental conditions.
(iii) For specimens A1–A4, the measured bulk resistivities ρ_{V1} and ρ_{V2} are closely matched, differing by only a few ohm (Fig. 2.8a–d).
(iv) Specimens A5–A9 show significant ρ_{V1} and ρ_{V2} discrepancies (Fig. 2.8e–i), with differences ranging 10^2–10^3 Ω m, correlated with the conductivity of surface coatings. Specimens A6–A9, which include MCNT (micro carbon nanotubes) interlayers, exhibit a monotonic decreasing in volume resistivity as the number of MCNT layers increases (see Fig. 2.9).

Combined with the equivalent circuit in Fig. 2.6, the above observations can be explained as follows:

(1) Specimens A1–A4 exhibit nearly identical R_{V1} and R_{V2}. This results from their highly conductive surface treatments, nonwoven fabrics or mesh coatings, that enforce near-equipotential conditions on the CFRP surface, thus minimizing the influence of surface resistance R_S. Moreover, the similar areas of measurement electrode B and guard electrode C contribute to the close agreement between ρ_{V1} and ρ_{V2}.
(2) Specimens A6–A9 show a progressive decrease in volume resistance as the number of MCNT interlayers increases. The embedded highly conductive MCNT layers enhance through-thickness electrical conductivity, thereby lowering the bulk resistivity.
(3) Specimens A5–A9 display significant differences between ρ_{V1} and ρ_{V2} due to several factors:

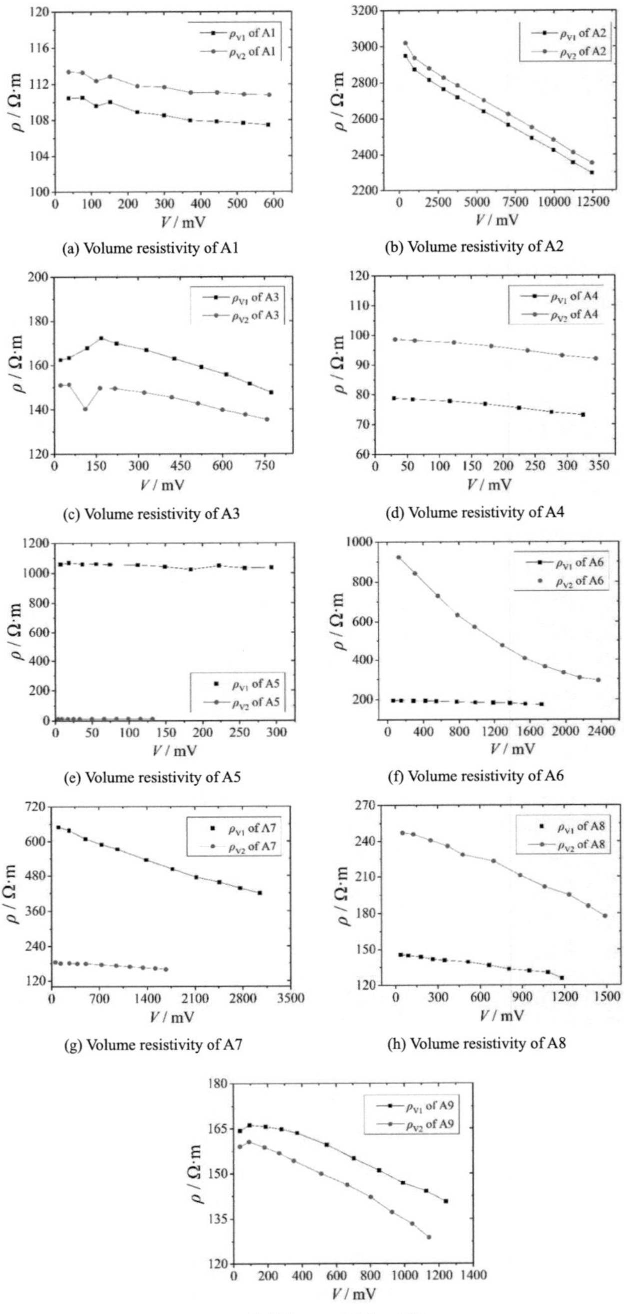

(a) Volume resistivity of A1

(b) Volume resistivity of A2

(c) Volume resistivity of A3

(d) Volume resistivity of A4

(e) Volume resistivity of A5

(f) Volume resistivity of A6

(g) Volume resistivity of A7

(h) Volume resistivity of A8

(i) Volume resistivity of A9

Fig. 2.8 DC volume resistivity of specimens A1–A9 measured by the improved method

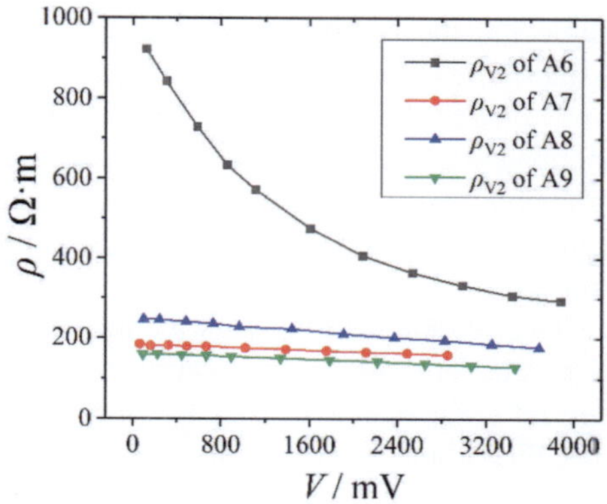

Fig. 2.9 Comparison of volume resistivity for specimens A6–A9

(i) In CFRP, each carbon fiber ply is bonded by resin, creating random point contacts between fiber. Carbon fibers are much more conductive than the resin matrix, so local contact resistances are much lower than in non-contact regions, producing a non-uniform resistivity distribution.

(ii) The low-conductivity surface layers of A5–A9 (short fibers or resin–fiber mixtures) introduce a non-negligible surface resistance R_S. The shunting through R_{V1} and R_{V2}, leading to over- or underestimated of these resistances. Smaller R_S. values amplify these errors.

(iii) The layered architecture of CFRP causes electric-field spreading, so the effective conduction area differs from the nominal electrode area. Cylindrical electrode B and annular electrode C produce different field distributions, further biasing the measured R_{V1} and R_{V2}. Therefore, explaining the discrepancy between R_{V1} and R_{V2} requires simultaneous consideration of both electric-field spreading and surface-resistance shunting in the CFRP equivalent circuit model.

2.2 Dynamic Electrical Conductivity of CFRP Composites Under Lightning Current

The in-plane and through-thickness conductivities of CFRP composites are directly related to the shunting behavior of lightning currents across layers and depth, determining lightning damage patterns and severity. The selection of initial conductivity parameters also critically influences the accuracy of simulations for lightning damage area and penetration depth.

Most experimental studies on CFRP lightning damage are performed under high-intensity lightning current pulses of 10 kA or greater [12, 13]. However, such currents cause irreversible damage to CFRP, preventing investigations into dynamic conductivity changes under lightning conditions. Consequently, existing thermo-electric coupled lightning damage models treat CFRP as a static, anisotropic resistor [12–14], neglecting the material's nonlinear transient and inductive responses during lightning strikes. The DC conductivity measurements in Sect. 2.1 indicate that the DC conductivity of CFRP composites is highly dependent on layup structure, processing, and the magnitude of the applied DC voltage. Therefore, in simulation of CFRP lightning damage, how to choose initial conductivity parameters, and whether using DC conductivity as the starting electrical parameters is valid, require further in-depth study.

This section first analyzes the intrinsic conductive mechanism of CFRP. It then introduces an experimental platform for non-destructive application of lightning-like current components. Through experiments on the dynamic electrical conductivity of CFRP composites, this section identifies factors causing variations in transient conductivity and proposes a measurement method tailored to these dynamic characteristics. It further derives the relationships between dynamic impedance and applied current parameters, and finally, based on experimental waveform data and the specimens' surface coating and internal sandwich structures, it develops an equivalent-circuit model for CFRP's through-thickness direction. This model explains the intrinsic conduction behavior of CFRP under lightning current components, laying the foundation for research on lightning damage mechanisms under both single, high-intensity and multiple sequential lightning current components.

2.2.1 Fundamental Theories of Conductivity in CFRP Composites

Classification of Conductive Materials

Materials can be classified by their electrical conductivity into conductors, semiconductors, and insulators. Conductors generally have resistivities $<10^{-7}$ Ω m, while insulators typically exceed 10^{8} Ω m. Semiconductors exhibit resistivities between 10^{-7}–10^{8} Ω m, occupying the intermediate range. The electrical behavior of these materials is explained by energy band theory. Figure 2.10 shows the band diagrams for each class.

In conductors, the conduction band contains a high density of free electrons that drift under an applied field, resulting in very low resistivity. In insulators, the conduction band lacks mobile electrons and the valence band is fully occupied. The bandgap is typically >6 eV, so thermal excitation of electrons into the conduction band is negligible, yielding minimal intrinsic conductivity (very high resistivity). Only under strong external excitation can a few electrons be promoted across the gap. Semiconductors have a similar band structure but with a smaller bandgap (usually <3 eV), allowing room-temperature thermal energy to excite enough electrons into the conduction band to produce intermediate conductivity. Given CFRP's microstructure, conductive carbon fibers embedded in an insulating polymer matrix, its bulk behavior falls within the semiconductor regime.

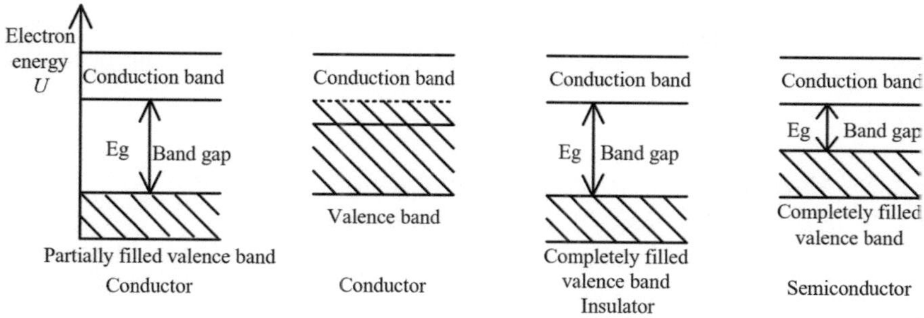

Fig. 2.10 Energy band diagrams of different material types

Conductive Theories of CFRP Composites

CFRP composites consist of carbon fiber embedded in an epoxy resin. The carbon fiber can exist in two configurations: direct fiber-fiber contact or separated by thin insulating resin layers. Two principal models explain their conductivity: percolation theory and quantum tunneling theory. These theories form the critical foundations for analyzing the nonlinear voltage-current behavior of conductive polymer composites as functions of filler concentration and applied electric field strength [15]. At low fiber volume fractions, insufficient direct contacts prevent a continuous conductive network, so the composite behaves as an insulator due to the resistive resin matrix. In the transitional region near the percolation threshold, some fibers form chainlike conductive networks; where thin polymer gaps remain, electrons can quantum-tunnel through the barrier under an applied field, producing a strongly nonlinear tunneling current [16]. As the fiber concentration increases beyond the threshold, direct percolation pathways emerge, causing an abrupt increases in composite conductivity by several orders of magnitude, after which Ohmic conduction dominates [17].

Compared to particulate fillers such as metal powders or graphene, high-aspect-ratio fibrous fillers promote the formation of percolating contact networks [18]. Carbon fiber can establish continuous conductive pathways in polymer matrices at very low volume fractions. For example, a percolation threshold of $\approx$1% for 5 mm short carbon fibers in cementitious composites [19]. In structural-functional CFRP, fiber volume fractions are typically high. Hirano et al. [20] reported that at 64% fiber volume, CFRP exhibits uniform conductivity along the fiber direction, with through-thickness conduction arising from occasional interlayer fiber contacts. Haider et al. [3] observed stable Ohmic behavior in CFRP with 55–60% fiber volume fraction under AC excitation.

Most studies of CFRP DC conductivity investigate the milliampere range, focusing on fiber content, distribution, and orientation effects on contact conduction, neglecting tunneling and other nonlinear effects. However, Rimska et al. [21] demonstrated frequency-dependent conductivity in CFRP under DC and AC fields, showing that high-frequency conductivity cannot be inferred from DC measurements. Given the vast differences in frequency, duration, and amplitude between lightning pulses and low-amplitude DC/AC currents, characterizing CFRP conductivity under lightning-like conditions is essential to accurately model its behavior during strikes.

2.2.2 Measurement Method for Dynamic Conductivity Characteristics of CFRP Composites

In non-destructive dynamic resistivity/conductivity measurements of laminated CFRP under lightning-current pulses, the specimen parameters are listed in Table 2.2.

To capture the dynamic conductivity of CFRP composites, the three-electrode recommended in the literature [6] and setup from Fig. 2.1 was initially employed, matching its electrode geometry. Figure 2.11 depicts the electrical schematic used for dynamic measurements under lightning currents.

In Fig. 2.11, appropriate capacitor *C*, resistor *R*, and inductor *L* values were selected to reproduce lightning current component A according to SAE APR 5412, targeting a peak time of $T_p = 30.2$ μs and a half-peak time of $T_t = 72.2$ μs, as shown in Fig. 2.12.

The dynamic conductivity test setup is shown in Fig. 2.13. Current-coil sensors, pulse dividers, and a Tektronix MDO 3012 oscilloscope were used to record both the

Table 2.2 Parameters of CFRP specimens for dynamic resistivity/conductivity experiments

No.	Dimensions	Structure and process characteristics
B1	100 × 150 × 5 mm	$[45/0/–45/90]_{4S}$, surface coated with micro carbon nanotubes (MCNT)
B2	100 × 150 × 5 mm	$[45/0/–45/90]_{4S}$, MCNT applied between 1/2 plies and surface
B3	100 × 150 × 5 mm	$[45/0/–45/90]_{4S}$, MCNT applied between 1/2, 2/3 plies and surface
B4	100 × 150 × 5 mm	$[45/0/–45/90]_{4S}$, MCNT applied between 1/2, 2/3, 3/4 plies and surface
B5	130 × 100 × 3 mm	$[0/90]_{3S}$, surface coated with low-conductivity short fibers

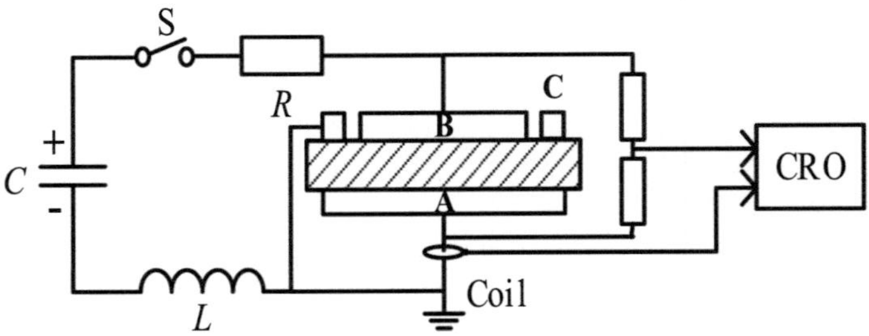

Fig. 2.11 Dynamic resistance/conductance measurement circuit for CFRP composites

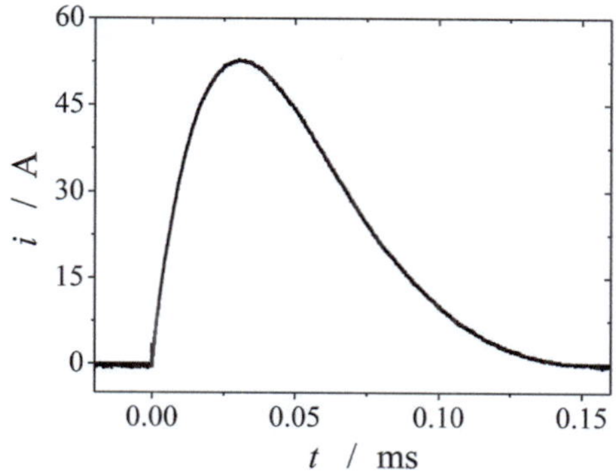

Fig. 2.12 Lightning current component A waveform

applied lightning current and the voltage across CFRP specimens, enabling analysis of how dynamic conductivity varies with lightning-current waveforms.

Lightning current impulse tests were conducted on CFRP specimen B1 using the circuit in Fig. 2.11, with electrode B tied to the high-voltage terminal, electrode A grounded, and electrode C in three configurations: grounded, floating, and connected to the high-voltage terminal. The measured voltage and current waveforms at a 2 kV discharge are shown in Fig. 2.14.

As shown in Fig. 2.14, the guard-electrode connection had negligible impact on the recorded waveforms. This is due to the ultra-short duration of lightning current component A (rise time ≈ 30 μs, half-peak time ≈ 70–80 μs), during which the 2 kV impulse is effectively applied across the CFRP in its high-resistance state. The equivalent circuit under these conditions is depicted in Fig. 2.15.

In Fig. 2.15, U denotes the discharge voltage. R_{C1}–R_{C32} represent the dynamic volume resistances of the 32 carbon-fiber plies through the thickness. G_1–G_{31} are

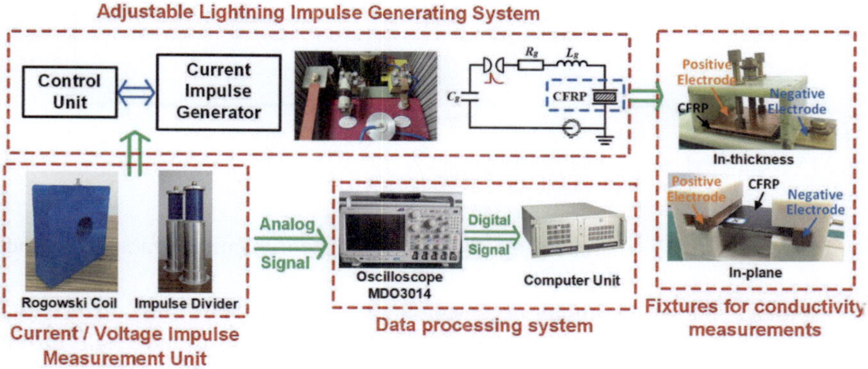

Fig. 2.13 Experimental platform for dynamic conductivity testing of CFRP composites under lightning currents

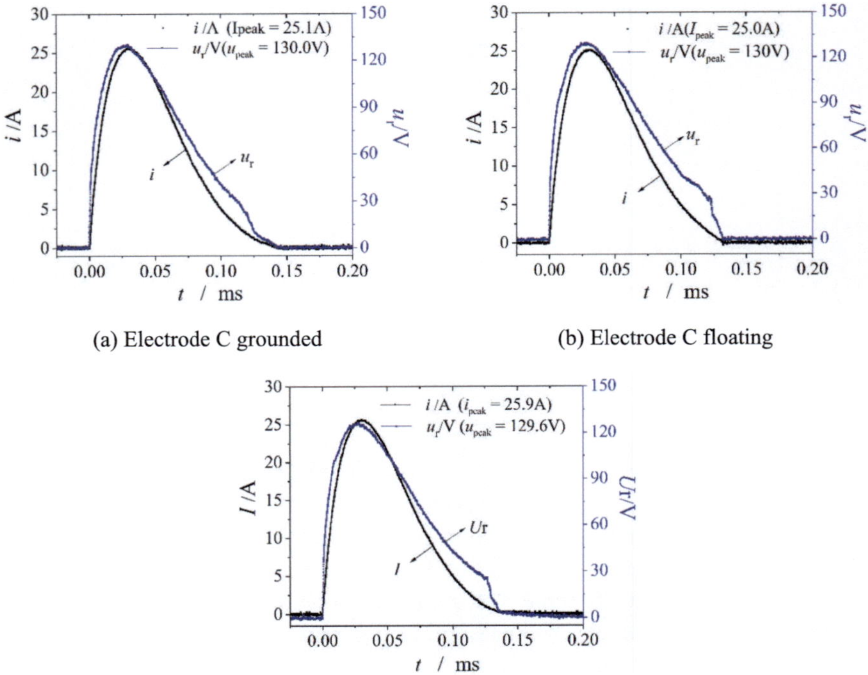

(a) Electrode C grounded

(b) Electrode C floating

(c) Electrode C connected to high-voltage terminal

Fig. 2.14 Experimental waveforms of specimen B1 under different guard-electrode configurations

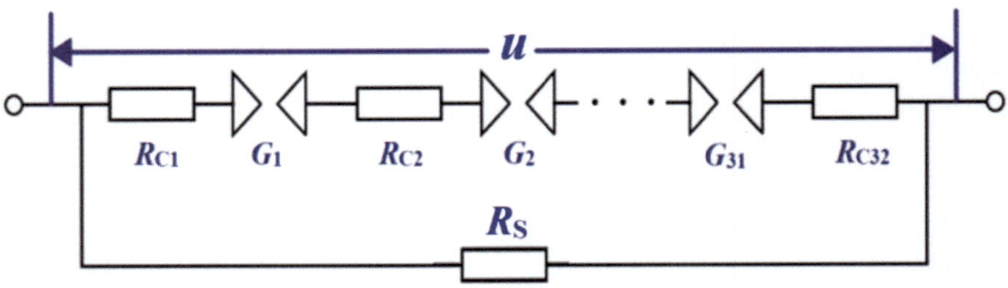

Fig. 2.15 Equivalent circuit of CFRP composites at discharge instant

the thin resin interlay between plies. R_S represents the dynamic surface impedance between electrodes B and C.

Due to manufacturing limitations, insulating residues inevitably remain on CFRP surfaces, keeping the surface-resistance branch in a high-resistance state during a lightning discharge. Furthermore, the short duration and moderate amplitude of the lightning component A do not inflict destructive damage on the specimens. As shown in Fig. 2.14, the experiments reveal that only a small fraction of the discharge current spreads through the composite plies, minimally affecting the total measured

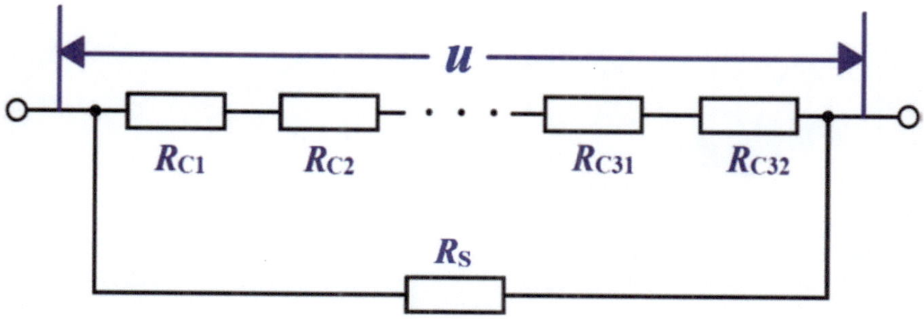

Fig. 2.16 Equivalent circuit diagram of CFRP composites after resin-layer breakdown

current. Therefore, surface charge dispersion and surface-path conduction in CFRP composites can be neglected in this context.

Additionally, random fiber-fiber contact points pre-condition the resin interlayer gaps (Fig. 2.15) into an easily ionizable state. Upon application of the lightning current pulse, component is applied, these interlayer gaps undergo dielectric breakdown, the specimen's equivalent circuit reconfigures to that shown in Fig. 2.16. The resulting decrease in dynamic volume resistance lowers the voltage drop across the specimen, further suppressing conduction via the high-resistance surface branch. Consequently, the electrical connection of electrode C does not significantly influence the test results.

For experimental simplicity, the reduced circuit in Fig. 2.17, retaining only electrodes A and B, can be used, where lightning current is injected at the specimen center and exit through the bottom surface. This setup allows investigation of the dynamic volume resistance or conductivity of CFRP composites under lightning current injection.

During testing, an aluminum-foil layer, matching the bottom electrode area, was placed at the electrode-specimen interface. Severe oscillatory noise appeared in the voltage waveforms, especially when the electrode and specimen surfaces were uneven, as shown in Fig. 2.18a. Applying a thin layer of silicone grease at the interface suppressed this noise. Figure 2.18b shows the significant reduction interference after greasing.

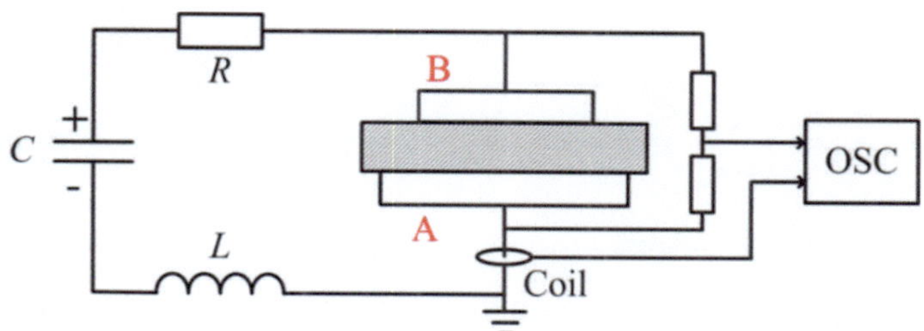

Fig. 2.17 Experimental circuit for dynamic impedance/conductance of CFRP composites

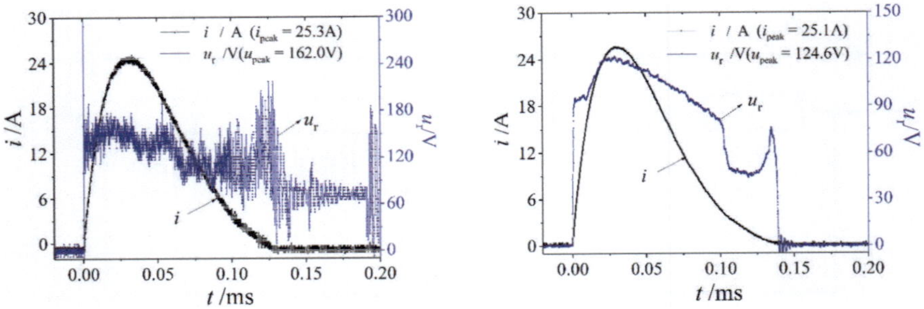

(a) Direct contact (b) Contact with silicone grease filling

Fig. 2.18 Dynamic conductive measurements for specimen B5 under different electrode contact conditions

Analysis of the specimen's surface morphology and silicone-grease properties indicated that the interference arises from high-frequency discharges in air gaps between the electrode and specimen. Silicone grease fills these gaps, preventing repeated breakdown and improving voltage-waveform fidelity. Although the grease may slightly alter impulse coupling, its thinness renders its impact in data interpretation. Specimen B5 displayed distinct dynamic conductivity under lightning current compared to other specimens. At the moment of current injection, the voltage across B5 spiked, while the current remained near zero, indicating a high-resistance initial state. Figure 2.19 presents lightning-current and response-voltage waveforms for braided CFRP under a 400 A impulse.

From Fig. 2.19, when the lightning current is injected through the thickness of the CFRP specimen, the voltage spikes to ≈180 V within microseconds of the pulse onset, twice the rise observed in Fig. 2.18 (~90 V). During this interval, the measured current is near zero, indicating that the CFRP is in "high-resistance state" at the moment of the external field applied.

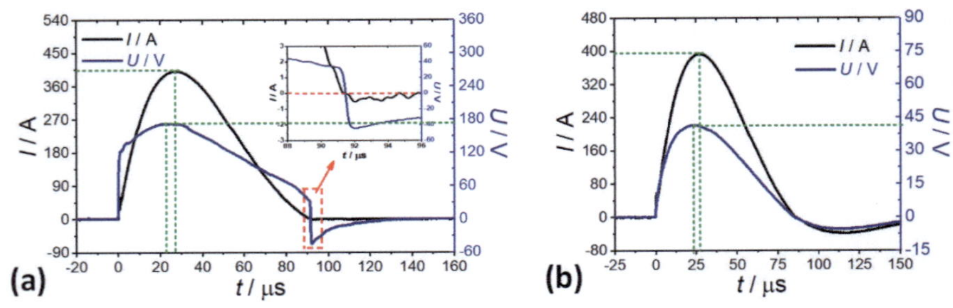

(a) Through-thickness (b) In-plane

Fig. 2.19 Lightning-current and response-voltage waveforms for specimen B5 under a 400 A impulse

As shown in Figs. 2.18 and 2.19, with increasing external electric field strength, the lightning current amplitude rises, while the response voltage exhibits a strong "clamping" behavior, i.e., its variation is much smaller than that of the current. This suggests that above a critical field, the CFRP transitions from a "high-resistance" to a nonlinear conductive state. As the current decays in the pulse tail, the CFRP material reverts to its initial "high-resistance" state, again produce high voltages even for small conduction currents. Furthermore, after the current zero-crossing, the response voltage undergoes reverse, polarity oscillations, abruptly switching from positive to negative at the zero point (see the amplified region in Fig. 2.19b). In planar conduction (not through-thickness), the CFRP response voltage follows the lightning-current waveform without clamping (Fig. 2.19b). In both conduction modes, the voltage pulse precedes the current pulses, revealing inductive effects in the composite's dynamic response [22–24].

2.2.3 Measurement Results and Analysis of Dynamic Conductivity Characteristics of CFRP Composites

Lightning current component A experiments were conducted on CFRP composites, with each specimen group containing three samples and each sample tested at least twice.

Experimental Results of Dynamic Conductivity Characteristics of CFRP Composites

CFRP composite specimens exhibited distinct response processes to lightning current components due to differences in manufacturing processes and layup structures, as shown in Fig. 2.20.

Analysis of Fig. 2.20 reveals that the response voltages at both ends of the CFRP composite specimens slightly precede the lightning-current peaks, indicating inductive behavior of CFRP under impulse excitation.

Nonlinear Conductivity Characteristics of CFRP Composites

Experimental results for CFRP under 50 and 100 A lightning pulses are shown in Fig. 2.21.

As shown in Fig. 2.21, the lead time of voltage over current decreases as the current amplitude increases, with the peak-time difference narrowing from 2.88 to 0.62 μs. This suggests that at kA-level currents, the voltage and current peaks become nearly synchronized. For simplicity in analysis and comparison, an impulse impedance R

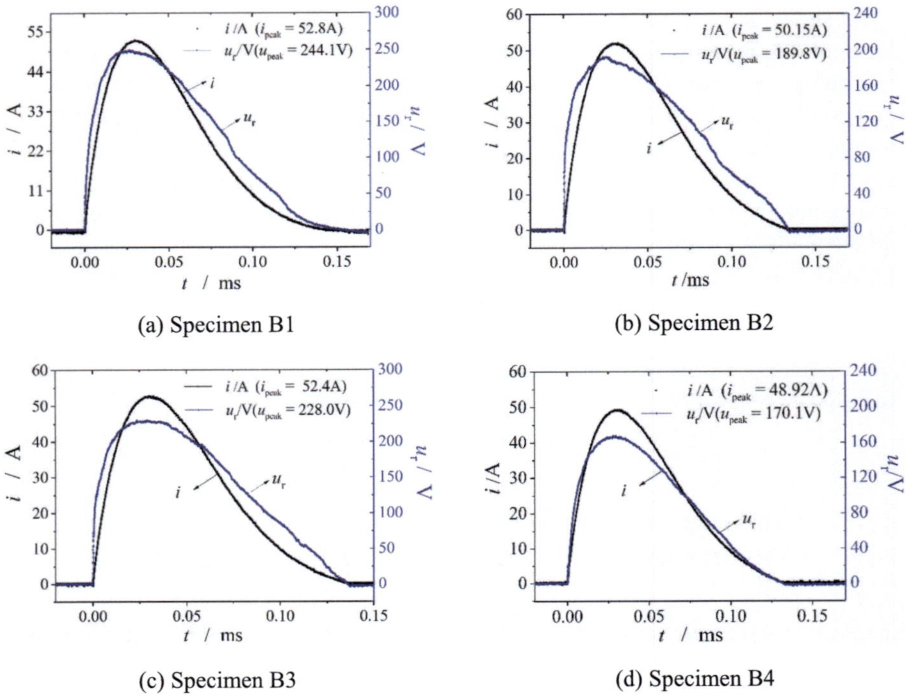

Fig. 2.20 Lightning-current and response voltage waveforms for CFRP specimens under component

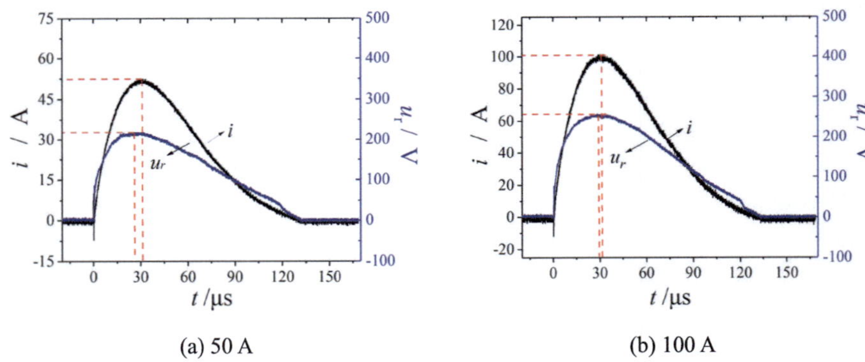

Fig. 2.21 Voltage leading current for specimen B2 under 50 and 100 A lightning pulses

$= u_{peak}/i_{peak}$, where u_{peak} and i_{peak} are the peak response voltage and peak lightning current, were defined, respectively. This resulting impulse impedance (and its reciprocal, impulse conductivity) as a function of current amplitude is plotted in Fig. 2.22.

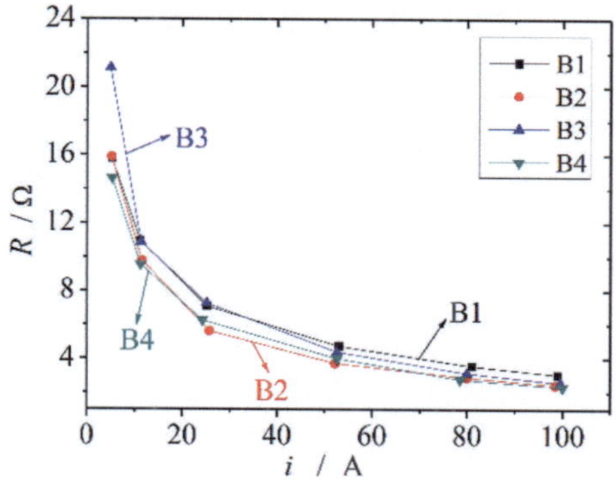

Fig. 2.22 Dynamic impedance characteristics of specimens B1–B4 under lightning current component A

Analysis of these results reveals that under lightning component A, the dynamic equivalent resistance of CFRP composites is on the order of 10^0 Ω, indicating good electrical conductivity. The dynamic impedance decreases with increasing pulse current, demonstrating strong nonlinear characteristics [21–23]. Moreover, specimens B1–B4 doped with varying numbers of high-conductivity micro-carbon nanotube (MCNT) interlayers, show negligible differences in through-thickness dynamic impedance,, implying that MCNT additions have minimal impact on dynamic conductivity characteristics under component A.

Further analysis of the single lightning component A transient yields the dynamic voltage-current characteristics plotted in Fig. 2.23.

From Fig. 2.23:

(i) As the lightning-current amplitude increases, the voltage rise across the specimen diminishes, confirming nonlinear impedance.
(ii) During both the rising and falling portions of the pulse, the voltage at the specimen ends differ, indicating a lagged response of CFRP to current changes.
(iii) The specimen's voltage peaks slightly before the current, revealing inductive effects in the material.

The nonlinearity dynamic impedance originates from CFRP's layered architecture: as the electric stress increases, breakdown of the resin interlayers expands, reducing overall impedance. The observed inductive behavior is tied to the internal sandwich structure, where conductive plies and dielectric layers form distributed inductive–resistive elements.

Dynamic Conduction Circuit Model for CFRP Composite Thickness Direction

CFRP composites are fabricated by arranging carbon-fiber tows at specified ply angles and consolidation them with prepreg resin via compression molding, resulting

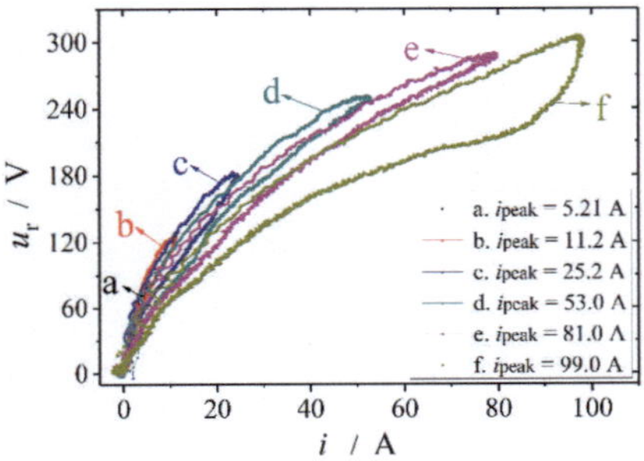

(a) Dynamic voltage-current of B1

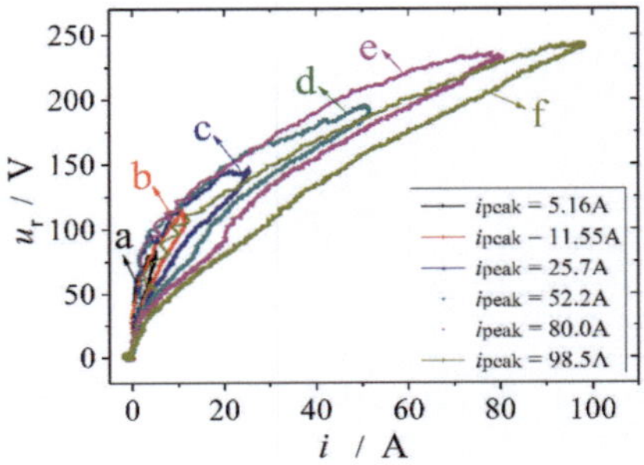

(b) Dynamic voltage-current of B2

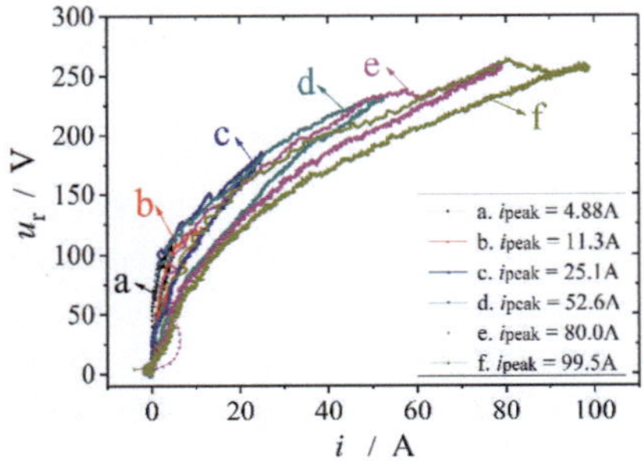

(c) Dynamic voltage-current of B3

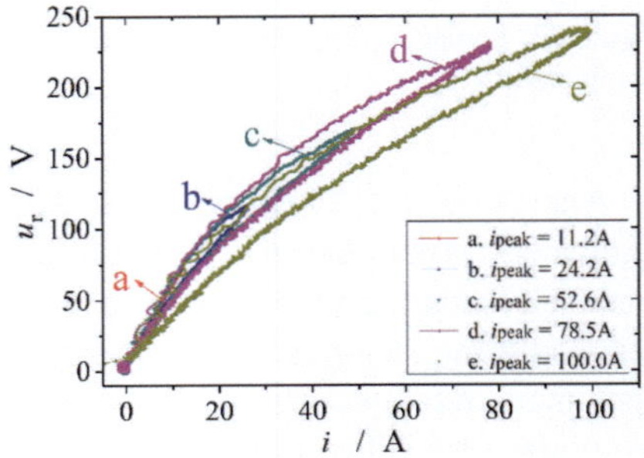

(d) Dynamic voltage-current of B4

Fig. 2.23 Dynamic voltage-current characteristics of CFRP composites

in a semiconductive fiber–resin network. From an equivalent-circuit perspective, each fiber tow can be represented by a series of small resistive segments. Under lightning-current conduction through the thickness, this corresponds to series connections of fiber-segment resistances (R_{ij}), inductances (L_{ij}), and internal discharge gaps (G_{ij}), as shown in Fig. 2.24 [22–24]. Each ply similarly forms a resistance-inductance network.

In Fig. 2.24, L_{ij}, R_{ij}, and G_{ij} denote the equivalent inductance, resistance, and discharge gap of resin layer between adjacent fiber tows, while r_{ij} denotes the resistance of a fiber-tow segment. Once the interlayer gaps G_{ij} undergo dielectric breakdown, the through-thickness equivalent circuit transforms as depicted in Fig. 2.25 [21–23]. Under a lightning impulse, the specimen voltage peaks slightly before the current, confirming inductive behavior. The aggregate inductive response arises from the series combination of each resin-layer inductances, and the observed voltage lag under pulse excitation is tightly linked to the inductive properties of both fibers and resin in CFRP materials.

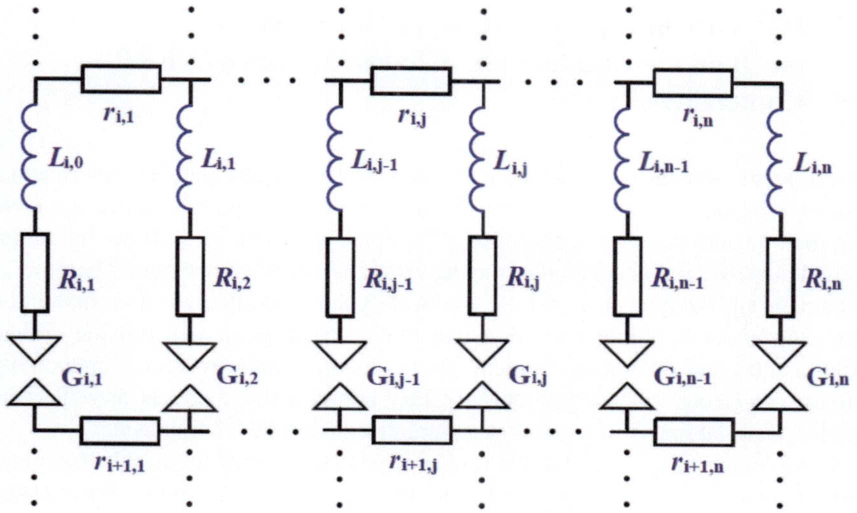

Fig. 2.24 Equivalent circuit diagram of stacked CFRP ply

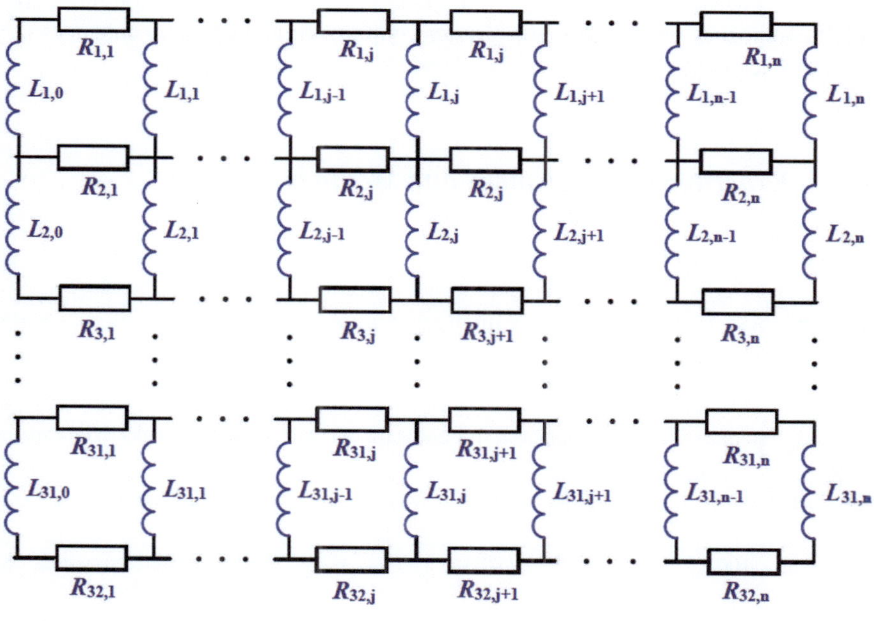

Fig. 2.25 Equivalent circuit diagram of CFRP composites after interlayer gap breakdown

2.3 Influence of Lightning-Current Parameters on Dynamic Conductivity Characteristics of CFRP Composites

Based on the dynamic-impedance measurement techniques and the experimental characterization of various composite architectures in the previous section, it is clear that the transient conductive response of carbon-fiber composites under lightning-current pulse differs substantially from their static resistive DC behavior. Therefore, a systematic and comprehensive investigation of dynamic conductivity and conduction modeling under lightning excitation is essential. Such work will provide critical experimental and theoretical foundations for incorporating dynamic conductivity into thermo-electric coupling models of CFRP lightning direct effects, as well as for guiding material optimization and structural design of CFRP components.

To this end, an experimental platform capable of generating adjustable lightning-current waveforms was established. Through dynamic-conductivity experimental on CFRP composites under varied lightning current parameters, and by correlating the results with internal layup architectures, the key factors that govern transient conductivity were identified. The development of a dynamic conduction model for CFRP composites underpins multi-physics coupling simulations and enhances the predictive accuracy of CFRP lightning damage assessments.

2.3.1 Experimental Specimens and Methods

Experimental Specimens

Three distinct CFRP composite specimen types were tested. Their characteristics are summarized in Table 2.3.

Table 2.3 Parameters of CFRP composite specimens for dynamic resistance/conductivity experiments

No.	Dimensions	Material grade	Fiber volume fraction (%)	Layup structure
C1	150 × 100 × 4 mm	CCF300/5228A	55	Braided structure, $[0/90F]_{16}$
C2	100 × 100 × 3.6 mm	T300/3021	60	Unidirectional fiber layup, $[0]_{24}$
C3	100 × 100 × 3.6 mm	T300/3021	60	Unidirectional fiber layup, $[45]_{24}$

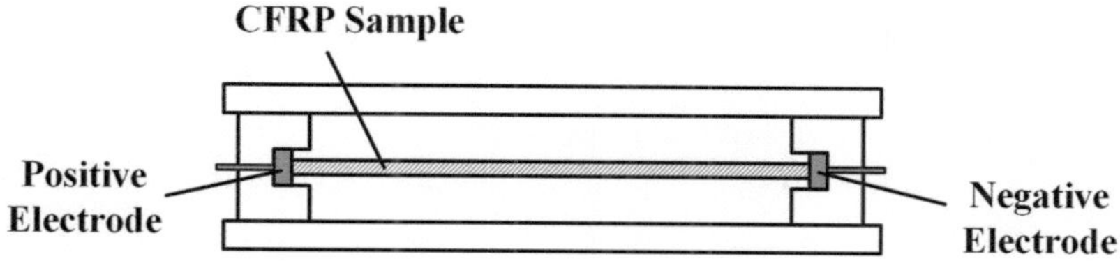

(a) Planar electrical conductivity measurement fixture

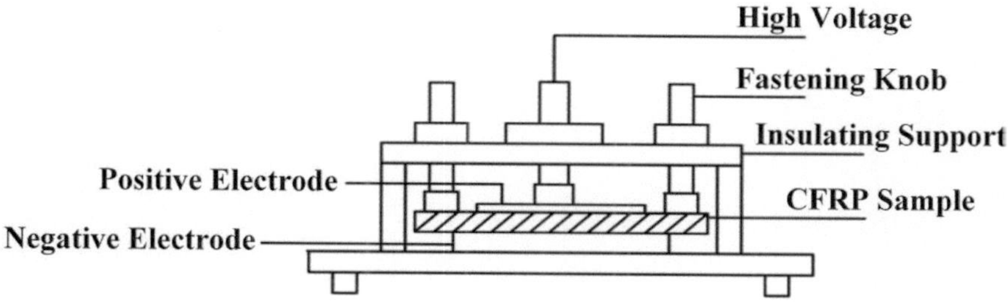

(b) Through-thickness electrical conductivity measurement fixture

Fig. 2.26 Experimental clamping devices for dynamic electrical conductivity measurements of CFRP composites

Experimental Fixtures

Experimental studies employed clamping along the planar length, planar width, and through-thickness directions of CFRP composites, evaluating dynamic conductivity for 0°, 45°, and braided layups. The electrode assemblies and clamping mechanism in Fig. 2.26 facilitate measurements in both planar and through-thickness orientations. The fixture consists of an insulating support, a clamping system, and copper measuring electrodes.

To enhance contact quality, copper electrode surfaces were polished with fine sandpaper and abrasive paste before each test to remove oxidation. Multiple layers of soft aluminum foil were placed between electrodes and specimens, and CFRP panels were firmly clamped to the electrodes, minimizing contact resistance artifacts.

Experimental Methods

To assess the effect of lightning-current waveform parameters on dynamic conductivity, three impulse shapes, 10/30, 20/60, and 40/120 μs, were generated. The generation circuit is shown in Fig. 2.11, and the component values for each waveform are provided in Table 2.4 (Fig. 2.27).

The waveforms and parameters of three lightning current components (10/30, 20/60, and 40/120 μs) are shown in Fig. 2.27.

Table 2.4 Circuit component values for different lightning-current waveforms

Lightning current component waveform parameters		C/μF	R/Ω	L/μH
Front time T_f(μs)	Half-peak time T_t (μs)			
10	30	14.0	1.0	9.9
20	60	14.0	2.0	39.5
40	120	28.0	2.0	79.1

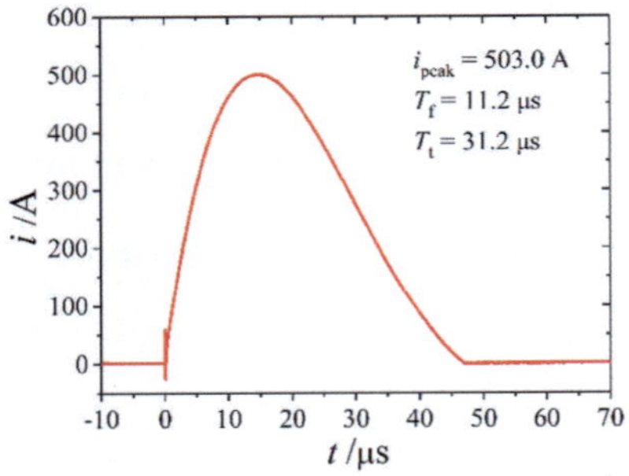

(a) 10/30 μs lightning current waveform

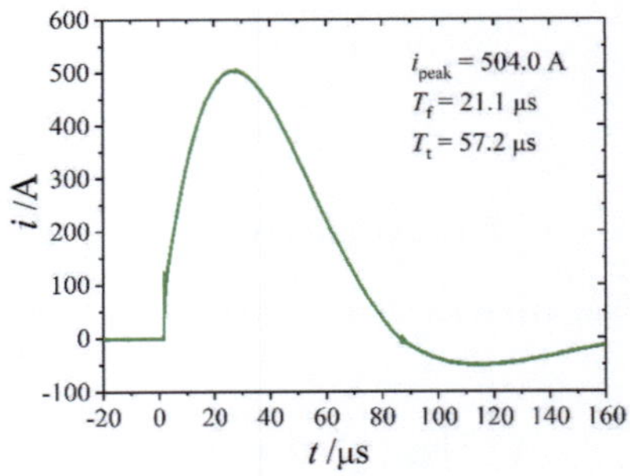

(b) 20/60 μs lightning current waveform

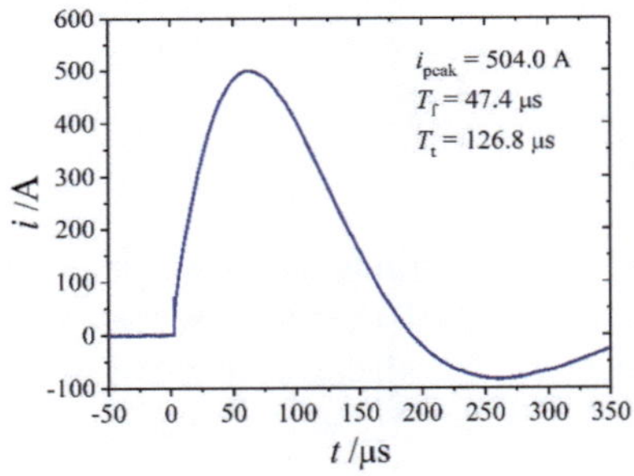

(c) 40/120 μs lightning current waveform

Fig. 2.27 Lightning current pulse waveforms and parameters for 10/30, 20/60, and 40/120 μs components

Lightning current through the CFRP specimen and response voltage were measured with a current coil, pulse divider, and Tektronix MDO3012 oscilloscope. Using the clamping fixtures from Fig. 2.26, dynamic conductivity characteristics under varying waveform shapes and peak currents were recorded. We then analyzed how dynamic impedance relates to layup architecture and waveform parameters, developing conduction models for CFRP. To ensure reproducibility, each test point was repeated three times.

2.3.2 *Influence of Lightning-Current Parameters on Dynamic Conductivity Characteristics*

Effect of Specimen Clamping Method on Dynamic Conductivity

Under different lightning-current parameters, CFRP composites with various layup structures exhibit distinct dynamic conductivity behaviors depending on the clamping orientation, while common trends emerge in the current–voltage relationship. For instance, specimen C1 under the 20/60 μs pulse shows the waveforms for planar length, planar width, and through-thickness clamping in Fig. 2.28 [22–25].

As shown in Fig. 2.28, the current waveforms are nearly identical for all clamping orientations. The voltage responses for planar length and width are also similar, but in both cases lightning current peak lags behind the voltage peak at the specimen ends. In contrast, for the through-thickness configuration, the voltage waveforms is markedly different, exhibiting a pronounced nonlinear conduction behavior. Further analysis of the transient response yields the dynamic voltage-current characteristics of through-thickness conduction. Figure 2.29 plots this curve for specimen C1 under the 20/60 μs lightning pulse.

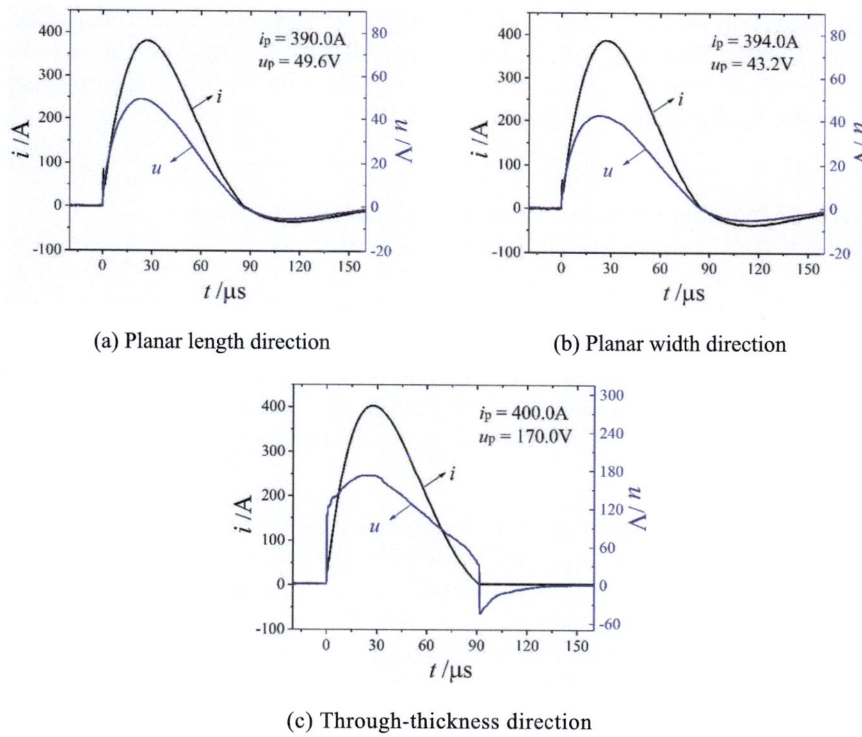

Fig. 2.28 Current-voltage waveforms of specimen C1 under different clamping configurations

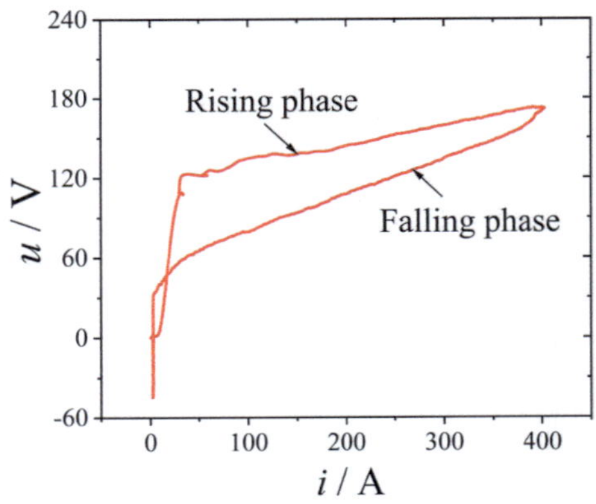

Fig. 2.29 Dynamic voltage-current characteristics of CFRP composites under through-thickness clamping

As shown in Fig. 2.29, the dynamic voltage-current curve displays significant hysteresis between the rising and falling phases of the current. During the falling edge, the specimen's voltage is substantially lower than during the rising stage at the same current level. When the current increases from 0 to ≈30 A, the voltage jumps from 0 to ≈110 V. As the current further increase to 400 A, the voltage rises by only ≈60 V. indicating that the CFRP has a high through-thickness impedance until the voltage exceeds a critical threshold, at which point the equivalent resistance transitions rapidly from a high to a low state.

Effect of Lightning-Current Waveform Parameters on Dynamic Conductivity Characteristics of CFRP Composites

To isolate the influence of waveform shape and peak amplitudes on CFRP conductivity, phase shifts between the lightning-current and response-voltage waveforms were neglected. The equivalent conductivity, σ, was defined as $\sigma = I_{peak}/U_{peak}$ representing the ratio of the current peak to the voltage peak, to quantify the composite's conductive performance. The variation of σ with lightning-current amplitude is plotted in Fig. 2.30.

(i) With increasing lightning-current amplitude, the equivalent conductivity of CFRP composites increases nonlinearly. For instance, as the current rises from ≈ 260 A to 1 kA, the through-thickness conductivity increases from 1.75×10^{-3} to 3.25×10^{-3} S/mm.

(ii) In the planar directions, conduction is governed by direct carbon-fiber pathways, yielding conductivities on the order of 100 S/mm whereas the through-thickness direction, with thicker resin interlayers, exhibits conductivities on the order of 10^{-3} S/mm, a three-order-of-magnitude difference.

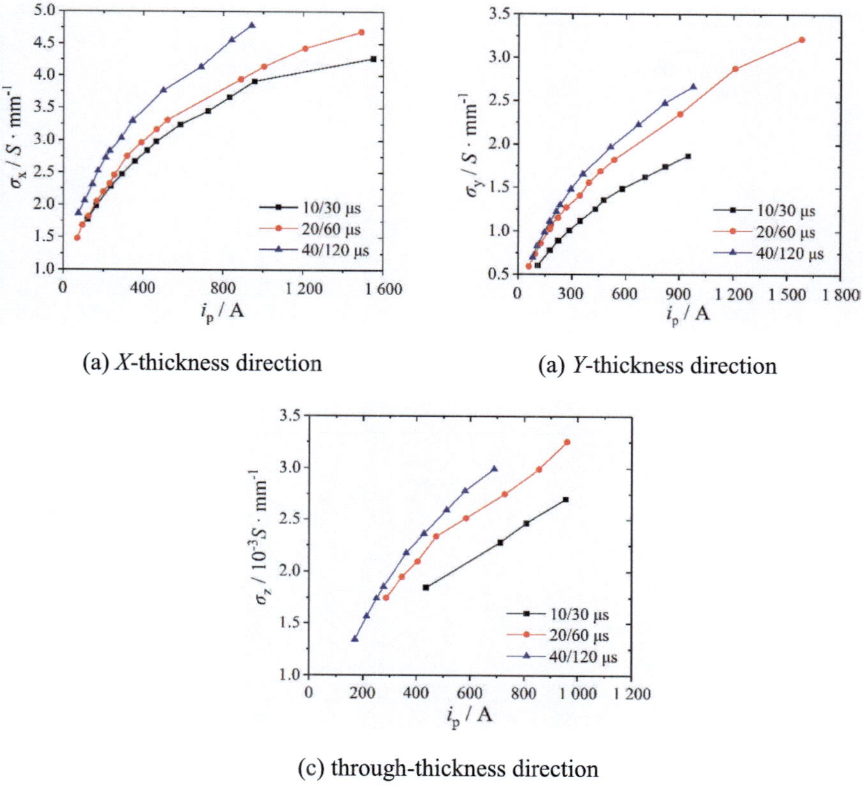

Fig. 2.30 Equivalent conductivity of CFRP composites versus lightning-current amplitude

(iii) Equivalent conductivity depends on the lightning current waveform parameters: slower rise times and longer durations produce higher conductivity. When the wavefront time increases from 10 to 30 μs, the planar-length conductivity rises from 3.25 to 4.10 S/mm. the planar-width conductivity increases from 1.50 to 2.25 S/mm, and the through-thickness conductivity grows from 2.25×10^{-3} to 3×10^{-3} S/mm.

Influence of Layup Structure on Dynamic Conductivity Characteristics of CFRP Composites

CFRP composites exhibit distinct dynamic responses to lightning-current components depending on manufacturing and layup architecture. For specimen C2 with a 0° layup, lengthwise clamping measures the 0° conductivity and widthwise clamping measures the 90° conductivity. Likewise, specimen C3's planar conductivity reflects its 45° ply orientation. Figure 2.31 compares the equivalent conductivity of specimens with different layups under three lightning pulses.

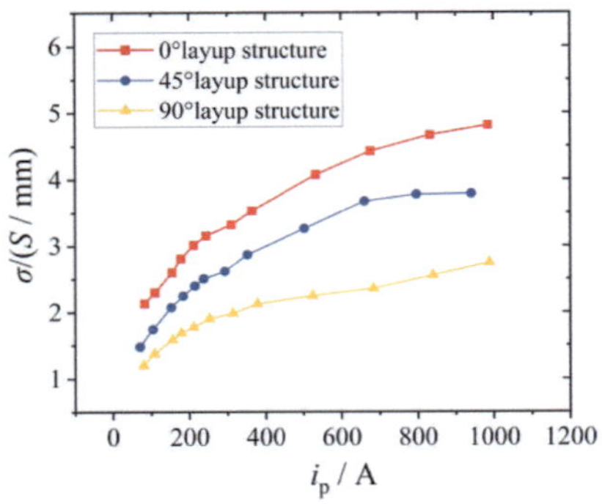

(a) 10/30 μs lightning current component

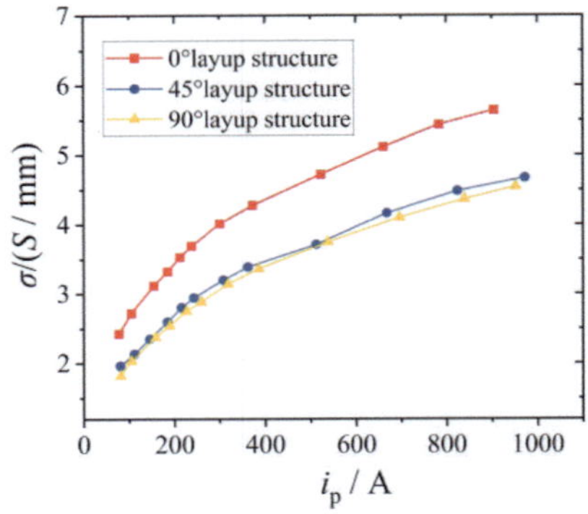

(b) 20/60 μs lightning current component

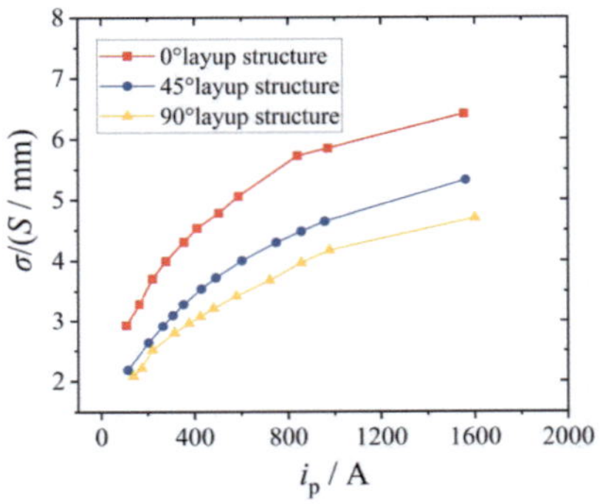

(c) 40/120 μs lightning current component

Fig. 2.31 Comparison of equivalent conductivity for CFRP composites with different layup structures

As shown in Fig. 2.31: conductive is highest along the fiber axial intermediate at 45°, and lowest in the transverse (perpendicular) direction. Notably, unlike the three-order-of-magnitude differences between planar and through-thickness conduction, the in-plane conductivity variations among 0°, 45°, and 90° orientations are within the same order of magnitude [22–25].

2.3.3 *Circuit Analysis of Dynamic Resistance in CFRP Composites*

Conduction Pathways in CFRP Composites Under Lightning Current

As demonstrated by the experimental in Sect. 2.3.2, the through-thickness conductivity of CFRP composites is orders of magnitude lower than their planar conductivity, and in-plane conductivity varies with fiber orientation. This behavior stems from the composites' internal architecture. CFRP parts are produced by stacking carbon-fiber prepreg plies, compressing, and curing in a mold. During this process, curved fibers create random filament contacts (as shown in Fig. 2.32).

Within each ply, carbon fibers exist either in direct contact or separated by thin resin layers. Planar conduction relies on networks of contacting fibers, which differ between plies. For axial conduction (0° layup), current flows along continuous fiber tows, yielding the highest conductivity, which can be estimated from fiber volume fraction. In off-axis directions (e.g., 45° or 90°), direct fiber-to-fiber contacts alone are insufficient to form continuous paths. Under the high voltages of lightning pulses, tunneling or partial dielectric breakdown of the thin resin films between fibers enhances inter filament conduction, effectively bridging adjacent fibers to complete the network. For example, the through-thickness conduction path in a 90° layup is illustrated in Fig. 2.33, with analogous paths existing in 45° laminates.

The lightning-current conduction across different plies of the composite differs primarily in path length, with planar conductivities showing no orders-of-magnitude variation compared to the 0° fiber-aligned layup. In contrast, through-thickness conductivity is substantially lower due to the thick, poorly conductive resin interlayers between plies. In most CFRP thermoelectric-coupling simulations, anisotropic conductivity parameters are directly adopted from DC static conductivity measurements. In these DC tests, the thin inter-fiber and inter-ply resin layers remain in high-resistance states under the low applied voltages, yielding extremely low measured

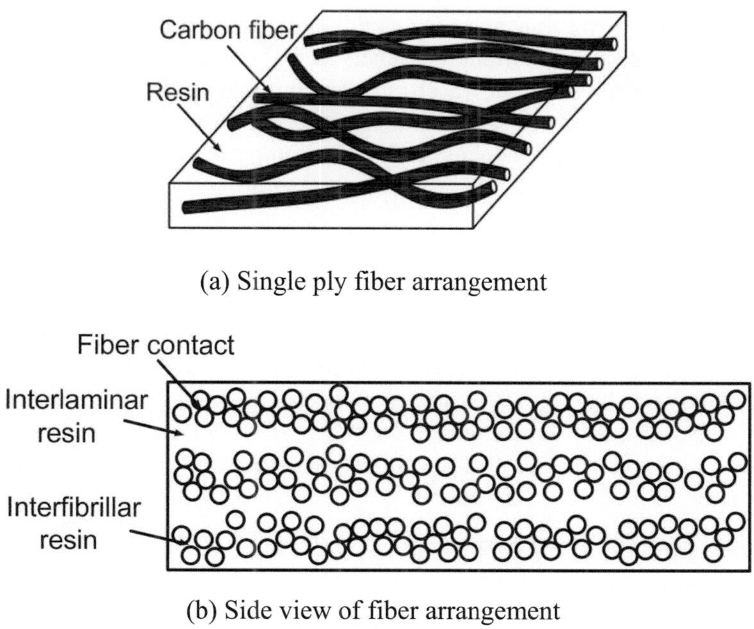

(a) Single ply fiber arrangement

(b) Side view of fiber arrangement

Fig. 2.32 Schematic of internal fiber arrangement in CFRP

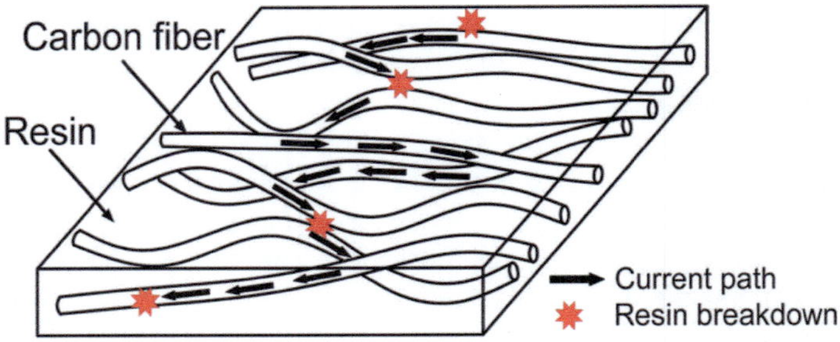

Fig. 2.33 Schematic diagram of current conduction path in 90° layup CFRP composites

DC conductivities for 90° layups and the through-thickness direction. However, Sect. 2.1 experiments show that both 0° and 90° layups exhibit conductivities on the order of 10^2 S/mm, and through-thickness conductivity on the order of 10^{-3} S/mm, which contrasts sharply with DC static values (~10^{-4} S/mm). Consequently, using DC conductivity parameters in thermoelectric-coupling models of CFRP lightning damage can introduce substantial errors in simulated damage extents and depths.

Dynamic Conductivity of CFRP Composites

During the application of a 20/60 μs lightning-current waveform, the through-thickness dynamic conductivity of CFRP composite specimen C1 is plotted in Fig. 2.34. As shown, the dynamic conductivity first follows the rising edge of the 20/60 μs pulse, increasing as current rises, then decreases during the pulse tail, closely mirroring the lightning-current waveform.. The through-thickness conduction path in CFRP can be modeled as a series combination of carbon-fiber filaments and epoxy resin. Therefore, the total conductivity G_l of the composite can be approximated by:

$$G_l = \frac{G_c G_r}{G_c + G_r} \tag{2.6}$$

where G_c is the conductance of the fibers and G_r is the conductance of the resin.

Given that carbon fibers exhibit excellent conductivity (~10^3 S/mm) and the epoxy resin is highly insulating (~10^{-15} S/mm), $G_c >> G_r$. Thus, $G_l \approx G_r$, indicating that the resin layer conductivity dominates the composite's dynamic impedance and its variations with applied voltage. The conduction process in CFRP under a lightning pulse can be divided into three stages:

(i) **Initial high-resistance stage**

The resin, as an insulator, maintains extremely low conductivity under low voltage and thus a high-resistance state. At the onset of the lightning pulse, the resin layer remains resistive, causing most of the discharge voltage to drop across the resin, while the through-thickness current is negligible.

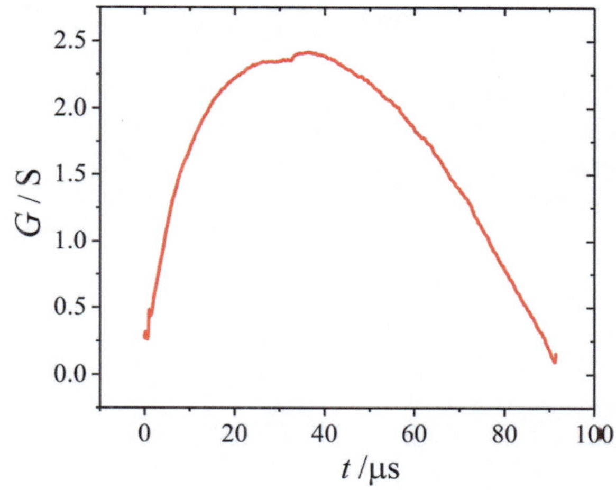

Fig. 2.34 Through-thickness dynamic conductance of specimen C1 under 20/60 μs lightning-current pulse

(ii) **Carrier-generation stage**

Continuous high-voltage application injects charge carriers into the composite. These carriers participate in conduction, progressively enhancing the resin's conductivity and increasing the current. As the pulse duration and carrier density grow, resin conductivity continues to improve, raising the overall conductance.

(iii) **Decay stage**

During the falling edge of the pulse, the electric field weakens below the level needed to sustain the injected carriers. Free electrons and other carriers recombine or escape, causing composite conductivity to decrease. The dependence of resin conductivity on pulse intensity underlies the nonlinear conductivity variation observed with different lightning-current amplitudes.

Inductive Effects in CFRP Composites

Under lightning-current components, the voltage waveform leads the current waveform, and the dynamic *V-I* curves for the rising and falling phases do not coincide. Moreover, slower rise rates and longer durations of the lightning-current components correspond to higher dynamic conductivities in CFRP composites, behavior attributable to their inductive impedance under high-speed excitation.

According to the expression for the current phasor:

$$\dot{I} = \dot{U}/(R_m + jX_m) \tag{2.7}$$

Assume the phase angles of voltage U and current I in Eq. (2.7) are φ_u and φ_i. According to experimental results, voltage across the CFRP composite leads the current. Thus, the phase angle of impedance $Z_m = R_m + jX_m$ can be expressed as:

$$\varphi_3 = \arctan(X_m/R_m) = \varphi_u - \varphi_i > 0 \tag{2.8}$$

This indicates that CFRP composites exhibit inductive impedance characteristics under lightning current. CFRP composites can be modeled as a series network of resistance and inductance, as shown in Fig. 2.35.

The instantaneous voltage across the specimen is:

$$U = i * R + L\left(\frac{di}{dt}\right) \tag{2.9}$$

During the rising and falling phases of the lightning pulse, the voltages are:

$$\begin{cases} U_1(t) = i_1 R + L\left(\frac{di_1}{dt}\right) \\ U_2(t) = i_2 R + L\left(\frac{di_2}{dt}\right) \end{cases} \tag{2.10}$$

where R and L represent the equivalent resistance and equivalent inductance of the CFRP composite respectively. When the current values are identical during rising and falling phases, the voltage difference between the two phases is:

$$U_1(t) - U_2(t) = (i_1 - i_2)R + L\left(\frac{di_1}{dt} - \frac{di_2}{dt}\right) \tag{2.11}$$

Considering that $\mathrm{d}i_1/\mathrm{d}t > 0$ during the rising and $\mathrm{d}i_2/\mathrm{d}t < 0$ during the falling phases of the current, there is $U_1 – U_2 > 0$. This explains why the voltage during the rising edge exceeds that during the falling edge, producing non-overlapping *V-I* curves. For lightning pulses with different waveform parameters, since voltage phase leads current phase, the current is still in the rising stage when the voltage across the CFRP composite reaches its peak. For all three lightning current components (10/30, 20/60, and 40/120 μs), $\mathrm{d}i/\mathrm{d}t > 0$. At identical current amplitudes, slower lightning current variation rates and longer durations result in smaller $\mathrm{d}i/\mathrm{d}t$. According to Eq. (2.9), lower $\mathrm{d}i/\mathrm{d}t$ leads to reduced voltage across the specimen and higher equivalent conductivity. Additionally, longer pulse durations inject more carriers into the resin, further boosting conductivity. Quantitatively,

$$\sigma_{40/120\,\mu\mathrm{s}} > \sigma_{20/60\,\mu\mathrm{s}} > \sigma_{10/30\,\mu\mathrm{s}} \tag{2.12}$$

Therefore, under a 40/120 μs pulse, CFRP exhibits the highest conductivity, followed by 20/60 μs, with the lowest under 10/30 μs components.

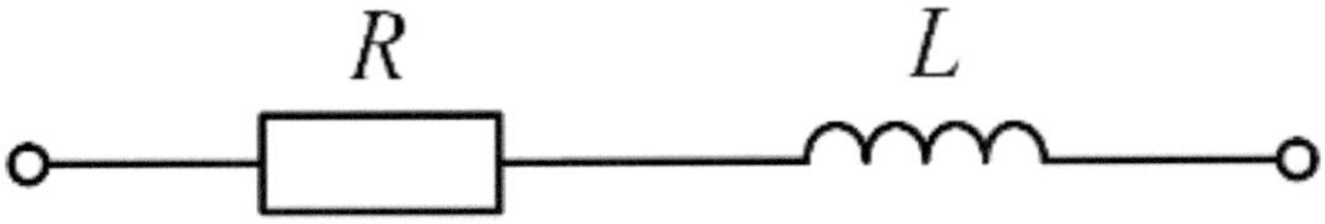

Fig. 2.35 Equivalent *R-L* circuit of CFRP composites

Source Analysis of Inductive Effects in CFRP Composites

Carbon fiber composites are composed of carbon fiber tows and polymer matrices such as epoxy resin. While planar conduction may traverse thin resin layers, the dominant current paths follow by interconnected carbon fiber. Therefore, inductive effects observed in CFRP originate primarily from the intrinsic inductance of the fibers. Each carbon fiber filament can be approximated as a straight wires, inherently possessing inductance. The inductance L of a straight conductor of length l and radius r is given by:

$$L = \frac{\mu_0 l}{2\pi}\left(\ln\frac{2l}{r} - 0.75\right) \tag{2.13}$$

where l is length of carbon fiber filament; r is fiber radius (carbon fiber diameter $\approx$ 7 μm); μ_0 is permeability of free space ($4\pi \times 10^{-7}$ H/m). From Eq. (2.13), the inductance per unit length of a single carbon fiber is approximately 2.5 μH/m. In a 100 × 100 × 3.6 mm CFRP panel, for planar conduction along the 90° layup, if we approximate the current path as 2 × 100 mm, then the inductance of one filament along that path is about 2.5 × 2 × 0.1 = 0.5 μH.

In practice, the planar inductance of CFRP arises from the parallel combination of many fiber filaments. Assuming roughly 50% of fibers participate in conduction, the number of parallel filaments in a composite with 60% fiber volume fraction (f_v) can be estimated by:

$$\frac{w \times h \times f_c}{d_c \times d_c} \times 50\% = \frac{100\ \text{mm} \times 3.6\ \text{mm} \times 60\%}{7\ \mu\text{m} \times 7\ \mu\text{m}} \times 50\% \approx 2.2 \times 10^6 \tag{2.14}$$

In the above equation, w and h are the width and thickness of the CFRP specimen, and the cross-sectional area per fiber is approximated as a square of side $2r$. Although parallelizing multiple carbon fibers reduces the net inductance, the reduction is not directly proportional to the number of fibers due to mutual inductance. From our engineering experience with parallel wire inductance, adding more conductors typically reduces the total inductance to about 60% of a single filament's inductance. Therefore, the equivalent planar inductance of the carbon-fiber network in CFRP is estimated at 0.5 × 60% ≈ 0.3 μH. Under a 20/60 μs lightning-current pulse of 384 A, the inductive voltage induced by this network is

$$\Delta U = L\frac{di}{dt} = 0.3\ \mu\text{H} \times \frac{384\ \text{A}}{20\ \mu\text{s}} = 11.5\ \text{V} \tag{2.15}$$

The measured voltage across the CFRP specimen is ≈31 V, so the inductive drop accounts for ≈12.4% of the total voltage. This demonstrates that the inductance of the carbon-fiber network significantly influences CFRP's dynamic conductivity under lightning excitation. As shown in Sect. 2.3.2, the resin-layer conductivity depends primarily on the voltage across the interlayers, producing a voltage-lead–current behavior that appears macroscopically as an inductive effect. Because the resin layers between plies are much thicker than in the planar direction, these inductive phenomena are even more pronounced in the through-thickness conduction mode.

2.4 Nonlinear Conductivity Mechanism of CFRP Composites

2.4.1 Microscopic Conductive Structure of CFRP Composites

Carbon fibers are manufactured from polyacrylonitrile (PAN) precursors via solid-phase pyrolysis at temperatures >1000 °C in inert atmospheres. Their excellent conductivity (carbon content > 90%) derives from graphitic microcrystalline domains. For example, Toray T300 and T700 fibers exhibit conductivities of 5 $\times$ 10^4 S/m. In contrast, epoxy resins are electrical insulators with conductivities on the order of 10^{-15} S/m, independent of applied field strength; any small resin-borne carriers arise from thermal excitation, quantum tunneling, or impact ionization. CFRP composites consist of conductive carbon-fiber tows embedded in an insulating polymer matrix. Figure 2.36 shows a micrograph of a 55% fiber-volume CFRP cross section. Within each ply, some fibers make direct contact, while others are separated by thin resin films; thicker resin-rich layers appear at ply interfaces.

Most existing models apply linear percolation theory to CFRP [19], asserting that above the percolation threshold, conductivity arises from Ohmic contacts between fibers, with resistance scaling linearly with filler volume fraction. The wide variation in measured longitudinal (fiber-direction) conductivity, σ_0, versus through-thickness conductivity supports these contact-based models [20]. However, such models assume that all charge carriers travel solely along the shortest, continuous

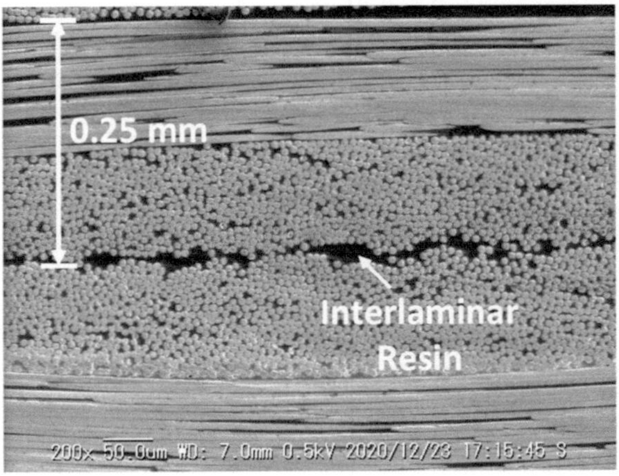

Fig. 2.36 Cross-sectional micrograph of 55% fiber-volume CFRP composite

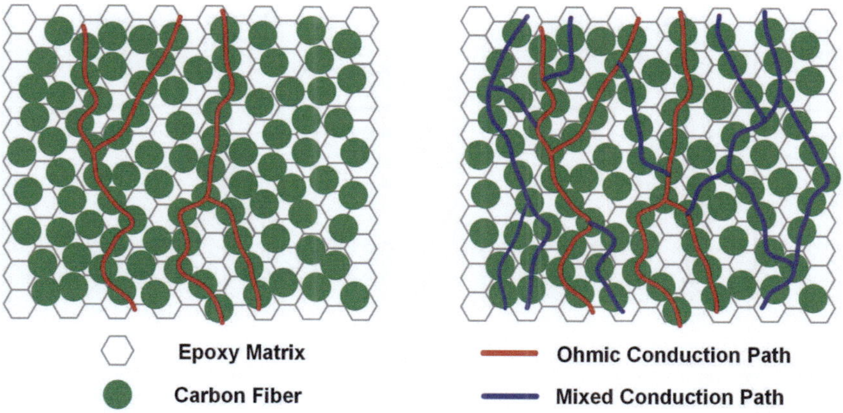

(a) Low-field (DC) conduction via Ohmic fiber-fiber contacts

(b) High-field conduction with additional fiber–resin interface paths

Fig. 2.37 Schematic of microscopic conduction paths in CFRP materials

fiber-contact paths, never detouring through resin. While valid under low-intensity currents, this assumption fails under high-intensity pulsed currents, where tunneling and resin breakdown become significant, leading to nonlinear conduction behavior.

In practice, CFRP is not a simple two-phase system of conductive fibers dispersed in an insulating matrix, but rather a multiphase composite containing extensive fiber–resin interfacial networks. Under small applied currents, charge transport occurs primarily through a percolation network of direct fiber–fiber contacts. At low electric field strengths, electrons cannot surmount the fiber–resin interface barrier, so "fiber–resin–fiber" paths remain inactive. Figure 2.37 illustrates the microscopic conduction mechanisms in CFRP composites.

Under low-amplitude currents, conduction in CFRP is confined to Ohmic fiber–fiber contacts (Fig. 2.37a), making the composite's conductivity field-independent. In contrast, lightning impulses with amplitudes of 50–1500 A far exceed the mA-range DC currents, overwhelming the charge-carrying capacity of direct contacts. Under these high-field conditions, the potential barriers at fiber-resin interfaces become modulated, enabling non-Ohmic conduction (e.g., quantum tunneling or dielectric breakdown) across the resin layers. Therefore, during high-intensity pulses, conduction involves both fiber–fiber Ohmic paths and fiber–resin interface paths, as depicted in Fig. 2.37b.

Barrier Structure of "Carbon Fiber-Resin-Carbon Fiber" Unit

Because carbon fibers and epoxy resin have different energy-band structures, charge transfer at their interfaces leads to space-charge barrier formation [26]. Figure 2.38 presents a rectangular-barrier model for this fiber–resin–fiber junction.

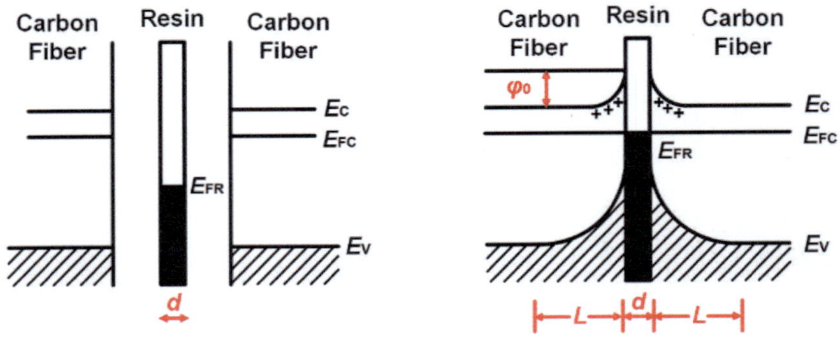

(a) Separate energy bands before contact (b) Band alignment after contact; carbon fiber

Fig. 2.38 Energy-level diagrams of the carbon fiber-resin-carbon fiber contact interface, E_{FC}: Fermi level of fiber; E_C: conduction band edge; E_V: valence band edge; E_{FR}: Fermi level of resin; L: depletion layer width; d: resin layer thickness; φ_0: Zero-bias barrier height ($U = 0$)

Upon contact, free electrons in the carbon fiber (due to its high Fermi level E_{FC}) migrate into the resin interface, becoming trapped in surface states and creating negative interface charge. This leaves positive donor ions at the fiber surface. Additional electrons from the fiber's graphitic domains then diffuse to the surface under the electrostatic attraction of these donor charges. These electrons can thermally surmount the barrier into the resin until the Fermi levels align $E_{FC} = E_{FR}$, at which point charge injection and trapping reach equilibrium. This process establishes a depletion region of width L in the fiber, producing band bending near the interface, a Schottky-type barrier. At low fields, this barrier creates a high-resistance layer. As the field increases, the barrier height and width modulate, driving the nonlinear *V-I* behavior observed in CFRP. Since barriers form on both fiber–resin interfaces, the system behaves like a back-to-back double Schottky barrier [26].

2.4.2 Nonlinear Conductivity Mechanism of "Carbon Fiber-Resin-Carbon Fiber" Unit

The height and width of the double Schottky barriers at fiber-resin-fiber interfaces change with the applied electric field, giving rise to CFRP's nonlinear conduction.

2.4.3 Low Electric-Field Region

As illustrated in Fig. 2.39a, when a voltage U is applied across the paired barriers, their band profiles become distorted. Conduction in this low-field regime follows a thermionic-emission mechanism: electrons absorb thermal energy (from both the electric field and ambient temperature), and when this energy exceeds the effective barrier height, they emit over the barrier into the resin.

In this regime, electrons from each fiber region surmount the barrier via thermionic emission. The thermal-activation energy on each side equals that barrier's height, and the current densities J_{L} and J_{R} injected into the resin from the left and right fibers are given by:

$$\begin{cases} J_L = J_0 \exp\left(-\frac{\varphi_1+\eta}{kT}\right) \\ J_R = J_0 \exp\left(-\frac{\varphi_2+\eta}{kT}\right) \end{cases} \tag{2.16}$$

where J_0 is a pre-exponential constant; $\varphi_1 = \varphi_0 - e_{\mathrm{U1}}$ and $\varphi_2 = \varphi_0 + e_{\mathrm{U2}}$ are the effective barrier heights on the left and right; k is Boltzmann constant. After electrons are captured in interface states, they can be thermally re-emitted into the right carbon fiber, yielding a back-emission current density J_{EM}:

$$J_{EM} = J_{EM0}\exp\left(-\frac{\varphi_1 + \xi + eU_S}{kT}\right) \tag{2.17}$$

where J_{EM0} is a constant. The net current density across the unit is proportional to the difference between the injected and reemitted electrons. Since $J_{\mathrm{R}} << J_{\mathrm{L}}$, the re-emission into the left fiber is negligible, so the total current density simplifies to [27]:

$$J = J_0\exp\left(-\frac{\varphi_0 - \beta F^{1/2}}{kT}\right) \tag{2.18}$$

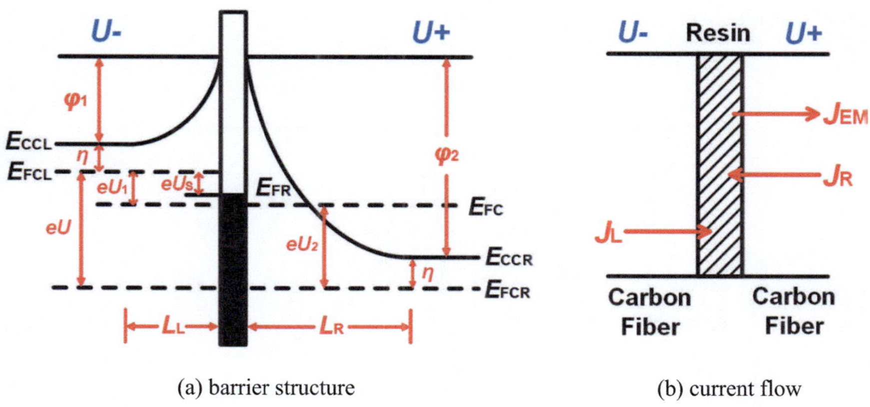

(a) barrier structure (b) current flow

Fig. 2.39 Barrier deformation under applied voltage U; E_{CCL} and E_{CCR}: conduction-band edges on left/right fiber, E_{FCL} and E_{FCR}: Fermi levels of left/right fiber, E_{FC} and E_{FR}: equilibrium Fermi levels of fiber and resin; L_{L} and L_{R}: depletion widths on each side; U_{L} and U_{R}: voltage drops across the left/right where $\beta = \sqrt{e^3/4\pi\varepsilon}$; ε is the dielectric constant; F: is the local electric-field strength. This expression shows that higher temperatures and stronger fields lower the apparent barrier, facilitating thermally assisted tunneling and increasing the current. Under ambient conditions and low fields, the carrier injection across the barrier is negligible, producing a saturated, high-resistance response akin to an insulator. Consequently, at low-amplitude DC excitation, conduction is dominated by Ohmic fiber-fiber contacts, explaining the linear conductivity observed in CFRP under small applied voltages.

Medium Electric-Field Region

As the applied voltage increases beyond the low-field regime, the U-I behavior of CFRP shifts into the medium-field domain, where quantum tunneling across the fiber–resin interface dominates conduction. Electrons with energy below the barrier height can penetrate the barrier in a wave-like fashion, generating a tunneling current. Because $U_1 << U_2$, the entire applied voltage drops across the reverse-biased barrier, whose profile is sketched in Fig. 2.40.

Using the WKB approximation, the transmission probability $D(E_x)$ for an electron traversing a potential barrier of height $V(x)$ into the carbon fiber-resin interface can be expressed as [28]:

$$D(E_x) = \exp\left\{-\frac{4\pi}{h}\int_0^d [2m(V(x) - E_x)]^{1/2} dx\right\} \tag{2.19}$$

where $E_x = mv_x^2/2$ is the Kinetic energy of incident electrons in the x-direction; m, e are electron mass and charge respectively; d is potential barrier width. The net number of electrons N traversing the barrier from left to right can be written as:

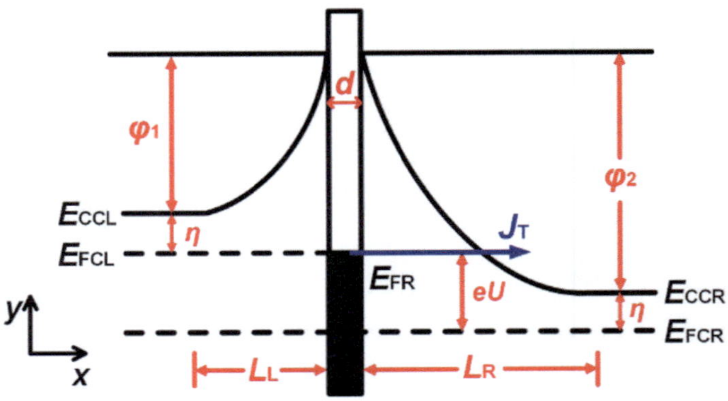

Fig. 2.40 Schematic of a fully depleted, reverse-biased Schottky barrier, I_T: tunneling current

$$N = \int_0^{E_m} D(E_x)dE_x \times \left\{ \frac{4\pi m^2}{h^3} \int_0^{\infty} f(E) - f(E + eU)dE_r \right\} \tag{2.20}$$

In the above equation, $f(E)$ is the Fermi-Dirac distribution function. Based on the calculation of electron tunneling probability by Simmons et al. [29], the tunneling current density J_T expression can be obtained:

$$J_T = \frac{4\pi me}{h^3} \left\{ eU \int_0^{\eta - eU} \exp[-A(\eta + \varphi_0 - E_x)^{1/2}]dE_x + \int_{\eta - eU}^{\eta} (\eta - E_x)exp[-A(\eta + \varphi_0 - E_x)^{1/2}]dE_x \right\} \tag{2.21}$$

where $A = (4\pi\gamma\Delta d/h)(2m)^{1/2}$; γ is the correction coefficient; Δd is the effective potential barrier width; φ_0 is the rectangular barrier height without applied electric field. Considering the differences between the generalized potential barrier and the ideal rectangular potential barrier, a correction term $\widetilde{\varphi} = \frac{1}{\Delta d} \int \varphi(x)dx$ representing the average barrier height was introduced in Reference [29]. After performing simplified integration on Eq. (2.21) and neglecting minor terms, the tunneling current density through the generalized potential barrier can be expressed as:

$$J_T = J_0\{\widetilde{\varphi} \exp(-A\widetilde{\varphi}^{1/2}) - (\widetilde{\varphi} + eU)\exp[-A(\widetilde{\varphi} + eU)^{1/2}]\} \tag{2.22}$$

where $J_0 = \frac{e}{2\pi h\gamma^2(\Delta d)^2}$, the tunneling current J_T can be viewed as consisting of two components: the current $J_L = J_0\widetilde{\varphi}\exp(-A\widetilde{\varphi}_0^{1/2})$ flowing from the left to the right of the barrier and the current $J_R = J_0(\widetilde{\varphi}_0 + eU)\exp[-A(\widetilde{\varphi}_0 + eU)^{1/2}]$ flowing from the right to the left. When no external voltage is applied ($U = 0$), the leftward and rightward currents generated by tunneling reach dynamic equilibrium. Under the interface barrier deformation shown in Fig. 2.40, Eq. (2.22) can be written as:

$$J_T = \frac{2e^3F^2}{8\pi h\varphi_0\gamma^2} \left\{ \exp\left[-\frac{4\pi\gamma}{eF} m^{1/2}\varphi_0^{3/2} \right] - \left(1 + \frac{2eU}{\varphi_0}\right) \exp\left[-\frac{4\pi\gamma}{eF} m^{1/2}\varphi_0^{3/2} \left(1 + \frac{2eU}{\varphi_0}\right)^{1/2} \right] \right\} \tag{2.23}$$

where $F = U/d$ is the electric field strength in the resin dielectric layer. When the applied voltage is sufficiently high such that the conduction band energy level of the right carbon fiber falls below the Fermi level of the left carbon fiber ($U > \eta/e$), electrons struggle to tunnel from the right electrode to the left electrode due to the absence of available empty energy levels in the left carbon fiber. For electrons moving from left to right, the abundant empty energy levels in the right carbon fiber facilitate tunneling. Under these conditions, the second term in Eq. (2.23) can be neglected, thereby obtaining:

$$J_T = \frac{2e^3F^2}{8\pi h\varphi_0\gamma^2}\exp\left[-\frac{4\pi\gamma}{eF}m^{1/2}\varphi_0^{3/2}\right] \tag{2.24}$$

Therefore, within the medium electric field conduction region, the tunneling current density is strongly influenced by electric field strength F, increasing rapidly with minor voltage variations and resulting in a high nonlinear coefficient for CFRP materials.

High Electric-Field Region

As the applied voltage continues to increase, the "fiber–resin–fiber" junction enters the high-field regime. In this region, the depletion layers collapse, rendering the interface barrier conductive. Consequently, the *U-I* behavior of CFRP no longer reflects Schottky-barrier characteristics. Instead, conduction is governed by the intrinsic conductivity of the carbon fibers and the resin matrix. The current density in this regime can be approximated by Ohm's law:

$$J = \frac{F}{\rho_R} \tag{2.25}$$

It follows that the material will exhibit near-linear resistance behavior. Once the electric field exceeds the critical threshold, the probability of electrons traversing the interface barrier approaches unity, and the fiber–resin interface behaves like an Ohmic contact. Wang et al. [18] highlight that Joule heating from high current densities raises the local temperature, slightly enhancing conductivity. Overall, the high-field conduction region acts as a recovery zone for CFRP's nonlinear resistance, where the nonlinear exponent decreases relative to the medium-field region, yielding weaker nonlinear behavior.

2.4.4 Equivalent Circuit Network of CFRP Composites Under Lightning-Current Pulses

Amplitude-Frequency Characteristics of CFRP Conductive Behavior

The dynamic equivalent circuit for CFRP composites is derived from experiments under lightning-current pulses. Prior studies [24, 25] show that CFRP's conductivity in both planar and through-thickness directions increases with rising pulse amplitude. Moreover, in a lightning environment, CFRP cannot be modeled as a pure resistor:

besides the inductive voltage-lead effect, its conductive ability actually decreases as the pulse rise rate (di/dt) increases. Thus, CFRP's conductance depends on both the current amplitude i and its rate of change di/dt, and its voltage–current response can be expressed as:

$$U = f\left(i, \frac{di}{dt}\right) \tag{2.26}$$

Equivalent Circuit Model

Based on the microscopic structure of fibers, resin, and fiber-resin-fiber interfaces, CFRP's semiconductive, nonlinear behavior stems from the I-V characteristics of resin inclusions and interface barriers. We thus first develop the microcircuit for the fiber-resin-fiber unit and then assemble these into a macroscopic, frequency- and amplitude-dependent equivalent circuit for CFRP under impulse excitation. The resulting model is shown in Fig. 2.41 [30].

The equivalent resistance of the carbon-fiber segments in a "fiber–resin–fiber" unit is defined as R_C. To account for the inherent inductive behavior of the fibers, R_C is placed in series with an equivalent inductance L_C forming the R-L branch that models the carbon-fiber network's dynamic response under impulse excitation. Within this same unit, the resin layer sandwiched between fibers contributes its own conduction mechanisms. First, space-charge-limited conduction occurs through the resin's bulk, which we model as a linear resistor R_{PL} proportional to the material's intrinsic conductivity. Second, interface-limited conduction arises at the fiber-resin interfaces, where field-dependent Schottky barriers modulate charge injection. These interface effects are captured by a nonlinear resistor R_{PN} whose resistance varies with the local electric field and carrier density. The series combination of R_{PL} and R_{PN} thus represents the resin bulk resistance plus the non-Ohmic interface resistance of the "fiber–resin–fiber" unit.

Considering the frequency-dependent conductive behavior exhibited by CFRP materials, an inductive component L_P is introduced and connected in series with the

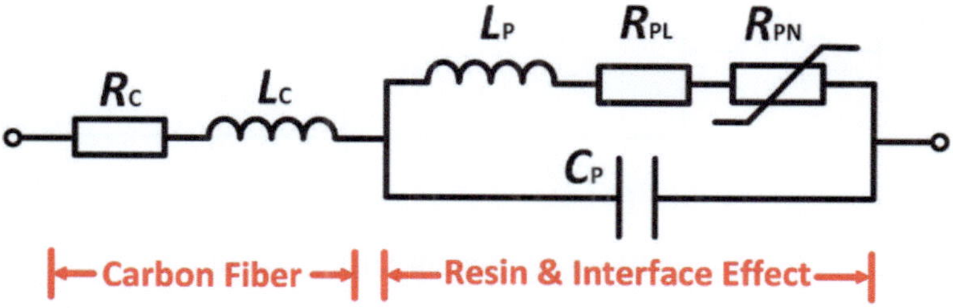

Fig. 2.41 Equivalent circuit model of the "carbon fiber-resin-carbon fiber" unit dependent on current intensity and frequency [29]

linear resistor R_{PL} and nonlinear resistor R_{PN}. The inductive reactance parameter $Z_L = j\omega L$, associated with the angular frequency of the lightning current, can regulate the electrical response of the circuit under different lightning current conditions. Additionally, the sandwich structure formed by highly conductive carbon fibers and high-permittivity resin (relative permittivity 3–5) serves as the origin of the intrinsic capacitive properties of CFRP materials within the "carbon fiber-resin-carbon fiber" microstructural unit, represented by a parallel equivalent capacitance C_P in the circuit model.

The conductive circuit of CFRP composites can be regarded as a three-dimensional network structure composed of numerous "carbon fiber-resin-carbon fiber" unit microcircuits connected in series and parallel. Therefore, the micro-unit circuit shown in Fig. 2.41 can be extended to the equivalent circuit of CFRP composites (Fig. 2.42a). The series branch of inductance L_0 and resistor R_0 represents the carbon fiber conductive network interconnected through Ohmic contacts; the series-parallel circuit of inductance L_L, linear resistor R_L, nonlinear resistor R_N, and capacitance C_1 models the non-Ohmic conductive network formed by "carbon fiber-resin-carbon fiber" units in CFRP composites [30].

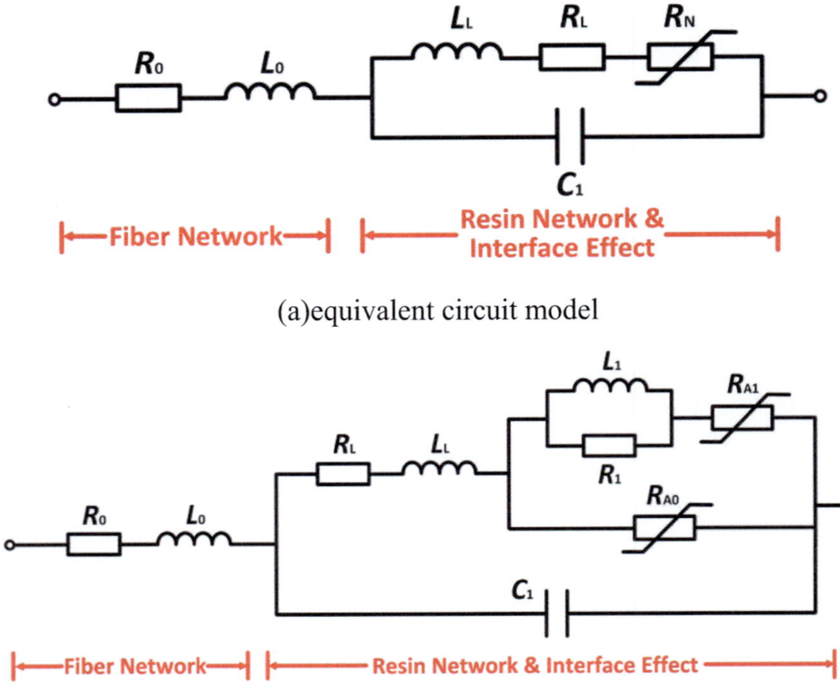

Fig. 2.42 Conduction models of CFRP composites under lightning current pulses

The circuit model in Fig. 2.42a is a universal equivalent circuit model capable of reflecting the microscopic conductive structure of CFRP materials. However, since the electrical response of CFRP materials depends on multiple factors such as current amplitude and frequency, and the circuit components, particularly the inductance L_L and nonlinear resistor R_N parameters, dynamically vary during different phases of lightning current pulse action, it is challenging to determine fixed circuit component parameters through numerical calculations or circuit simulations. o enhance the practicality of the circuit model, certain improvements have been made to the circuit model in Fig. 2.42a, as shown in Fig. 2.42b.

In the improved circuit, R_{A0} and R_{A1} are nonlinear resistors with distinct resistance values and volt-ampere characteristics. Under the same current intensity, the resistance of R_{A0} is greater than that of R_{A1}. The added parallel network of inductor L_1 and resistor R_1 separates the nonlinear resistors R_{A0} and R_{A1}, acting as a low-pass filter to regulate the material's volt-ampere characteristics under pulse currents with varying rise rates. Specifically:

(i) Under low-frequency lightning currents, the inductive reactance ωL_1 of inductor L_1 is small. Compared to the nonlinear resistors R_{A0} and R_{A1}, the total impedance of the parallel branch $R_1//L_1$ is negligible, causing minimal impact on circuit current division. At this stage, the nonlinear portion of the circuit approximates the parallel combination of R_{A0} and R_{A1}.

(ii) As the lightning current rise rate increases, the pulse frequency rises, leading to a corresponding increase in ωL_1. This amplifies the total impedance of the $R_1//L_1$ branch, forcing more lightning current to divert into the R_{A0} branch. Since $R_{A0} > R_{A1}$, higher voltages develop across the circuit under high-frequency lightning currents, consistent with the experimentally observed phenomenon that shorter lightning current wavefronts correspond to lower equivalent conductivity of the material. By appropriately selecting parameters for the nonlinear resistors R_{A0}, R_{A1}, and the parallel branch $R_1//L_1$, the equivalent circuit in Fig. 2.42b can effectively simulate the electrical response of CFRP materials under lightning current conditions.

Circuit Simulation Analysis and Results

Based on the established pulse equivalent circuit model for CFRP materials, the SIMULINK running in the Matlab environment was used to develop the circuit simulation model shown in Fig. 2.43 [30]. This model was employed to analyze the dynamic conductive characteristics of CFRP under lightning currents and validate the effectiveness of the proposed dynamic equivalent circuit model of CFRP. Linear circuit components in the simulation model were from the PowerSystem Library of the software, while nonlinear resistors R_{A0} and R_{A1} were configured through custom resistor modules. Post-processing of loop current and voltage signals was completed by adding measurement and data processing units.

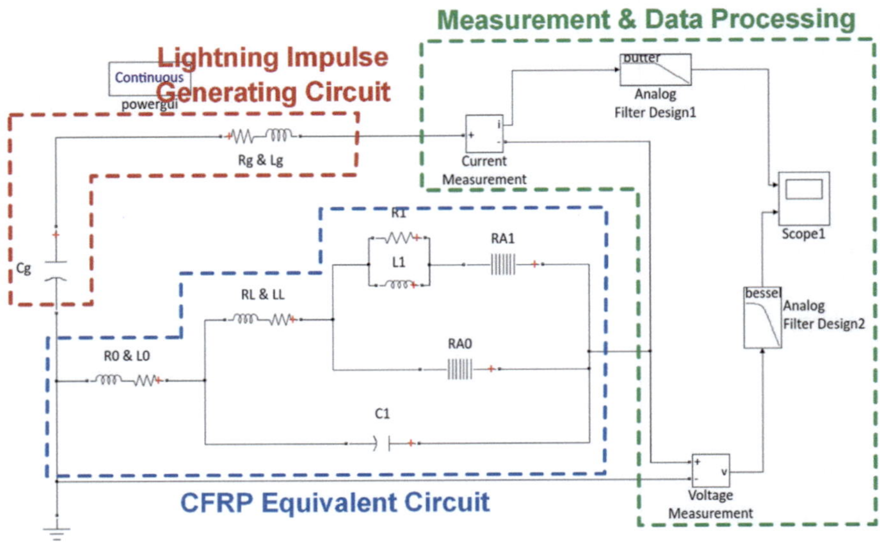

Fig. 2.43 Circuit simulation model for dynamic electrical conductivity characteristics of CFRPs

Through analysis of the internal conductive structure of CFRP composites and extensive experimental simulations, a set of circuit component parameters as shown in Tables 2.5 and 2.6 were selected.

Under the condition of a current peak $I_{\text{peak}} = 346$ A, the established circuit simulation analysis model can be used to calculate the pulsed current and response voltage waveforms at a specified intensity. A comparison between simulation and experimental results is shown in Fig. 2.44. Based on simulation calculations of the single lightning current application process and CFRP conductive parameters within a certain intensity range, the effectiveness and accuracy of the simulation circuit model were validated. This indicates that the equivalent circuit model (Fig. 2.44) can satisfactorily predict the dynamic conduction behavior of CFRP materials under lightning current Component A.

Table 2.5 Component parameters for through-thickness conductive equivalent circuit model

<table>
<tr><td>R_0/Ω</td><td>$L_0/\mu\text{H}$</td><td>C_1/nF</td><td>R_1/Ω</td><td>$L_1/\mu\text{H}$</td><td>R_L/Ω</td><td>$L_L/\mu\text{H}$</td></tr>
<tr><td>0.02</td><td>0.1</td><td>0.3</td><td>0.1</td><td>10.0</td><td>0.14</td><td>1.0</td></tr>
<tr><td colspan="5">R_{A0}</td><td colspan="2">R_{A0}</td></tr>
<tr><td colspan="5">$U = k_i I^{\alpha_i}$
$\begin{cases} k_1 = 61.9,\ \alpha_1 = 0.125\ I \le 600 \\ k_2 = 95.3,\ \alpha_2 = 0.098\ 600 < I \le 1500 \\ k_3 = 7.31,\ \alpha_3 = 0.409\ I > 1500 \end{cases}$</td><td colspan="2">$U = kI^{\alpha}$
$k = 46.5, \alpha = 0.146$</td></tr>
</table>

Table 2.6 Component parameters in the planar electrical conductivity equivalent circuit model

R_0/Ω	$L_0/\mu H$	C_1/nF	R_1/Ω	$L_1/\mu H$	R_L/Ω	$L_L/\mu H$
0.01	0.1	0.01	0.05	1.0	0.01	0.25
R_{A0}					R_{A0}	
$U = k_i I^{\alpha_i}$ $\begin{cases} k_1 = 36.1,\ \alpha_1 = 0.163 & I \le 600 \\ k_2 = 52.8,\ \alpha_2 = 0.103 & 600 < I \le 1500 \\ k_3 = 5.18,\ \alpha_3 = 0.421 & I > 1500 \end{cases}$					$U = kI^{\alpha}$ $k = 45.3, \alpha = 0.442$	

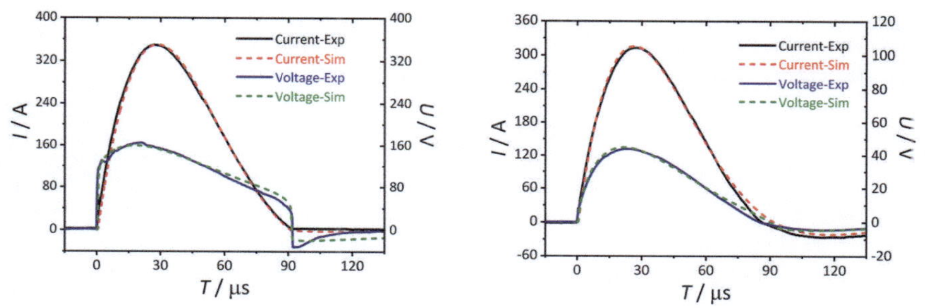

(a) through-thickness conductivity (b) planar conductivity

Fig. 2.44 Comparison of simulation and experimental results for lightning current and response voltage

By comparing the simulation and experimental results, it can be seen that the pulsed current and response voltage waveforms obtained from the circuit simulation model closely match the experimental measurements. In the circuit simulation model, adjusting the initial voltage of the energy storage capacitor C_g allows simulation of material conduction under different current intensities. By comparing the simulation and experimental results of peak equivalent resistance varying with current intensity (Fig. 2.45), the simulation results were found to be highly consistent with measured values, with planar and through-thickness simulation prediction deviations not exceeding 3%.

Based on this, a three-dimensional equivalent circuit model of CFRP materials as shown in Fig. 2.46 can be constructed [30].

In the figure, Z_{PX} and Z_{PY} represent the equivalent impedances of CFRP in planar X and Y directions respectively, while Z_T denotes the through-thickness equivalent impedance.

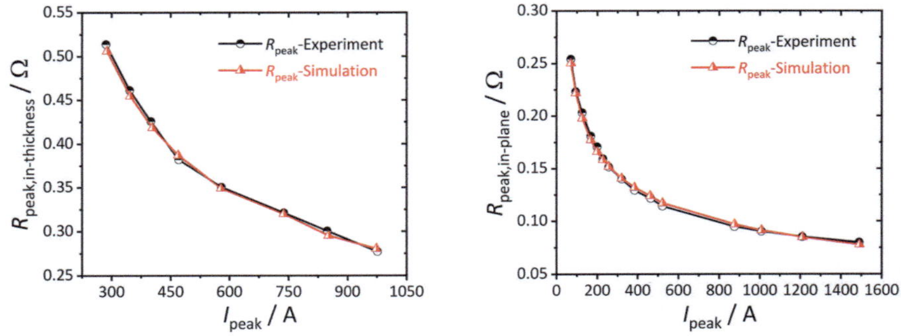

(a) through-thickness conductivity (b) planar conductivity

Fig. 2.45 Comparison of simulation and experimental peak equivalent resistance

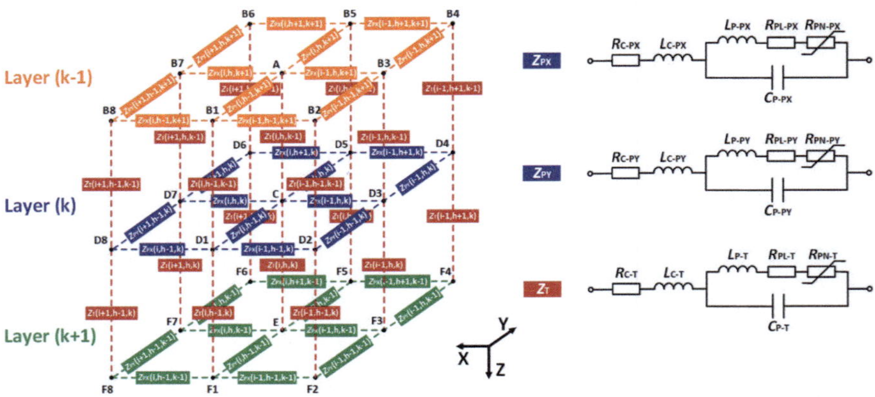

Fig. 2.46 Microscopic three-dimensional conductive circuit model of CFRP materials

Mathematical Characterization of CFRP Conductive Behavior

When a high-intensity lightning current traverses CFRP, the Joule heating raises the local temperature. Once this exceeds a critical threshold, the resin matrix undergoes pyrolysis, leading to damage modes such as delamination, cracking, and debonding. Finite-element analysis (FEA) is a powerful tool for quantitatively predicting lightning-induced damage in CFRP. In developing thermo-electrical coupled models, the conductivity parameters are vital for simulation fidelity. However, CFRP's nonlinear, dynamic conduction under pulsed currents complicates the selection of these parameters. Accordingly, extracting effective resistance or conductance values that incorporate both electrical and thermal effects from dynamic-response data is essential for building accurate and robust thermal-electrical coupled lightning-damage simulations for CFRP.

Using the dynamic equivalent circuit model, the electrical response and conductive characteristics of CFRP materials under specific lightning strike conditions can be predicted. Based on simulation results of lightning current and response voltage, and analysis of current Joule heating effects, the effective resistance R_{eff} of CFRP materials is defined by Eq. (2.27):

$$\sum_{i=1}^{n} I_i U_i \Delta t_i = R_{eff} \sum_{i=1}^{n} I_i^2 \Delta t_i \tag{2.27}$$

where t_{i}: is the sampling time dynamically selected by SIMULINK software within an appropriate range during simulation; I_{i} is pulsed current at time t_{i}; U_{i} is response voltage at time t_{i}; Δt_{i} is Time interval between adjacent sampling points, $\Delta t_{\text{i}} = t_{\text{i}} - t_{\text{i-1}}$. Through simulation analysis and subsequent calculations, the effective resistance R_{eff} reflecting Joule heating effects in CFRP materials can be obtained for different magnitudes of lightning current component A. The variation patterns of through-thickness and planar effective resistances with lightning current amplitude are shown in Fig. 2.47.

As the lightning current amplitude increases, the effective resistance $R_{\text{eff-T}}$ of CFRP materials gradually decreases for the through-thickness conductive state. When the intensity of lightning current component A exceeds 20 kA, $R_{\text{eff-T}}$ values stabilize. Numerical fitting reveals that the variation of through-thickness effective resistance $R_{\text{eff-T}}$ with the logarithm of lightning current amplitude I_{peak} follows an exponential function:

$$R_{eff\text{-}T} = 57.34 \times e^{-\ln I_{peak}/1.164} + 0.1398 \tag{2.28}$$

According to numerical fitting results, $R_{\text{eff-T}}$ asymptotically approaches 0.1398 Ω with increasing lightning current intensity, representing the "limiting" conduction

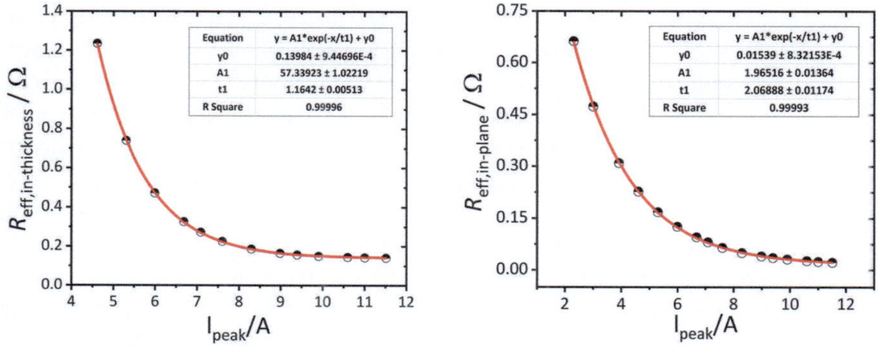

(a) through-thickness conductivity (b) planar conductivity

Fig. 2.47 Variation of effective resistance R_{eff} with peak pulse current capability of CFRP materials in the through-thickness direction. From a microscopic conduction mechanism perspective, as applied lightning current intensity increases, the nonlinear conductive characteristics dominated by "carbon fiber-resin-carbon fiber" interface barriers weaken, and the linear volume resistance of resin materials in the conduction path becomes the determining factor for overall CFRP conductivity. For planar conduction, a similar functional relationship between effective resistance $R_{\text{eff-P}}$ and lightning current amplitude I_{peak} is obtained:

$$R_{eff\text{-}T} = 1.965 \times e^{-\ln I_{peak}/2.069} + 0.01539 \tag{2.29}$$

Notably, the calculation of effective resistance values is based on current Joule heating effects, reflecting the Joule heat generation efficiency during specific lightning current-CFRP interactions. While effective resistance does not correspond to the material's instantaneous actual conductive state, its magnitude still reflects the material's conduction capability, as evident in the result that the limiting value of $R_{\text{eff-P}}$ (0.01539 Ω) is significantly smaller than that of $R_{\text{eff-T}}$ (0.1398 Ω).

2.5 Summary

This chapter presents experimental studies on the direct-current (DC) static conductivity and dynamic conductivity of CFRP composites under non-destructive lightning-current pulses. An improved three-electrode method for measuring anisotropic DC conductivity was proposed, and the DC volume resistivity of CFRP composites was quantified. A dynamic conduction model under lightning pulses was established, and the impacts of pulse waveform parameters and ply-stacking structures on dynamic conductivity were investigated. The key findings are as follows:

(1) To overcome the inaccuracy of traditional three-electrode methods (designed for insulators) in measuring CFRP volume resistance, we introduced a sampling resistor between the guard electrode and the low-voltage system terminal. Experimental results show that through-thickness conductivity of CFRP lies between that of conductors and insulators, resistivity of 10^2–10^3 Ω m, and decreases with increasing DC voltage, exhibiting nonlinear behavior. Volume resistivity also varies significantly with manufacturing process, ply stacking, and surface coatings.

(2) Under lightning-current excitation, the response voltage of CFRP leads the current peak, and the voltage profiles during the rising and falling edges do not overlap, indicating inductive effects. For currents of 10^0–10^2 A, the equivalent impedance ranges from 10^0 to 10^1 Ω, with through-thickness resistivity of 10^0–10^1 Ω m, demonstrating a dramatic conductivity enhancement compared to DC values. Through-thickness resistance shows strong nonlinearity, decreasing

with increasing current. The equivalent circuit comprises a network of resistors, inductors, and discharge gaps.

(3) For in-plane conduction (along or across fiber axes), the conductive network is dominated by carbon-fiber tows, connected via direct fiber contacts or thin resin films. Through-thickness conduction is limited by thick, low-conductivity resin layers, resulting in anisotropic behavior. Under pulses of 10^0–10^2 A, in-plane conductivity (10^3 S/m) far exceeds through-thickness conductivity (10^0 S/m). Dynamic conductivity depends on layup structure, current paths, and pulse waveform: slower rise rates and longer durations yield higher conductivities, a trend closely tied to the inductive properties of individual carbon fibers.

References

1. Todoroki A, Tanaka M, Shimamura Y. Measurement of orthotropic electric conductance of CFRP laminates and analysis of the effect on delamination monitoring with an electric resistance change method[J]. Composites Science and Technology, 2002, 62: 619–628.
2. Cheng J, Ji H, Qiu J, et al. Role of interlaminar interface on bulk conductivity and electrical anisotropy of CFRP laminates measured by eddy current method[J]. Ndt & E International, 2014, 68: 1–12.
3. Haider MF, Majumdar PK, Angeloni S, et al. Nonlinear anisotropic electrical response of carbon fiber-reinforced polymer composites[J]. Journal of Composite Materials, 2018, 52(8): 1017–1032.
4. Park JB., Hwang TK, Kim HG. Et al. Experimental and numerical study of the electrical anisotropy in unidirectional carbon-fiber-reinforced polymer composites[J]. Smart Materials and Structures, 2007, 16(1): 57–66.
5. Todoroki A, Ohara K, Mizutani Y, et al. Lightning strike damage detection at a fastener using self-sensing TDR of composite plate[J]. Composite Structures, 2015, 132: 1105–1112.
6. IEC 60093, Methods of test for volume resistivity and surface resistivity of solid electrical insulating materials[S]. Geneva, Switzerland: International Electro-technical Commission, 1980.
7. Yao X, Chen J, Sun J. Two-electrode measurement method and device for DC small current volt-ampere characteristics of carbon fiber composite material[P]. China National Intellectual Property Administration, 2018. Patent No. ZL201510451148.1.
8. Yao X, Chen J, Sun J. Measurement method and device for small current volt-ampere characteristics of carbon fiber composite material[P]. China National Intellectual Property Administration, 2018. Patent No. ZL201510454287.X.
9. Yao X, Chen J, Sun J. Three-electrode measurement method and device for small current volt-ampere characteristics of carbon fiber composite material[P]. China National Intellectual Property Administration, 2017. Patent No. ZL201510451277.0.
10. Yao X, Sun J, Chen J, et al. Direct current resistance testing methods of carbon fiber reinforced polymer[J]. Materials Research Innovations, 2015, 19(5): 64–69.
11. Sun J, Yao X, Xu W, et al. DC volume resistance measurement method for carbon fiber composites with conductive layup[J]. High Voltage Engineering, 2018, 44(2): 561–567.
12. Yao X, Guo C, Sun J, Chen J. Damage simulation and experiment of carbon fiber reinforced polymer under lightning current[J]. High Voltage Engineering, 2017, 43(5): 1400–1408.
13. Guo Y, Dong Q, Chen J, et al. Comparison between temperature and pyrolysis dependent models to evaluate the lightning strike damage of carbon fiber composite laminates[J]. Composites: Part A, 2017, 97: 10–18.

14. Yin J, Chang F, Li S, et al. Lightning strike ablation damage influence factors analysis of carbon fiber/epoxy composite based on coupled electrical-thermal simulation[J]. Applied Composite Material, 2017, 24(5): 1089–1106.
15. Lu P, Qu Z, Wang Q, et al. Conductive switching behavior of epoxy resin/micron-aluminum particles composites[J]. e-Polymers, 2018, 18(1): 85–89.
16. Roldughin VI, Vysotskii VV. Percolation properties of metal-filled polymer films, structure and mechanisms of conductivity. Progress Org Coatings 2000;39:81–100.
17. Moseenkov S I, Zavorin A V, Ishchenko A V, et al. Using Current-Voltage Characteristics to Control the Structure of Contacts in Polyethylene Based Composites Modified by Multiwalled Carbon Nanotubes[J]. Journal of Structural Chemistry, 2020, 61(4): 628–639.
18. Wang J, Yu S, Luo S, et al. Investigation of nonlinear I-V behavior of CNTs filled polymer composites[J]. Materials Science and Engineering: B, 2016, 206: 55–60.
19. Chen B, Wu K, Yao W. Conductivity of carbon fiber reinforced cement-based composites[J]. Cement and Concrete Composites, 2004, 26(4): 291–297
20. Hirano Y, Yamane T, Todoroki A. Through-thickness electric conductivity of toughened carbon-fiber-reinforced polymer laminates with resin-rich layers[J]. Composites Science and Technology, 2016, 122: 67–72.
21. Rimska Z, Křuesálek V, Spacsek J. AC conductivity of carbon fiber-polymer matrix composites at the percolation threshold[J]. Polymer composites, 2002, 23(1): 95–103.
22. Sun J, Yao X, Xu W, et al. Dynamic characteristics of carbon fiber composites under non-destructive lightning current component A[J]. Journal of Xi'an Jiaotong University, 2016, 50(6): 130–135.
23. Cao C, Sun J, Xu W, et al. Conductive characteristics of carbon fiber composites under impulse current[J]. Journal of Xi'an Jiaotong University, 2018, 52(2): 103–109.
24. Sun J, Yao X, Xu W, et al. Dynamic characteristics of carbon fiber reinforced polymer under nondestructive lightning current[J]. Polymer Composites, 2018, 39(5): 1514–1521.
25. Sun J, Li X, Tian X, et al. Dynamic electrical characteristics of carbon fiber-reinforced polymer composite under low intensity lightning current impulse[J]. Advanced Composites Letters, 2020, 29:1–13.
26. Tucci V, Vitelli M. On the effect of anisotropy in nonlinear composite materials for stress grading applications-a numerical study[J]. IEEE Transactions on Dielectrics and Electrical Insulation, 2000, 7(3): 387–393.
27. Eda K. Conduction mechanism of non-Ohmic zinc oxide ceramics[J]. Journal of Applied Physics, 1978, 49(5): 2964–2972.
28. D. Bohm, Quantum Theory (Prentice-Hall, Inc., Englewood Cliffs, New Jersey, 1951), p. 275.
29. Simmons J G. Generalized formula for the electric tunnel effect between similar electrodes separated by a thin insulating film[J]. Journal of applied physics, 1963, 34(6): 1793–1803.
30. Jinru Sun, Xiangyu Tian, Yafeng Li, Yi Wu, Yugang Duan, Jingliang Chen, Gerhard Ziegmann, Mingzhe Rong, Xueling Yao, Ben Wang (2021) Lightning strike-induced dynamic conduction characteristics and damage behavior of carbon fiber-reinforced polymer composites. Composite Structures, 275(2021)114391:1–14.

Chapter 3
Lightning Damage of Carbon Fiber Composites Under Single Lightning Current Component

3.1 Research on Lightning Damage of Carbon Fiber Composite Materials

The interaction of lightning with CFRP composites is a multiphysics phenomenon involving arc attachment, high-pressure shock waves, electromagnetic forces, and Joule heating. The direct effects of lightning current on CFRP depend on the material's formulation and processing, the lightning-current waveform parameters, and the mode of strike.

A review of experimental research reveals that studies on lightning-induced damage in CFRP began in 2007 with a collaboration between Prof. Paolo Feraboli (University of Washington) and Yoshiyasu Hirano (JAXA). The first experimental results on CFRP lightning damage were published in 2009, followed by additional findings [1], followed by subsequent related research findings [2–4]. Early experiments examined the effects of Component D pulses with peak currents of 10 kA, 40 kA, and 50 kA on fastener-type and flat-type CFRP panels [1]. Damage was quantified by measuring residual tensile and compressive strengths, and ultrasonic and microscopic inspections were used to map damage area and depth. Results showed that fastener-reinforced CFRP exhibited more localized but deeper damage, often through-thickness, whereas flat panels displayed damage concentrated in the second to fourth surface plies. As illustrated in Fig. 3.1, fiber fracture, resin ablation, and delamination extended through the full thickness in fastener-type samples.

Reference [2] compared residual strength, damage area, and damage depth of CFRP after mechanical impact versus lightning-current damage. Mechanical impacts of 5, 15, and 25 ft lb (6.8, 20.4, and 44.9 J) were matched to lightning-pulse Component D peaks of 40, 50, and 70 kA. It found that, for equivalent overall damage, mechanical impacts produced larger damage areas and greater strength loss than lightning strikes. Reference [3] examined three lightning-current waveforms, 2.6/10.5, 4/20, and 7/150 μs, across 10–40 kA on graphite/epoxy laminates, establishing how peak current, charge transfer, and action integral govern damage depth and area,

J. Sun et al., *Lightning Effects on Carbon Fiber-Reinforced Polymer Composites*,
https://doi.org/10.1007/978-981-95-3196-7_3

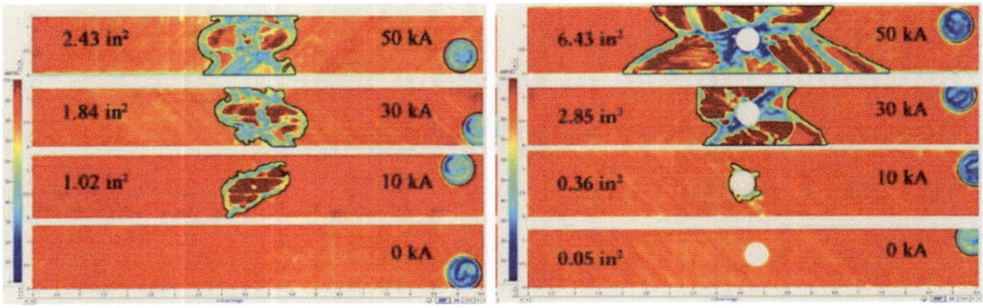

(a) Flat CFRP composites (b) CFRP composite materials with fasteners

Fig. 3.1 Comparison of lightning damage areas of CFRP composites without fasteners and with fasteners

respectively. It also determined that panel size, thickness, and waveform shape have minimal influence on the overall damage extent.

The severity of CFRP lightning damage depends not only on current parameters, but also on the test environment and the composite's ply architecture. Reference [5] investigated glass-fiber/carbon-fiber woven laminates under dry and hygrothermal conditions, showing that moisture degrades fiber-matrix bonding and exacerbates lightning damage after aging, as illustrated in Fig. 3.2.

Between 2009 and 2014, numerous researchers worldwide built upon the work of Hirano [2] and Feraboli [3] on CFRP lightning damage using single lightning current components, specifically 4/20, 8/20, and 7/150 μs. The characteristics of these current components closely approximate the waveforms of lightning-induced and switching overvoltages encountered in power engineering. In power systems, overhead transmission lines above 35 kV employ static wires (overhead ground wires) that are struck by lightning, with the surge current conducted through down conductors into the

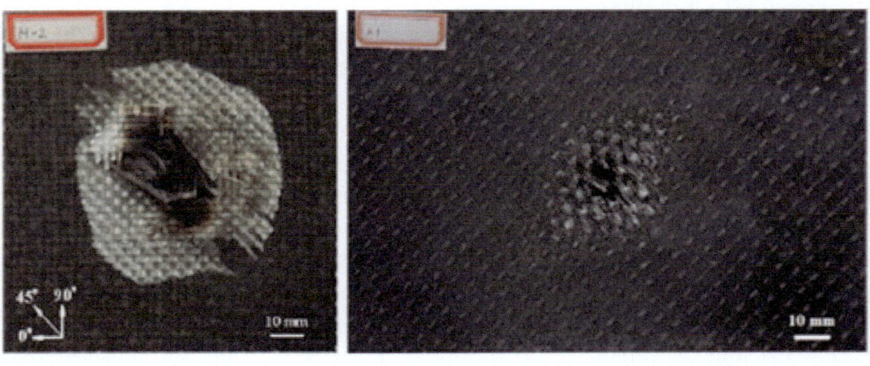

(a) hot and humid environment (b) CFRP with fasteners in dry environment

Fig. 3.2 Comparison of lightning damage of CFRP composites under hygrothermal and dry environment tower grounding grid. Consequently, power system lightning protection design and research focus mainly on mitigating lightning- and switching-induced overvoltages arising from frequent circuit breaker operations, with direct strike effects and damage mechanisms seldom addressed. In contrast, aerospace applications must contend with electromagnetic environments involving both cloud-to-ground and cloud-to-cloud lightning discharges. Standards such as MIL STD 464 [6] and SAE APR 5412 [7] define natural lightning characteristics, specifying current components A, B, C, and D for direct strike effects, whose parameters differ significantly from those adopted in power engineering. Therefore, evaluating CFRP lightning damage in aerospace contexts requires selecting test waveforms that reflect these distinct parameters rather than those typical of power transmission systems.

Research on lightning-induced damage in CFRP composites in China started comparatively late. Since 2013, Xi'an Jiaotong University, in collaboration with Shandong University and the Air Force Engineering University, has conducted experimental investigations on CFRP lightning damage using current component A (6.4/69 μs) [8–16]. Ultrasonic imaging systems were used to identify damage areas and penetration depths in CFRP, while electron microscopy provided high-resolution observation of the microscopic morphology of lightning-induced damage. This work quantified the relationship between damage severity in CFRP samples and the applied current parameters. References [12–14] measured and analyzed mechanical properties, notably the residual mechanical strength of CFRP samples post-impact, revealing that lightning strikes of varying intensities led to different degrees of mechanical performance degradation. References [15, 16] investigated CFRP composites incorporating fasteners, producing results consistent with those in Ref. [1].

3.2 Experimental Design of Lightning Damage of Carbon Fiber Composites Under Lightning Current Component A

3.2.1 Lightning Current Component A and the Experimental Fixture

The multiple successive discharges of lightning current component A were selected as the injection waveform for the damage tests, and experiments were conducted under at least four peak-current levels from 10 to 200 kA. To accommodate direct-strike tests on CFRP composites of varying sizes, diversion configurations, and specific fasteners (e.g., bolts, rivets), an experimental fixture (Fig. 3.3) was designed.

The fixture comprises insulating supports, cover plates, copper connection terminals, and lightning current injection electrodes. The left and right insulating vertical plates mount on the support base plate, with the right plate adjustable via sliding guide rails. Grooves accommodate elongated metal electrodes used for current diversion

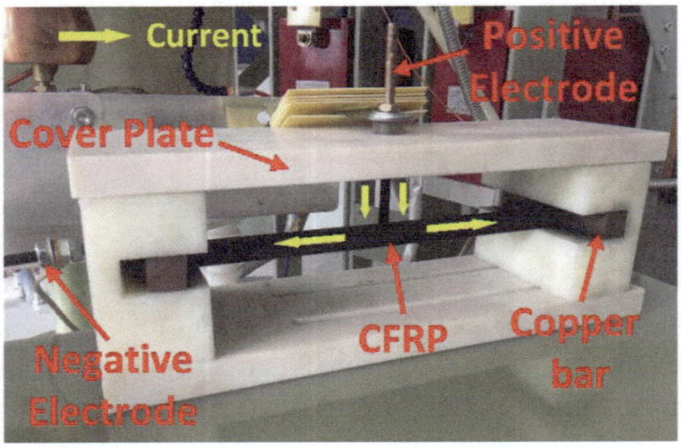

(a) Lightning damage experimental fixture

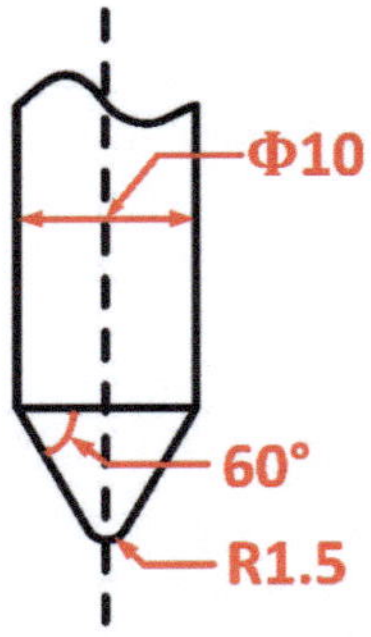

(b) Injection electrode assembly

Fig. 3.3 Lightning damage test fixture for CFRP composites

in the mid-section of both plates. The CFRP specimen is clamped between symmetrically arranged copper rod electrodes and secured by nuts on the vertical plates. Lightning current component A is delivered through a copper screw-rod electrode, which is fixed to the insulating cover plate with high-temperature rubber sleeves and nuts. Its conical end face sits directly above the specimen center, and the electrode-to-specimen distance is adjustable per test requirements. Both sides of the specimen are electrically connected to the return path via side copper rod electrodes.

During testing, one terminal of the current generation circuit injects through the top copper rod electrode, while the opposite terminal connects to the two elongated electrodes in the grooves of the vertical plates, simulating a two-end diversion scenario. To emulate a four-end diversion configuration, two additional insulating vertical plates and fixed metal electrodes can be added to the fixture's front and rear.

3.2.2 Lightning Damage Detection Method of Carbon Fiber Composite Materials

Lightning Strike Process Observation

A high-speed camera and infrared thermal imager were deployed to monitor the lightning strike process on CFRP specimens Fig. 3.4). The high-speed camera (Vision Research Phantom Miro M410) was positioned to face the test area, offering a resolution of 126 × 126 pixels and capturing at 200,000 fps. The Swedish FLIR T440 infrared thermal imager recorded surface temperature in real time during direct strikes on the composite. Its key specifications include a maximum measurable temperature of 1200 °C, thermal sensitivity <0.045 °C, and an image capture rate of 60 Hz.

(a) Phantom Miro M410 high-speed camera (b) FLIR T440 infrared thermal imager

Fig. 3.4 Lightning strike monitoring instruments

Lightning Direct Damage Detection of CFRP Composites

Nondestructive evaluation was performed using an American Sonoscan D9500 ultrasonic microscope and a YXLON Y. Cheetah micron X-ray 3D imaging system (Fig. 3.5). The ultrasonic microscope provides a scanning resolution of 9 μm, while the X-ray system produces high-resolution three-dimensional images of internal damage.

The microscopic morphology of the internal lightning-induced damage regions in CFRP composites was examined using a Hitachi SU-8010 field-emission scanning electron microscope (Fig. 3.6). Prior to testing, specimens were sectioned as required. Under a 1 kV operating voltage, the microscope achieves a resolution of 1.4 nm.

3.2.3 Design of Carbon Fiber Composite Specimen Under Single Lightning Current Component

The CFRP composite materials used in large-aircraft structural components fall into unidirectional laminate and two-dimensional woven architectures. To investigate the

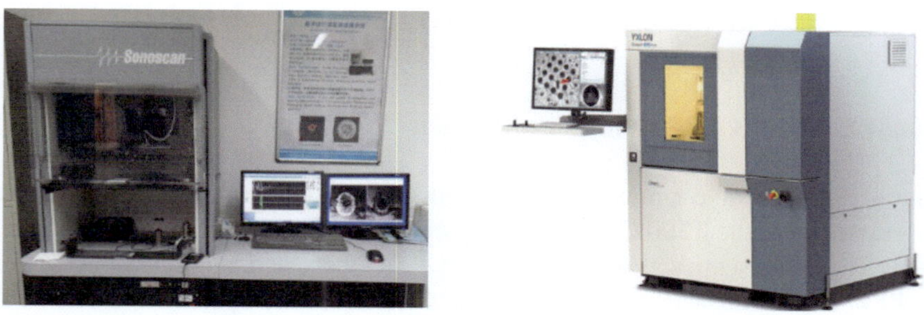

(a) Sonoscan D9500 Ultrasonic Microscope (b) YXLON X-ray 3D imaging system

Fig. 3.5 Nondestructive testing systems for lightning damage

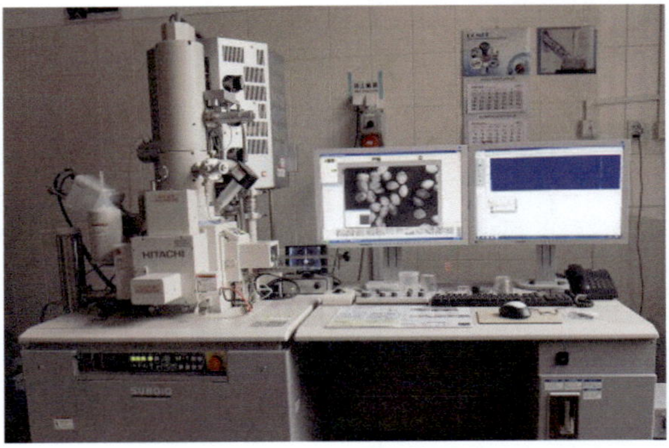

Fig. 3.6 HITACHI SU-8010 field emission scanning electron microscope

direct lightning damage modes and mechanisms of these different CFRP configurations, specimens were fabricated as follows: unidirectional laminate CFRP composites, two-dimensional woven CFRP composites, CFRP composites incorporating a single fastener, and unidirectional laminate composites with copper-mesh shielding.

Unidirectional Laminate CFRP Composites (Type S1)

The unidirectional laminate carbon-fiber composite utilizes the T400/4021 material system, with a carbon-fiber volume fraction of 60%. The specimen dimensions are 150 mm × 100 mm × 4.6 mm, and the layup configuration is $[45/-45/0/90]_{3S}$, comprising 24 layers at 0.15 mm per layer.

Two-Dimensional Woven CFRP Composites (Type S2)

As shown in Fig. 3.7b, the two-dimensional woven composite employs the CCF400/5228A material system, with a carbon-fiber volume fraction of 55%. The specimen dimensions are 150 mm × 100 mm × 4 mm, consisting of 16 layers at 0.25 mm thickness each, with a layup sequence of $[0/90]_F$.

Single Fastener CFRP Composite Speimen (Type S3)

Figure 3.8 shows the laminated CFRP composite specimen with a single fastener, fabricated from the T300/C08 material system at a carbon-fiber volume fraction of 70%. The layup sequence is $[45/-45/0/90]_{2S}$, and the specimen dimensions are 150 × 40 × 2 mm. The fastener is positioned at the center of the specimen, with both the hole and fastener diameters 4 mm.

In the lightning damage experiments using single lightning current component A, specimens were clamped at the ends, and the arc injection gap was set to 2 mm. The

(a) unidirectional ply structure (b) two-dimensional woven structure

Fig. 3.7 Carbon fiber composite specimens

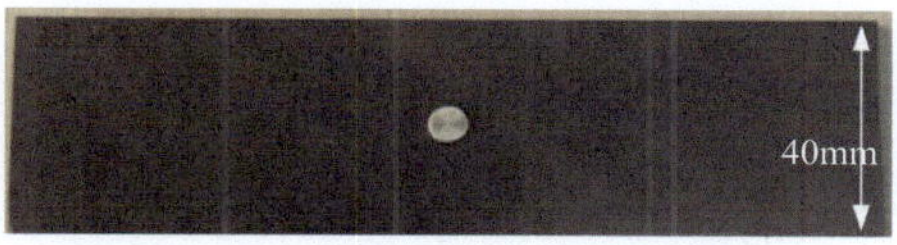

(a) Sample physical map (150 × 40 × 2 mm)

(b) Sample physical map (300 × 40 × 2 mm)

Fig. 3.8 CFRP composite specimen with a single fastener

test conditions for CFRP composites with different configurations are summarized in Tables 3.1 and 3.2 (Table 3.3).

Table 3.1 Lightning damage experimental conditions of Type S1 sample

Sample number	Current peak I_p (kA)	Front waveform time T_f (μs)	Half-peak time T_t	Charge transfer Q_C (C)	Action integral W_A (A^2s)
Type S1-1	21.4	17.4	67.6	1.54	19208
Type S1-2	41.9	17.9	98.9	4.52	108475
Type S1-4	56.4	17.5	142.0	7.74	260205
Type S1-4	94.8	26.0	117.5	11.17	628494

Table 3.2 Lightning damage experimental conditions of Type S2 specimen

Sample number	Current peak I_p (kA)	Front waveform time T_f (μs)	Half-peak time T_t (μs)	Charge transfer Q_C (C)	Action integral W_A (A^2s)
Type S2-1	20.0	16.4	70.5	1.54	17412
Type S2-2	40.4	17.4	77.4	4.41	76619
Type S2-4	61.7	17.2	99.8	7.70	252110
Type S2-4	100.5	12.7	74.5	8.46	510805

Table 3.3 Lightning damage experimental conditions of Type S3 specimen

Sample number	Current peak I_p (kA)	Front waveform time T_f (μs)	Half-peak time T_t (μs)	Charge transfer Q_C (C)	Action integral W_A (A^2s)
Type S3-1	20.0	16.8	72.6	1.58	22104
Type S3-2	32.3	17.0	78.4	3.52	56813
Type S3-4	48.3	17.2	99.6	7.70	152273
Type S3-4	102.0	22.7	119.6	11.62	666807

3.3 Lightning Damage of Ply Carbon Fiber Composites Under Lightning Current Component A

3.3.1 The Process of Lightning Strike on Composite Materials

The lightning strike on CFRP composites comprises stages including leader development, conductive channel formation, current injection, arcing and combustion, and specimen cooling. Real-time recording of this process was performed using a high-speed digital camera (Phantom Miro M310).

The leader formation begins with air ionization induced by the high-intensity electric field near the injection electrode. As one leader branch advances toward the specimen, it initiates a connecting leader from the CFRP surface, visible as arc jets between the electrode and laminate (Fig. 3.9(a1)). Upon intersection with the oppositely propagating leader, a continuous conductive channel spans the gap (Fig. 3.9(a2)). When the lightning current propagates through this channel, an intense white flash is emitted (Fig. 3.9(a3)). The composite's temperature then escalates rapidly, causing ablation and combustion of the CFRP, as evidenced by the red flames in Fig. 3.9(a4). After the strike, the surface temperature distribution of the specimen is captured (Fig. 3.10).

The surface temperature distribution of CFRP composite material specimens was measured using an infrared thermal imager (FLIR T460) operating at 30 Hz with a recording resolution of 320 × 240 pixels. Since the lightning current component

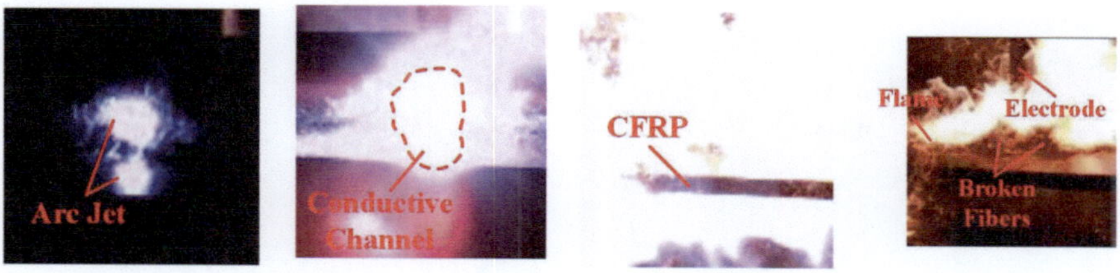

(a1) $t = 4.52$ μs (a2) $t = 9.04$ μs (a3) $t = 363.2$ μs (a4) $t = 726.4$ μs

Fig. 3.9 Lightning strike process recorded by high-speed camera

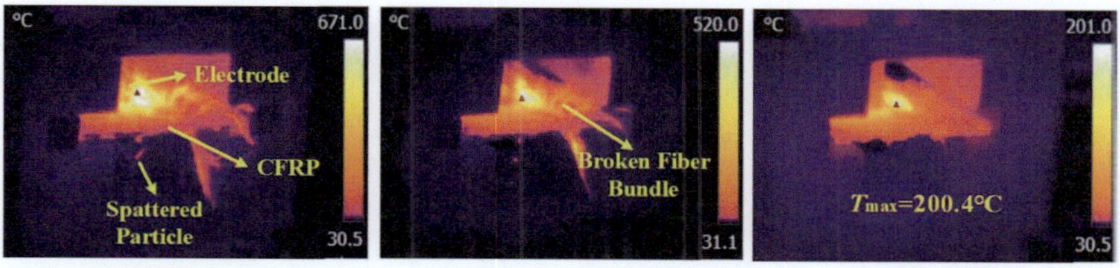

(a) $t \approx 0.1$ s (b) $t = 1.233$ s (c) $t = 7.701$ s

Fig. 3.10 Surface temperature distribution on the CFRP specimen after lightning strike

A lasts less than 400 μs, which is much shorter than the imager's 0.033 s frame interval, the infrared thermal imager only capture the post-strike cooling of the specimen surface. From the thermal image distribution on the CFRP specimen surface (Fig. 3.10), it can be observed that shortly after the lightning strike, high-temperature regions are concentrated around the lightning attachment points. At $t \approx 0.1$ s, the maximum surface temperature exceeds 670 °C, and fiber fragments are ejected outward from the damaged area. At $t \approx 1.233$ s, fractured fiber bundles at elevated temperatures appear on the surface, matching the observed damage morphology. By $t \approx 7.701$ s, the maximum surface temperature remains at 200.4 °C, indicating ongoing resin pyrolysis and progressive damage expansion. At $t \approx 7.6$ s, resin matrix pyrolysis persists, driving further expansion of the damaged region.

3.3.2 Morphology and Mode of Lightning Damage of Unidirectional Ply Carbon Fiber Composites

Lightning Damage Morphology of Unidirectional Laminate CFRP Composites

Under the lightning current component A conditions in Table 3.1, the surface damage morphology and damage-zone dimensions for unidirectional laminate CFRP specimens are shown in Fig. 3.11 [18].

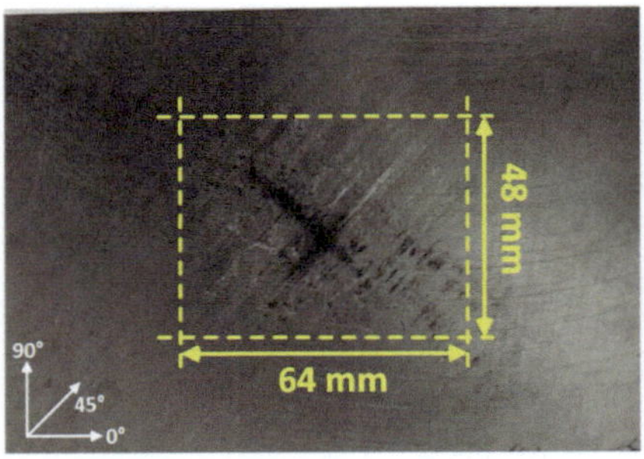

(a) $Ip = 20$ kA

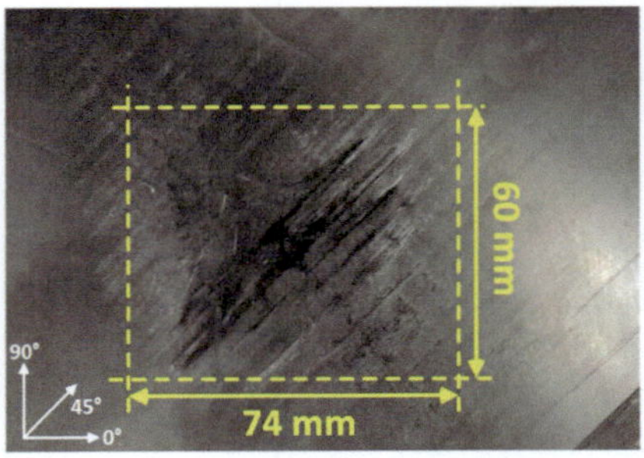

(b) $Ip = 40$ kA

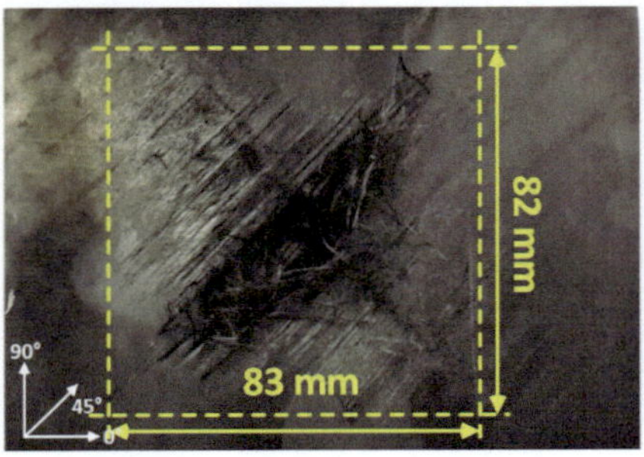

(c) $Ip = 60$ kA

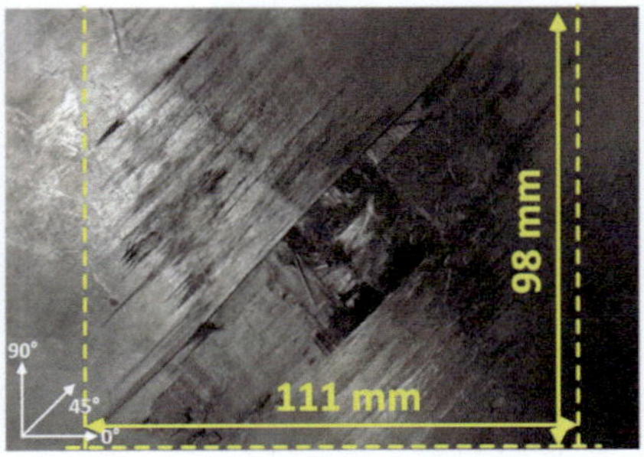

(d) $Ip = 100$ kA

Fig. 3.11 Surface damage morphology of CFRP composites

From these results, the following damage characteristics emerge:

(i) At $I_p = 20$ kA (Type S1-1), fiber bundle fractures at the lightning injection point and propagate outward along the surface fiber orientation (45°). Secondary fractures align along the –45° direction of the second ply. Delamination appears near the strike area in the first two plies due to fiber bulging and internal pyrolysis gas pressure. Surface fiber debris is scattered, as highlighted in Fig. 3.11a.

(ii) As I_p increases (Fig. 3.11b–d), heat from high-temperature arcs and Joule heating intensifies. Deeper and larger ablation pits form at the injection point. Fiber fractures and outward warping along the 45° direction become more severe, while deeper plies (−45°, 0°) exhibit bundle fractures, warping, and progressive delamination around the fractured bundles.

Lightning Damage Mode Analysis of Unidirectional Laminate CFRP Composites

Figure 3.12 shows the X-ray 3D reconstruction of the damage zone in Type S1-4. Figure 3.13 presents its top and side views before and after trimming warped fibers. A close-up of the damage zone is given in Fig. 3.13c.

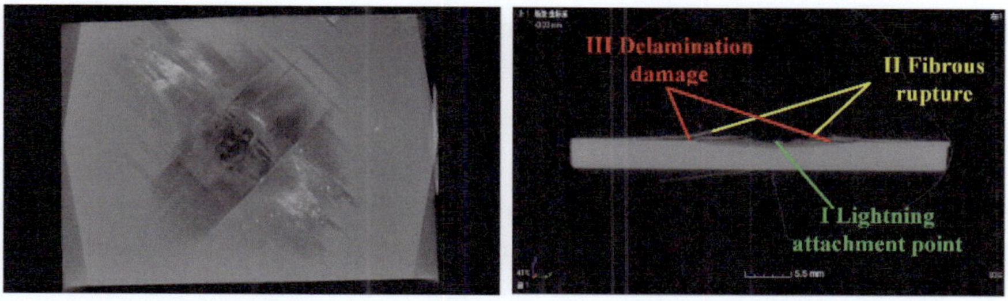

(a) top view (b) side view

Fig. 3.12 X-ray 3D image of the lightning damage area of Type S1-4

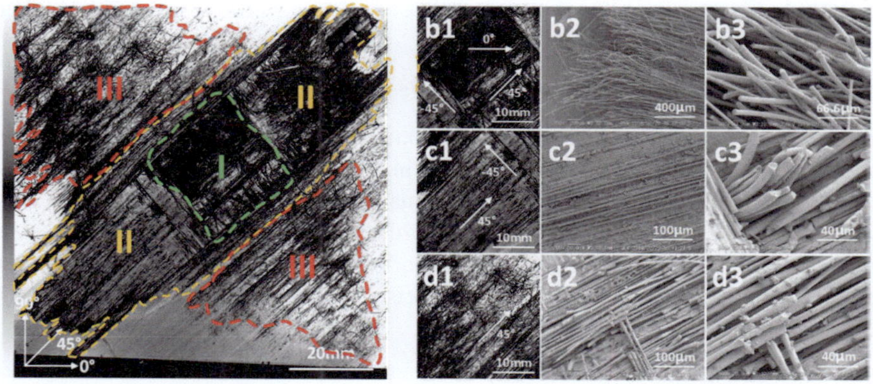

(a) overall view (b) microstructural view

Fig. 3.13 Surface damage observed by ultrasonic scanning: (b1) Zone I; (c1) Zone II; and (d1) Zone III; SEM images: (b2) and (b3) Zone I; (c2) and (c3) Zone II; and (d2) and (d3) Zone III

Combining Figs. 3.12 and 3.13, surface plies fracture in discrete bundles, with damage zones symmetrically flanking each fracture. Based on distinct damage features, the zone is divided into three regions (Fig. 3.13c) [7].

(i) Lightning attachment Zone I

Zone I shows extreme damage, characterized by resin pyrolysis, fiber fracture, and carbon sublimation. Here, the arc directly contacts the composite surface, where plasma temperatures exceed 30,000 K, pressure reaches 10^6 Pa, and energy transfer approaches 39.55×10^3 J/Ω under 40 kA conditions [10]. Resin pyrolysis begins at 200–600 °C, while carbon fiber sublimation near 3000 K, causing rapid ablation of both resin and fibers. Explosive gas releases from pyrolysis ruptures adjacent plies, at least the 45°, −45°, and 0° layers fully fracture, as shown in Fig. 3.13(b1). Fiber bundles are ablated and scattered by the arc's extreme temperature and pressure, as shown in Fig. 3.13(b2), and matrix material largely vanishes (Fig. 3.13(b3)). Visual inspection of Zone I reveals complete resin removal, leaving only exposed, fractured fiber ends, and this region exhibits the maximum damage depth.

(ii) **Carbon fiber fracture damage Zone II**

Zone II exhibits interlaminar delamination and fiber bundle fracture, forming an elongated damage region extending from the attachment Zone I along the surface fiber orientation. To clarify the morphology, protruding fibers in the delamination region were trimmed. On the exposed second ply, minor ablation along 45° and −45° fiber directions is visible, yet the fiber architecture remains ordered, and the bottom ply shows no significant degradation, as shown in Fig. 3.13(c1). This suggests the delamination arises from mechanically induced interlaminar cracks. Micrographs of the second ply, as illustrated in Fig. 3.13(c2) and (c3), reveal much milder damage than Zone I, with clean, flat fracture surfaces indicating brittle fiber failure rather than plastic deformation.

(iii) **Resin pyrolysis delamination damage Zone III**

Zone III is a fan-shaped region flanking Zone II, marked by more extensive fiber breaks and matrix decomposition, as shown in Fig. 3.13a. Layered fiber protrusions align with the 45° orientation, and the resin matrix pyrolyzes, leaving surface cracks along fiber lines. Microstructures in Fig. 3.13(d2) and (d3) show deeper ablation pits, greater matrix loss, and increased fiber exposure compared to Zone II. Unlike the brittle fractures of Zone II, fiber ends here display rough, irregular tips, indicating advanced thermal degradation.

Analysis of Internal Damage and Microstructure of CFRP Composites

To elucidate the lightning damage sequence, cross-sectional CT images from the YXLON Y. Cheetah 3D X-ray system were analyzed. A conical ablation pit at the cross-section center, corresponding to Zone I, shows multiple plies bending outward (Fig. 3.14a), indicating strong interlaminar forces. DBecause CFRP's in-plane conductivity far exceeds its through-thickness conductivity, and high-temperature arc ablation further reduces through-thickness current paths, most lightning current flows along surface plies, producing rapid localized heating. The resin matrix pyrolyzes, releasing CO_2 and H_2O vapor. The extreme heating zone then propagates outward from the central pit, implying that damage initiates at the ablation site. Figure 3.14b–e schematically depict the inferred delamination procession across Zones I-III based on these observations.

As shown in Figs. 3.13 and 3.14, the damage modes in Zone II arise from the following sequence. Heat from high-temperature arcs and Joule heating within the CFRP raises specimen temperature, initiating resin pyrolysis. Confined by the outer plies, pyrolysis gases accumulate between layers, generating upward pressure on the surface plies. When this pressure exceeds the mechanical strength of the upper fiber layers, carbon fiber bundles fracture. In regions farther from the strike, where the resin matrix is not fully pyrolyzed, the surface plies are stripped by explosive gas

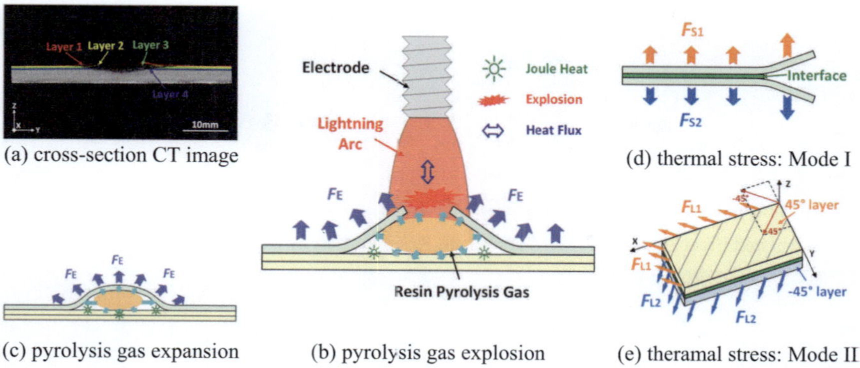

Fig. 3.14 Schematic diagram of internal damage process

pressure F_E and inertial forces, producing the flake-like tears characteristic of Zone II. Microstructural analysis of the damage zones was performed using a field-emission scanning electron microscope, with representative images shown in Fig. 3.15.

From Fig. 3.15, the three lightning damage zones in unidirectional CFRP laminates exhibit the following microstructural characteristics:

(i) In the lightning attachment zone, carbon fiber bundles undergo ablation, sublimation, and fracture under the high-temperature and high-pressure field generated by the lightning arc, accompanied by intense pyrolysis of inter-fiber resin. As a result, numerous exposed and fractured carbon fiber ends remain, as shown in Fig. 3.15a. Higher magnification reveals irregular particles of pyrolyzed and carbonized resin adhering to the CFRP surface, as shown in Fig. 3.15b.

(ii) In the fiber fracture zone, microstructure of the exposed second ply shows granular residues on fiber surfaces and in fiber gaps. Fiber alignment remains relatively intact, and the underlying resin matrix shows minimal degradation, as observed in Fig. 3.15c, d. Fracture surface magnification in Fig. 3.15e reveals brittle fractures that are perpendicular to the fiber axis, with minimal plastic deformation and flat fracture ends, supporting the inference that fiber breakage result from explosive pressure of interlaminar pyrolysis gases.

(iii) Representative features of the resin pyrolysis delamination zone are shown in Fig. 3.15f–h. Compared to Zone II, this zone exhibits flake-like pyrolyzed, resin, melt pits in the resin filler layers, and a notably higher exposure of carbon fiber bundles, as shown in Fig. 3.15f and g. Unlike the brittle fractures in Zone II, fibers near the fracture tips in this region often show swelling and thickening (Fig. 3.15h), attributed to disruption of carbon chain crystalline regions during thermal decomposition.

(iv) Through combined macro- and microstructural analyses, the lightning damage in CFRP specimens can be categorized into three primary mechanisms: (i) direct arc-induced thermal damage, (ii) mechanical damage from explosive gas pressure impacting fiber plies, and (iii) interlaminar delamination caused by pyrolysis gas expansion.

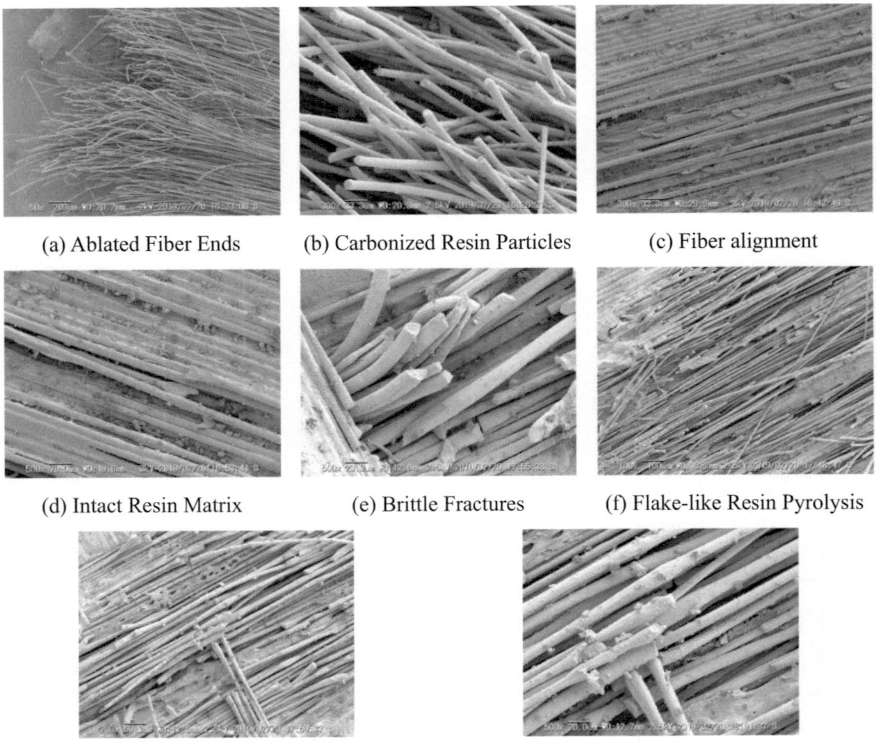

(a) Ablated Fiber Ends (b) Carbonized Resin Particles (c) Fiber alignment

(d) Intact Resin Matrix (e) Brittle Fractures (f) Flake-like Resin Pyrolysis

(g) Exposed Fiber Bundles (h) Swollen and Thickened Fibers

Fig. 3.15 Microstructure of lightning damage in Type S1-4 regions of the specimen

In summary, under the action of lightning current component A, unidirectional CFRP laminates exhibit resin pyrolysis and carbonization, fiber bundle ablation and fracture, and interlaminar delamination. The dominant mechanisms include thermal effects from arc and Joule heating, and mechanical effects from pyrolysis gas expansion.

3.3.3 Analysis of Lightning Damage Results for Unidirectional Laminate CFRP

Scanning Results of Lightning Damage in Ply-Structured CFRP

The lightning damage ultrasonic scan results for unidirectional laminate CFRP specimens subjected to lightning current component A at various peak currents are shown in Figs. 3.16 and 3.17. These scans quantify damage progression: surface disruption,

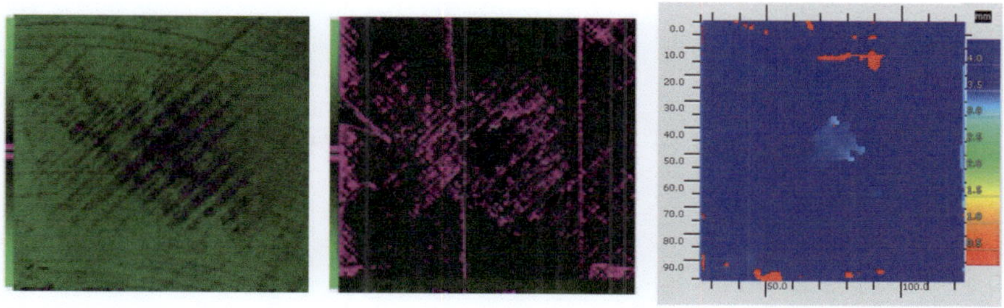

(a) surface scan (b) transmission scan (c) depth profile

Fig. 3.16 Ultrasonic scanning results of Type S1-1 (20 kA)

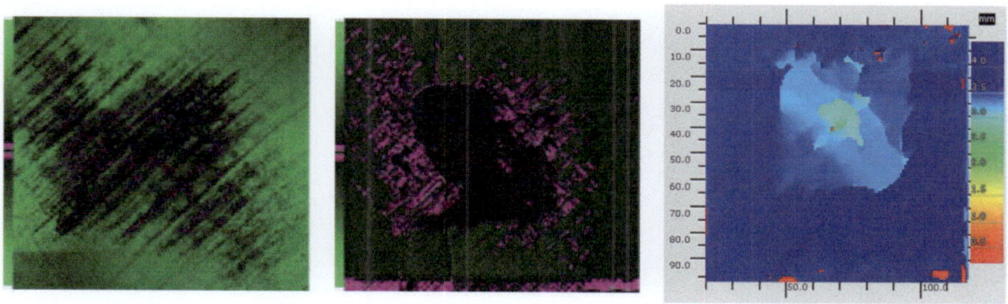

(a) surface scan (b) transmission scan (c) depth profile

Fig. 3.17 Ultrasonic scanning results of Type S1-2 (40 kA)

internal delamination, and maximum penetration depth all increase with peak current, demonstrating the correlation between lightning intensity and damage severity.

Lightning Damage Area and Depth of Ply-Structured CFRP Composites

Based on the ultrasonic scans in Figs. 3.16, 3.17, 3.18 and 3.19, a pixel-counting method was used to quantify the lightning damage area (S_Z) and damage depth (D_d) for unidirectional laminate CFRP under lightning current component A. During injection, Joule heating induces resin pyrolysis and fiber sublimation, weakening interlaminar bonding and causing internal delamination. Concurrently, pyrolysis gases accumulate; once their expansive pressure exceeds interlaminar strength, gas eruptions produce brittle fractures and outward warping of surface plies. In ultrasonic transmission mode, the total damage area represents the combined delamination and fiber-bursting regions.

Reflecting Sect. 4.3.2's morphological insights, S_Z comprises two parts:

(i) S_1: Fiber-fracture area, where bundles break and warp under explosive gas forces, even if the resin matrix is partially intact.

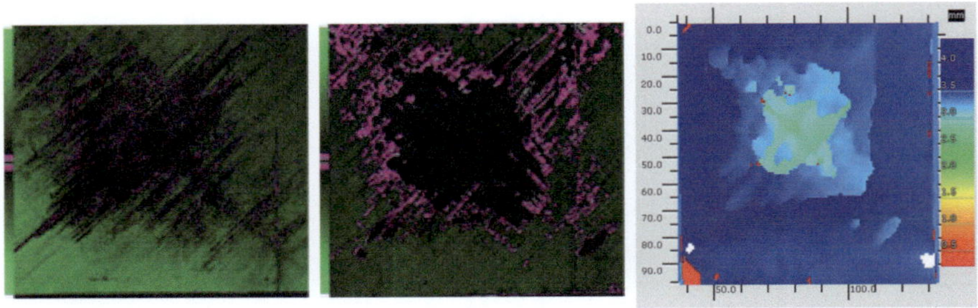

(a) surface scan (b) transmission scan (c) depth profile

Fig. 3.18 Ultrasonic scanning results of Type S1-3 (60 kA)

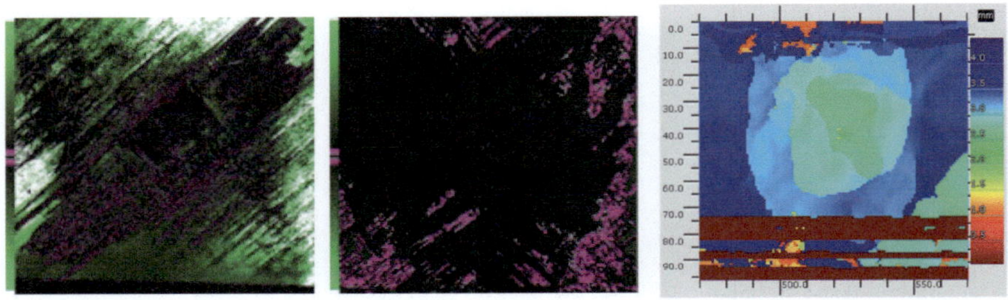

(a) surface scan (b) transmission scan (c) depth profile

Fig. 3.19 Ultrasonic scanning results of Type S1-4 (100 kA)

(ii) S_2: Delamination area, resulting from resin pyrolysis and fiber ablation driven by arc and Joule heating.

With a fixed 2 mm gap, the electrical parameters of current component A and corresponding damage metrics (S_Z, S_1, S_2, D_d) are summarized in Table 3.4.

Table 3.4 Lightning damage metrics for unidirectional laminate CFRP under component A

No	I_p (kA)	Rising rate R_i (kA/μs)	Q_C (C)	W_A (A^2s)	S_Z (cm^2)	S_1 (cm^2)	S_2 (cm^2)	D_d (mm)
Type S1-1	21.4	1.224	1.54	19208	5.24	1.26	4.98	0.70
Type S1-2	41.9	2.441	4.52	108475	16.5	7.41	9.16	1.02
Type S1-4	56.4	4.217	7.74	260205	29.2	14.7	15.5	1.27
Type S1-4	94.8	4.646	11.17	628494	61.2	45.7	25.5	1.58

Fig. 3.20 Ultrasonic scanning results of lightning damage areas

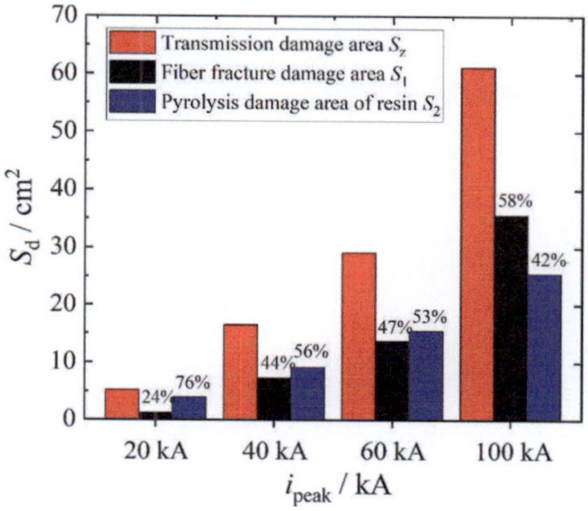

Figure 3.20 compares the total transmission damage area (S_Z), fiber-fracture area (S_1), and resin-pyrolysis area (S_2) of unidirectional laminate CFRP composites [17].

As shown in Table 3.4 and Fig. 3.20, with increasing peak current of component A, the fraction of damage attributable to fiber fractures (S_1) driven by trapped pyrolysis gases and explosive pressures grows, eventually dominating the total damage. Higher peak currents penetrate deeper, causing pyrolysis farther into the resin matrix. These pyrolysis gases become more confined beneath intact surface plies, and coupled with enhanced thermal effects from the intense current, generate stronger explosive impacts that amplify brittle fractures in the surface fibers. Figures 3.21 and 3.22 present curve fits illustrating the relationships between peak lightning current: (i) Total transmission damage area S_z and resin-pyrolysis damage area S_2, in ply-structured CFRP (Fig. 3.21). (ii) Damage penetration depth D_d in unidirectional laminates (Fig. 3.22).

From Table 3.4 and Figs. 3.21 and 3.22, lightning damage in unidirectional laminate CFRP follows these patterns:

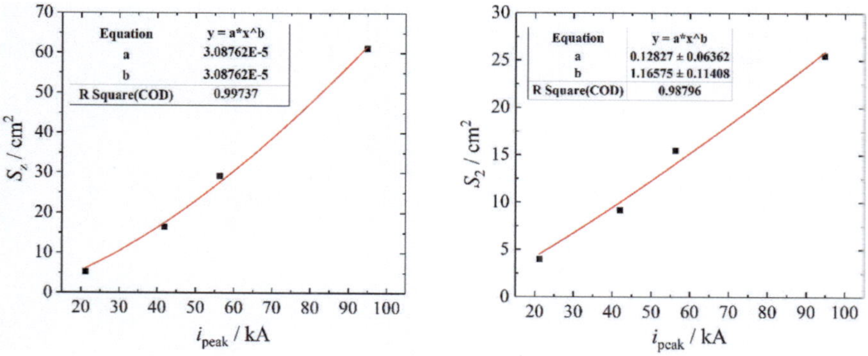

(a) transmission damage

(b) pyrolysis damage

Fig. 3.21 Data fitting of lightning damage areas in ply-structured CFRP

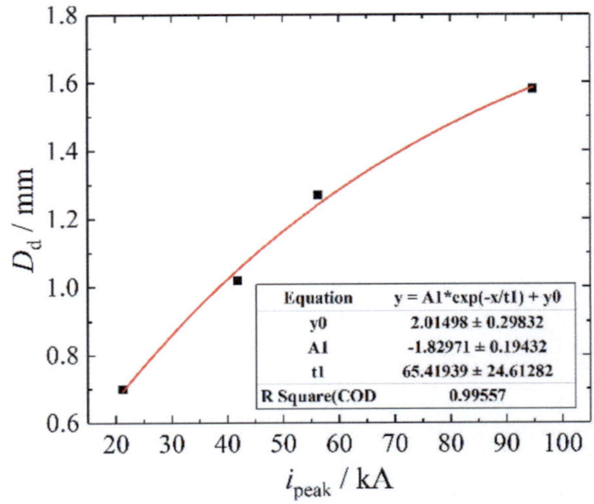

Fig. 3.22 Data fitting of damage depth in unidirectional laminate CFRP

(i) S_z and S_2 of scale with the peak current I_p by a power law, because higher I_p increases transferred charge and action integral, amplifying Joule heating. The resulting higher temperatures intensify resin pyrolysis and fiber sublimation, expanding the damage areas.

(ii) Damage depth D_d grows exponentially with I_p, as larger peaks produce higher temperatures and faster heating rates at the strike point. Severe pyrolysis between the first and second plies forces deeper layers to conduct current earlier, increasing energy penetration and exacerbating internal damage.

(iii) Within the 100 kA range, the power-law fit of S_2 versus I_p approaches linearity, as shown in Fig. 3.22b, while S_2 versus I_p retains a distinctly nonlinear power-law trend, as presented in Fig. 3.22a. Thus, fiber-fracture damage from explosive pyrolysis gases escalates faster than resin delamination, increasing its proportion of total damage, consistent with Fig. 3.20.

3.4 Lightning Damage of Two-Dimensional Woven CFRPs Under Lightning Current Component A

3.4.1 Lightning Damage Morphology and Mode of Two-Dimensional Woven CFRP Composites

Lightning Damage Morphology of Two-Dimensional Woven CFRP

The lightning damage of CFRP composites depends on fiber architecture and fabrication. Figure 3.23 shows surface damage under lightning current component A at various peaks.

Observations from Fig. 3.23 reveal the following damage characteristics of two-dimensional woven CFRP composites under lightning current component A:

(i) Damage footprint differs markedly from unidirectional laminates: it is smaller and approximately circular, reflecting the quasi-isotropic conductivity of the woven architecture.
(ii) At 20 kA (Type S2-1, ~1.54 C), damage is minimal, limited to small ablation pits and fiber protrusions from resin gas pressure around the injection point.

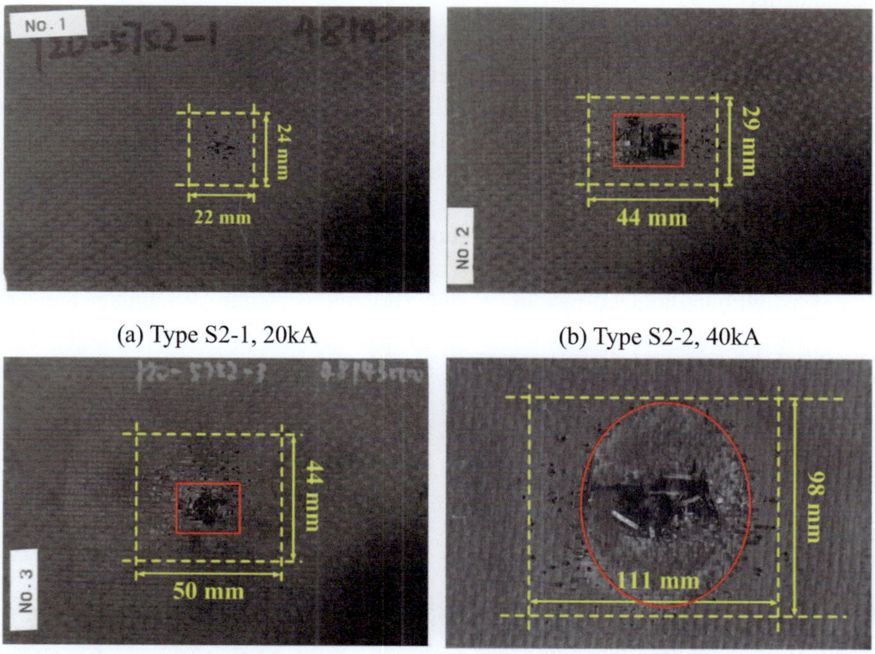

(a) Type S2-1, 20kA
(b) Type S2-2, 40kA
(c) Type S2-4, 60kA
(d) Type S2-4, 100kA

Fig. 3.23 Surface damage morphology of two-dimensional woven CFRP composites

(iii) At 40 and 60 kA, specimens (Type S2-2 and S2-4) show notable fiber bursting and protrusions at the strike center, with localized delamination adjacent to the impact site, though damage remains centrally concentrated.

(iv) At 100 kA, Type S2-4 exhibits significantly enlarged damage, with distinct delamination. The red circle in Fig. 3.23d marks the first-to-second ply boundary: the first ply within this zone is completely removed by explosive forces. Maximum damage, fiber fractures and resin ablation, occurs at the injection center, with satellite ablation spots in the periphery.

Lightning Damage Mode of Two-Dimensional Woven CFRP Composites

The damage zone of specimen Type S2-4 was imaged by the YXLON Y. Cheetah micro-CT system, as shown in Fig. 3.24.

To distinguish damage mechanisms, a magnified, annotated region is shown in Fig. 3.25.

Combined insights from Figs. 3.24 and 3.25 reveal four primary damage modes [17]: fiber fracture/warping, resin pyrolysis/ablation, delamination, and ablation pitting. Accordingly, the damage is partitioned into three concentric regions: Zone I (innermost, blue ellipse), Zone II (between blue and green), and Zone III (between green and red), each corresponding to distinct damage characteristics and mechanisms.

(i) Lightning damage Zone I corresponds to the attachment epicenter, characterized by elliptical ablation pits penetrating at least the fourth ply. This zone exhibits resin pyrolysis, fiber sublimation/fracture, and delamination from explosive pyrolysis gas expansion, all driven by the arc's extreme temperature and pressure, which simultaneously trigger thermal, chemical, and mechanical damage processes.

(ii) Lightning damage Zones II and III both expose the second ply, but mechanisms differ. Zone II's dark gray surface results from delamination via resin pyrolysis, with carbonized residues visible. In contrast, Zone III's lighter appearance reflects brittle fiber fractures and ply separation caused by explosive

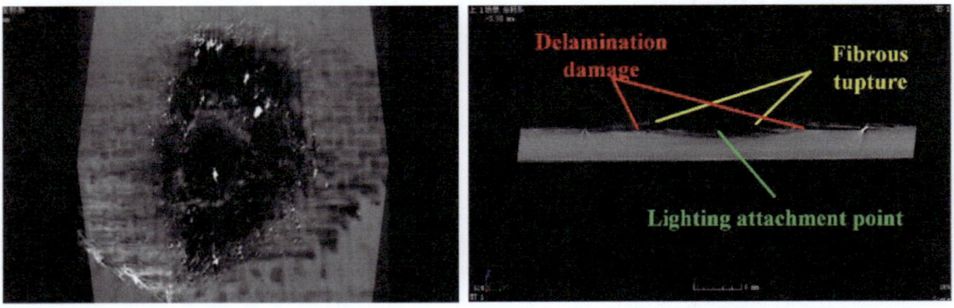

(a) vertical view (b) side view

Fig. 3.24 X-ray 3D imaging of lightning damage in Type S2-4

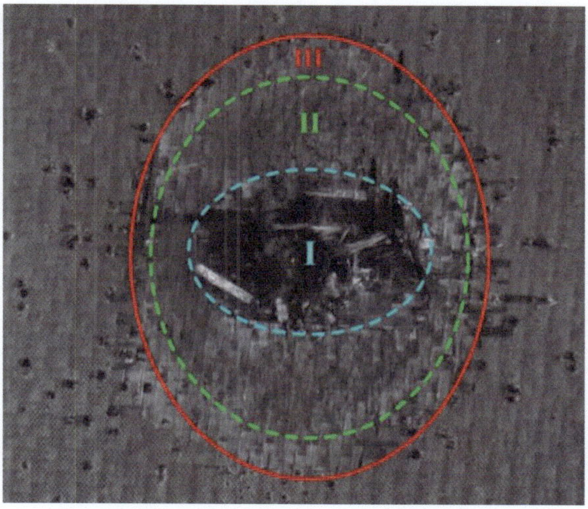

Fig. 3.25 Magnified view of the Type S2-4 lightning attachment point and surrounding damage

gas impacts, which is a mode akin to the flake-tearing seen in unidirectional laminates.

(iii) Significant differences exist in lightning damage characteristics between woven and unidirectional CFRPs: (i) In unidirectional laminate (Type S1-4), fiber fracture constitutes 58% of total area, with 42% from resin delamination. (ii) In two-dimensional woven (Type S2-4), fiber fracture drops to 26%, while resin delamination rises to 74%. (iii) Below 100 kA, Fig. 3.23a–c show no distinct fiber-fracture zones in woven specimens; only resin delamination around the strike site, indicating that pyrolysis gas pressures remain below the bundle tensile strength.

These differences reflect woven architecture's mutual fiber constraints, which limit bursting fractures to cases where current peaks generate sufficient pyrolysis gas to exceed ply tensile strength. Under 100 kA, such critical conditions are reached only locally, confining fiber-tear damage in woven CFRP.

Microstructure Analysis of Lightning Damage Zone of Two-Dimensional Woven CFRP Composites

Samples from Zone II (Fig. 3.26a) and III (Fig. 3.26a and c) of specimen Type S2-4 were examined by scanning electron microscopy to assess damage in the woven architecture. In darker Zone II, distinct resin-ablation grooves appear between fiber bundles, with exposed fibers forming sheet-like patches and an uneven laminate surface. Conversely, lighter Zone III shows orderly fiber strands and intact resin matrix, confirming that Zone II suffers delamination from resin pyrolysis, whereas

(a) Zone II delamination region (b) Zone III impact region (c) magnified fiber fracture in Zone III

Fig. 3.26 SEM micrographs of lightning damage in Type S2-4

Zone III endures mechanical impact damage from internal gas explosions. High-magnification images of Zone III, as presented in Fig. 3.26c reveal smooth, flat fracture surfaces, indicative of brittle fiber failure due to explosive gas pressure.

3.4.2 Analysis of Lightning Damage Results of Two-Dimensional Woven CFRPs

Scanning Results of Lightning Damage of Two-Dimensional Woven CFRP Composites

The lightning damage scanning results of two-dimensional woven CFRP composites under lightning current component A at different peak levels are shown in Figs. 3.27, 3.28, 3.29 and 3.30 [17].

Combining the lightning damage morphology of two-dimensional woven CFRP composites shown in Fig. 3.23 with the ultrasonic scanning results in Fig. 3.27, 3.28, 3.29 and 3.30, he following observations are made:

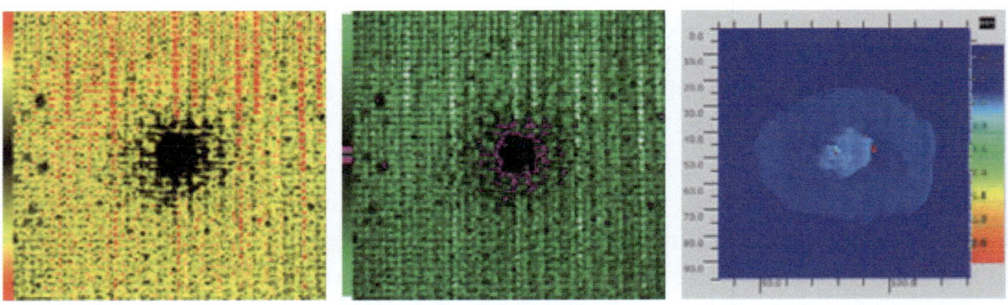

(a) Surface scan (b) Transmission scan (c) Depth scan

Fig. 3.27 Type S2-1 lightning damage scanning results (20 kA)

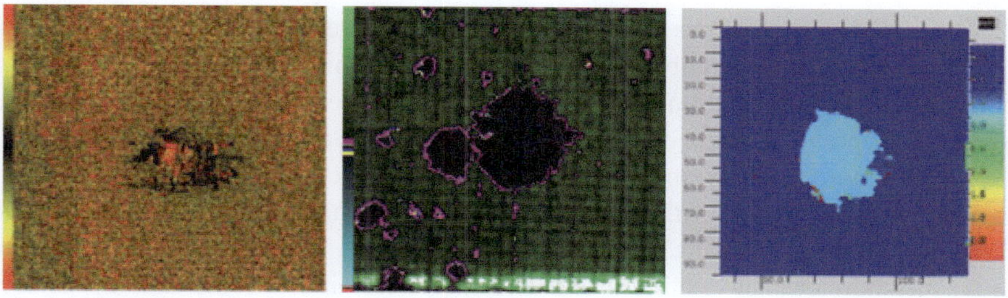

(a) Surface scan (b) Transmission scan (c) Depth scan

Fig. 3.28 Type S2-2 lightning damage scanning results (40 kA)

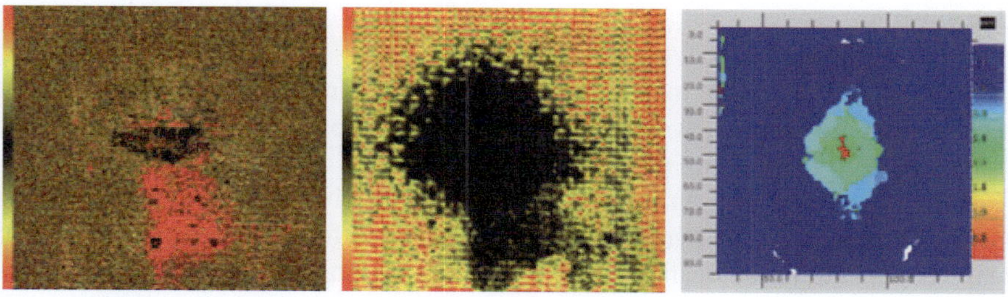

(a) Surface scan (b) Transmission scan (c) Depth scan

Fig. 3.29 Type S2-2 lightning damage scanning results (60 kA)

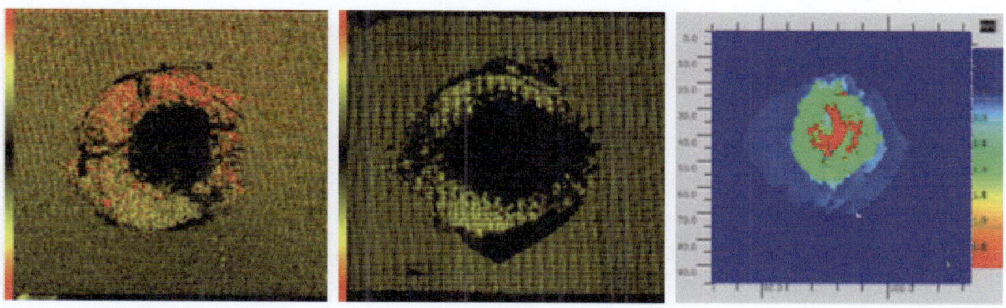

(a) Surface scan (b) Transmission scan (c) Depth scan

Fig. 3.30 Type S2-2 lightning damage scanning results (100 kA)

(i) Under lightning current peaks of 20, 40, and 60 kA, there is no evident fiber fracture zone. The resin pyrolysis delamination region nearly coincides with the total damage area shown in transmission scans.
(ii) Compared to unidirectional laminate CFRP composites, the fiber fracture damage area ratio is significantly lower in two-dimensional woven CFRP specimen Type S2-1.

The underlying cause lies in the interlaced architecture of two-dimensional woven laminates. Radial and weft fiber interlacing enables lightning current to disperse along both directions, forming a nearly circular damage pattern. During resin pyrolysis, interlaminar pressure gradually increases as gas accumulates and expands. However, for fiber fracture to occur, internal gas pressure must exceed not only the interlaminar bonding strength and fiber bundle tensile strength, but also overcome multi-directional fiber constraints. Thus, fiber fracture areas remain much smaller than those in unidirectional laminates under identical conditions, with visible damage regions mostly matching pyrolysis-induced delamination zones.

Lightning Damage Area and Depth of Two-Dimensional Woven CFRP Composites

Experimental data on the total transmission damage area (S_Z) and damage depth (D_d) of two-dimensional woven CFRP specimens subjected to lightning current component A (with a 2 mm gap) are listed in Table 3.5.

Based on the previous analysis of damage modes, fiber fracture is minimal, and surface damage is mainly pyrolysis delamination. Therefore, pyrolysis trends are not analyzed separately but are instead represented by the total transmission damage area (S_Z). Figure 3.31 illustrates the relationship between S_Z and D_d versus the lightning current peak [17].

Table 3.5 and Fig. 3.31 indicate the following damage characteristics:

(i) With increasing peak current, the damage area expands markedly. This is because higher current peaks increase charge transfer and energy deposition, leading to elevated temperatures and enhanced resin pyrolysis, thereby enlarging the damaged region.

(ii) Higher lightning current also leads to greater penetration depth, damaging deeper layers of the composite. Specifically, elevated peak currents raise surface temperatures and heating rates at the lightning strike point, which accelerates resin decomposition between the first and second plies. This allows deeper layers to conduct current earlier, resulting in a progressive increase in damage depth.

Comparison of lightning damage between two-dimensional woven and unidirectional laminate CFRP composites (Fig. 3.32) [18].

Table 3.5 Lightning damage parameters of two-dimensional woven CFRP composites

No	I_p (kA)	Rising rate R_i (kA/μs)	Q_C (C)	W_A (A^2s)	S_Z (cm^2)	D_d (mm)
Type S2-1	20.0	1.220	1.54	17412	4.05	0.48
Type S2-2	40.4	2.416	4.41	76619	7.79	1.05
Type S2-4	61.7	4.587	7.70	252110	14.29	1.46
Type S2-4	100.5	7.914	8.46	510805	22.90	1.97

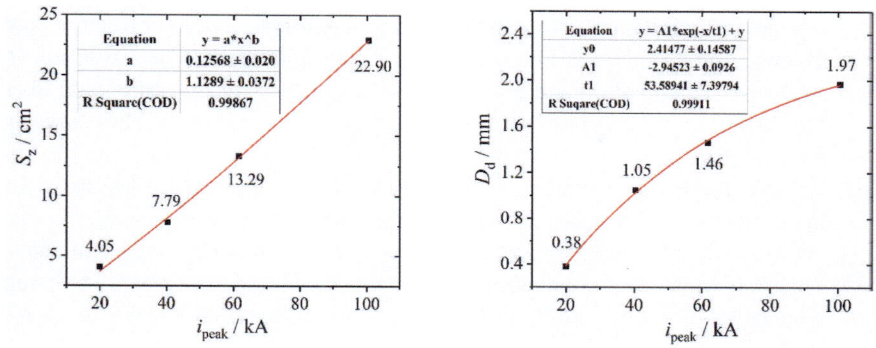

(a) Transmission damage area

(b) Lightning damage depth

Fig. 3.31 Fitted curves of lightning damage area and depth versus peak current in two-dimensional woven CFRP composites

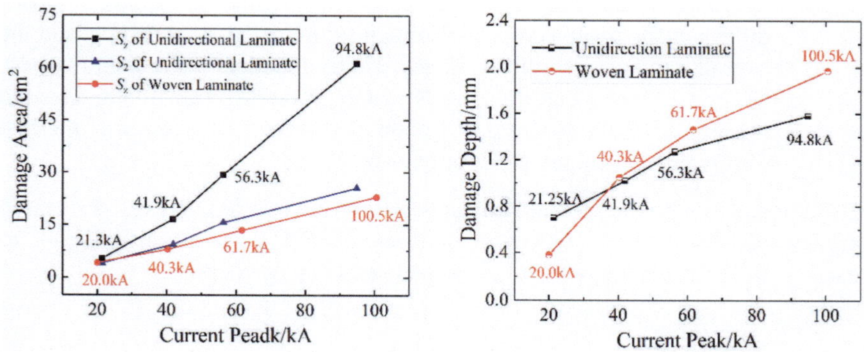

(a) Transmission damage area

(b) Lightning damage depth

Fig. 3.32 Comparison of lightning damage between unidirectional laminate and two-dimensional woven CFRP composites

Experimental results under lightning current component A reveal distinct damage characteristics across CFRP structures:

(i) The lightning damage area of two-dimensional woven composites is slightly smaller than the resin pyrolysis area in unidirectional laminates. This is because current disperses along both radial and weft fibers in the woven structure, while in unidirectional laminates, current concentrates along a single fiber orientation, leading to localized Joule heating and higher pyrolysis efficiency, thus producing a larger damaged area.

(ii) The total transmission damage area in unidirectional laminates significantly exceeds that of two-dimensional woven composites, with this difference becoming more pronounced at higher current amplitudes. In unidirectional laminates, total damage consists of both resin pyrolysis and fiber fracture, and higher currents increase internal pressure from pyrolysis gases, resulting in explosive delamination and brittle fiber damage. In contrast, orthogonal fiber interlacing in woven composites restricts gas-driven expansion, thus limiting fiber fracture regions.

(iii) At lower lightning currents, damage depth in woven composites is less than that in unidirectional laminates. However, as the current peak increases, the damage depth in woven structures gradually exceeds that in unidirectional laminates. This is because dual conduction paths along radial and weft fibers in woven composites initially confine current to surface layers. As current increases, internal plies are progressively engaged, deepening the damage.

(a) Figure 3.32a shows that in unidirectional laminates, fiber fracture area increases faster than resin pyrolysis area as current amplitude rises. This leads to significantly larger total damage areas under high-current conditions.

(b) In contrast, due to structural constraints, lightning damage in two-dimensional woven composites remains dominated by pyrolysis, and the overall damage area grows slowly. Under identical test conditions (40, 60, and 100 kA), current spreads widely in unidirectional laminates but remains spatially confined in woven composites, concentrating thermal energy and enhancing penetration depth.

Additionally, in unidirectional laminates, part of the lightning energy is released via gas explosion, while in woven composites, restricted gas expansion leads to greater thermal dissipation, which further increases damage depth.

3.5 Lightning Damage of CFRP Composites with Fasteners Under Lightning Current

Fastener assemblies in large aircraft introduce high contact resistance at the metal-composite interface, causing localized energy absorption during a strike. This leads to surface ablation, interface degradation, loss of mechanical strength, and ignition hazards (e.g., to fuel tanks). Thus, investigating direct lightning effects on fastener-CFRP structures is of critical theoretical and practical importance.

3.5.1 Lightning Strike Process of CFRP Composites with Fasteners

A high-speed camera (120 kfps, 128 × 128 pixels) recorded the attachment and propagation of a 48 kA lightning arc in a fastener-CFRP assembly (Fig. 3.33).

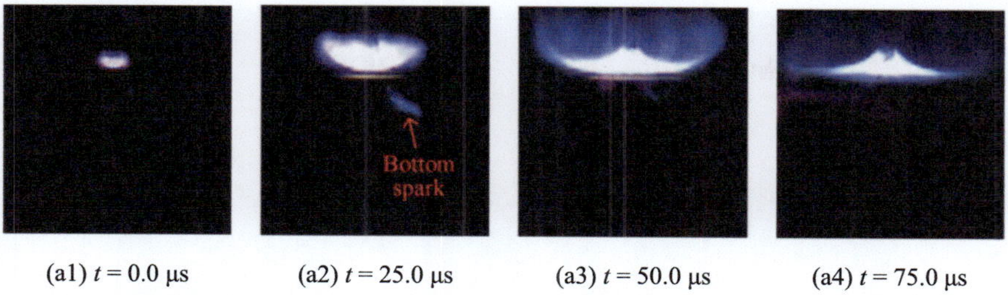

Fig. 3.33 Lightning strike process of CFRP composites with fasteners

As the arc forms at 25 μs, symmetrical fan-shaped channels emerge beside the fastener, with intense spark discharge and gas ejection at the interface. By 33.3 μs, these channels expand, sustaining sparks and sputtered plasma at the metal–composite junction. At 75 μs, as the current decays, spark and gas ejection cease. The observed electrical and thermal sparks can be attributed to:

(i) Electrical ignition sources: dielectric breakdown in micro-gaps around the fastener and at sharp edges.
(ii) Thermal ignition sources: Joule heating at the interface, and high-temperature gas jets containing molten fastener metal and pyrolyzed composite particles expelled from assembly gaps.

These combined phenomena underscore the complex ignition risk posed by fastener interfaces under lightning strike conditions.

3.5.2 *Lightning Damage Morphology and Analysis of CFRP Composites with Fasteners*

Damage Morphology of CFRP Composite Fasteners

The 48 kA lightning damage on fastener-CFRP assemblies is shown in Fig. 3.34.

Upon injection, the arc initially anchors at the fastener before spreading through the plies. The resulting damage zoning comprises:

(i) Damage Zone I: The attachment region at the fastener-composite interface. High contact resistance here causes charge accumulation and intense Joule heating, leading to severe resin pyrolysis, fiber fracture, and ablation pits.
(ii) Damage Zone II: The conduction region where current propagates along fiber orientations from the fastener. This yields resin pyrolysis and fiber ablation, producing ply fractures along conductive paths.
(iii) Damage Zone III: The expansion region where current diverges both along and across fiber plies. Here, interlaminar resin pyrolysis, fiber uplift, and gas-driven separation damage the composite between fibers and plies.

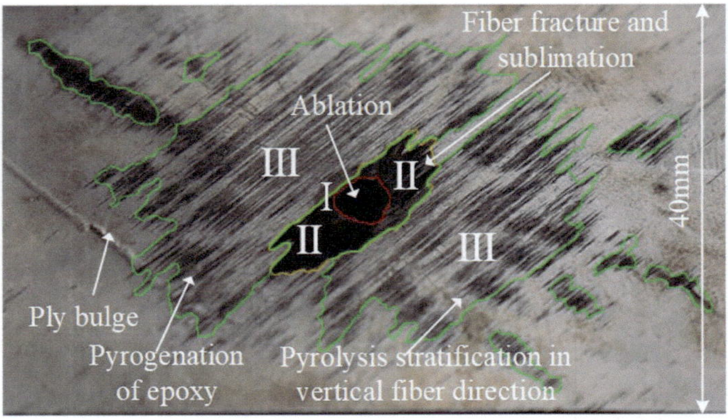

Fig. 3.34 Lightning damage of CFRP with fasteners under 48 kA lightning current

Damage distribution is symmetrical and fan-shaped, centered on the fastener and aligned with the 45° ply orientation, forming approximately circular zones. Integrating this morphology with surface conduction, Fig. 3.35 depicts the diversion paths of lightning current in fastener-CFRP structures.

When lightning strikes fastener regions, the current diverts from fastener components into successive CFRP plies. Unlike the fastener's continuous metal paths, the thin, discontinuous resin between carbon filaments impedes charge diffusion, causing initial charge accumulation near the fastener hole. High contact resistance at the fastener-CFRP interface then produces intense Joule heating, leading to localized damage. As charge builds, dielectric breakdown occurs at the interface, triggering fiber ablation, resin pyrolysis/carbonization, and gas evolution. The volume of pyrolysis gases directly correlates with interlaminar delamination in the composite.

(i) First-ply damage in fastener-CFRP assemblies depends on both internal current paths and the surface arc propagation patterns.

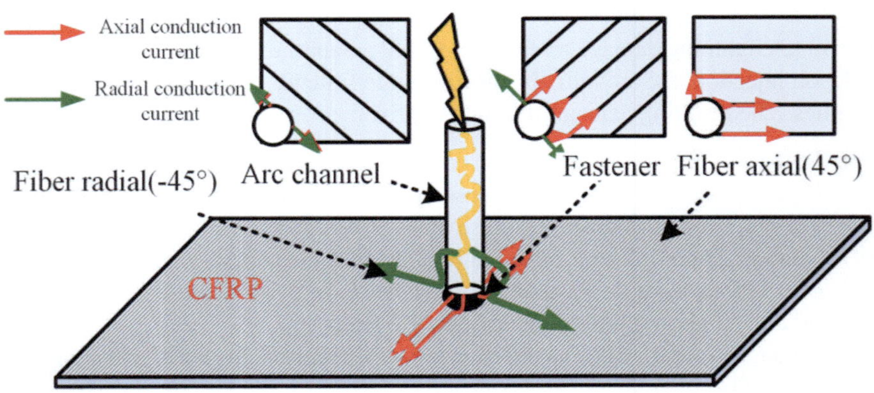

Fig. 3.35 Lightning current diversion mechanism in fastener-CFRP assemblies

(ii) Lightning current travels primarily along fiber axes, generating heat that pyrolyzes inter-fiber and inter-ply resin. The resulting spike in radial and through-thickness conductivity enables 3D current diversion, establishing an interconnected conductive network, as shown in Fig. 3.36 [18, 19] (Fig. 3.37).

Figure 3.38 illustrates the lightning interaction process with fastener-CFRP composites [20]. During lightning attachment, metallic fasteners conduct charge, preventing its buildup. However, intense arc heating melts and vaporizes the fastener head while residual charges concentrate at micro-gaps in the fastener-CFRP interface. When the local electric field exceeds breakdown thresholds, secondary discharges occur between plies or at interface gaps, as depicted in Fig. 3.38b, releasing intense

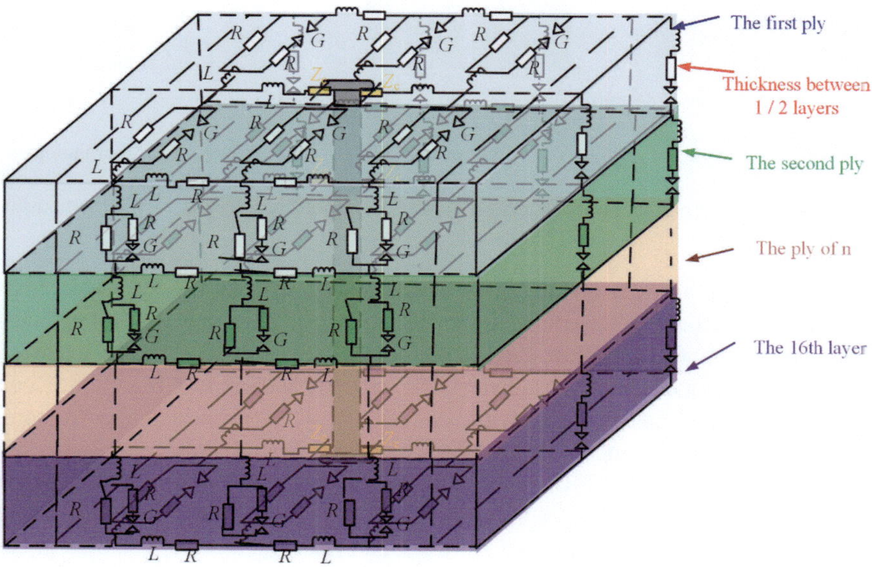

Fig. 3.36 Conduction network in a laminated CFRP composite under lightning current

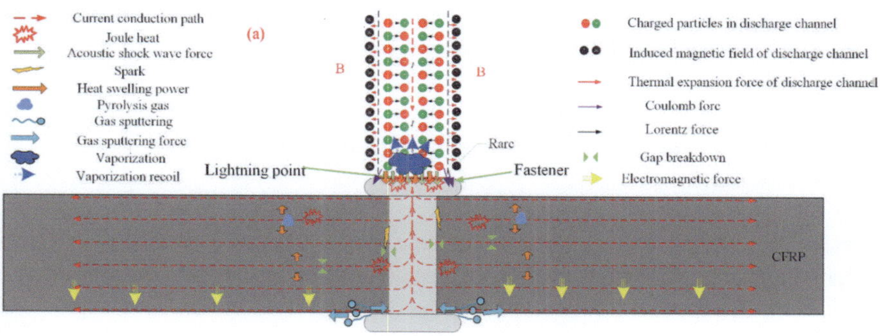

Fig. 3.37 Lightning interaction process with fastener-CFRP composites

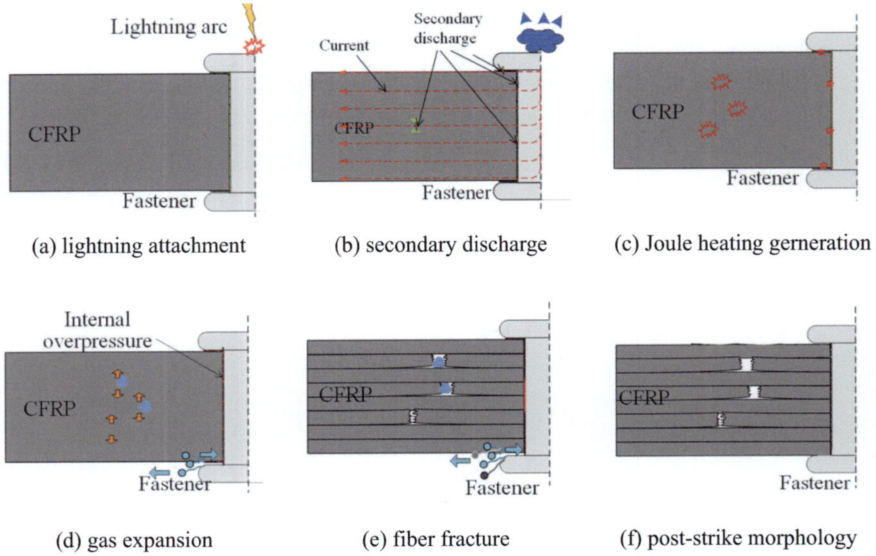

Fig. 3.38 Lightning current diversion and gas expansion phenomena in fastener-CFRP composites

Joule heat that causes resin melting, vaporization, and ablation, alongside fiber combustion, fracture, sublimation, and ablation pit formation in Zone I After interface breakdown, the arc propagates along fiber axes, inflicting axial extension damage. Simultaneously, currents diverge through the ply thickness, producing Zone II (conduction) and Zone III (expansion) damage via ply pyrolysis and fiber fracture.

As the peak current increases, pyrolysis gas volume rises, expanding along the fastener base and within CFRP layers, as shown in Fig. 3.38d. The confined space between the fastener and hole traps gas, causing gas sputtering, including arc plasma sparks, pressurized gas jets, ionized particles, and fiber fragments. Sustained injection sustains overpressure and gas shockwaves, triggering internal gas explosions that yield severe fiber fractures and deformation Fig. 3.38e. Pyrolysis-weakened interply bonds delaminate, exacerbated by trapped gases. Upon cooling, residual gases escape, leaving fractured fibers protruding at the interface, deformed fastener holes, and visible CFRP delamination, shown in Fig. 3.38f.

Microstructure Analysis of Lightning Damage in Fastener-CFRP Composites

Higher lightning currents produce complex surface and internal damage via two primary mechanisms. Localized ablation at attachment points absorbs thermal energy through arc heating and Joule effects, generating pyrolysis gases. Gas explosions intermittently interrupt current paths, concentrating charges in conduction zones, which slows pit growth but intensifies resin pyrolysis. Simultaneously, gas from

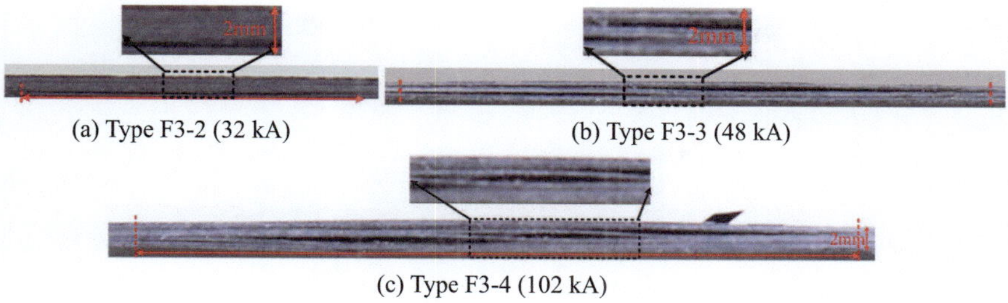

Fig. 3.39 Side delamination morphology of CFRP composite fasteners

fiber sublimation and resin breakdown triggers thermal explosions, expanding the fastener-CFRP interface. Delamination is a critical failure mode in fastener-CFRP assemblies. Figure 3.39 shows side-view delamination under varying currents [20].

At 32 kA, delamination is minor, limited to a short, single layer. At 48 kA, delamination lengthens and occurs across two regions through the thickness, indicating severe internal splitting and a large damage area. At 102 kA, extensive delamination spans approximately 62 mm, covering nearly the full thickness with noticeable deformation. Figure 3.40 plots the correlation between delamination length and lightning current peak in fastener-CFRP composites.

Compared to the 48 kA case, the 102 kA delamination length increases by 34.8%, and three distinct delamination zones emerge. The reasons are:

(i) Surface pyrolysis consumes part of the arc energy, enlarging ablation pits and rupturing the confined fastener-CFRP interface space via gas explosions at the attachment point.

Fig. 3.40 Delamination length versus lightning current peak for fastener-CFRP composites

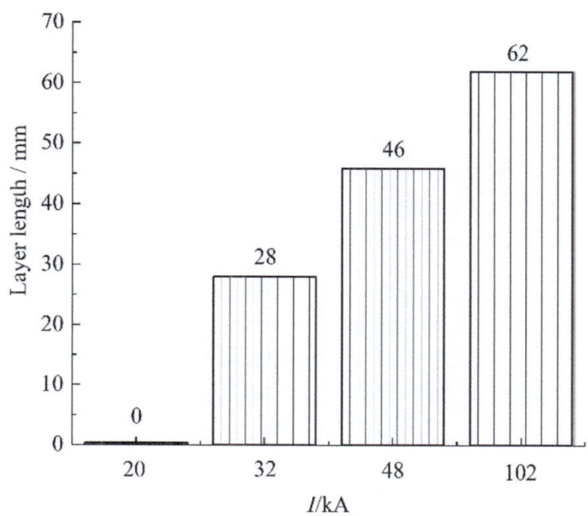

(a) Overview of Zone II

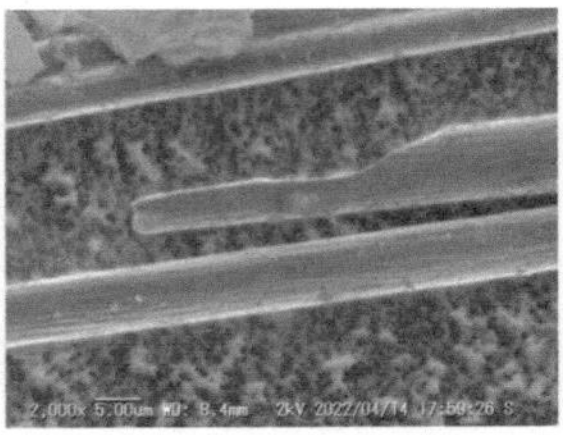

(b) Magnified view of Zone II

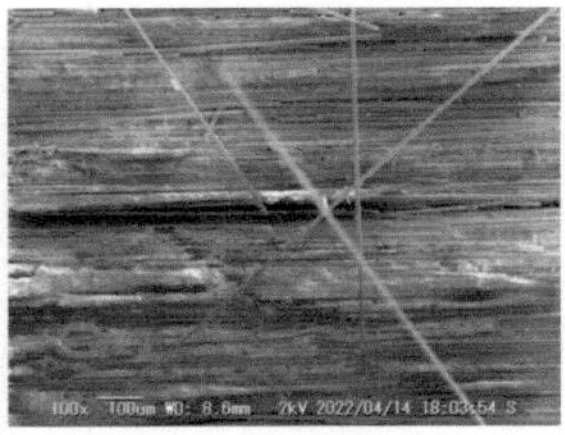

(c) Overview of Zone III

(d) Magnified view of Zone III

(e) Microstructure in expansion zone

(f) Reduced pyrolysis and milder fractures.

Fig. 3.41 Lightning damage microstructure of CFRP composite fasteners

(ii) Instantaneous high pressure and pyrolysis gas buildup at the interface cause abrupt interfacial gas pressure spikes, propagating delamination in structurally vulnerable areas.
(iii) Expanded delamination zones enable gas diffusion between plies, reducing explosive force and limiting further delamination growth.

Electron microscopy of surface plies under 102 kA (Fig. 3.41) reveals:

(i) Zone II (Fig. 3.41a, b) shows severe fiber ablation and sublimation, with "bamboo-shoot" tip-shaped fiber ends and extensive resin pyrolysis.
(ii) Zone III (Fig. 3.41c, d) exhibits partial resin pyrolysis, fiber ablation, fiber-bundle fractures, and uneven ply surfaces. Magnified cross-sections reveal fiber fractures, bundle delamination along fiber directions, and outward fiber uplift.
(iii) In distal expansion zones (Fig. 3.41e, f), resin pyrolysis is reduced and fiber fractures are less severe.

Figure 3.42 presents micro-damage at the second-ply fastener hole, where complete resin pyrolysis exposes randomly oriented carbon fibers coated with carbon particles. Fiber bundles display uneven thickness and "bamboo-shoot" tip morphology at their fracture ends.

(a) Overview (b) Magnified view

Fig. 3.42 Microstructure at the second-ply fastener hole

3.5.3 Analysis of Lightning Damage Results of CFRP Composites with Fasteners

Influence of Lightning Current Peak on CFRP Damage

Ultrasonic C-scan and photographic results for fastener-CFRP specimens under varying lightning peaks, as presented in Table 3.3, are shown in in Figs. 3.43 and 3.44.

From these results in Figs. 3.43 and 3.44, it can be concluded that:

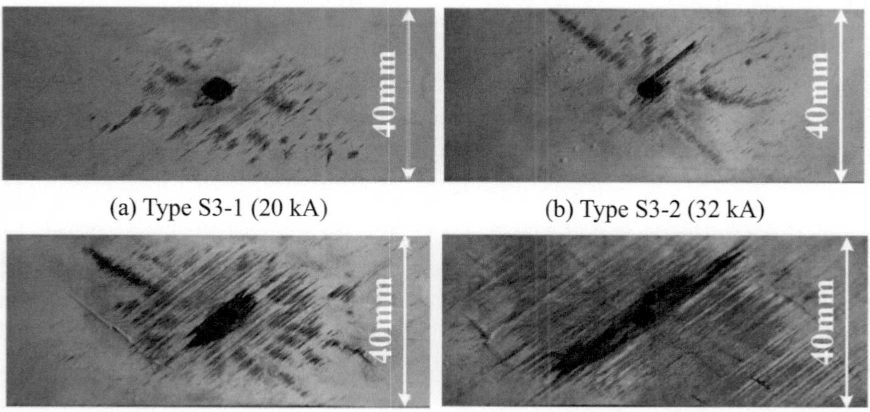

(a) Type S3-1 (20 kA) (b) Type S3-2 (32 kA)

(c) Type S3-3 (48 kA) (d) Type S3-4 (102 kA)

Fig. 3.43 Experimental surface damage of CFRP composite fasteners

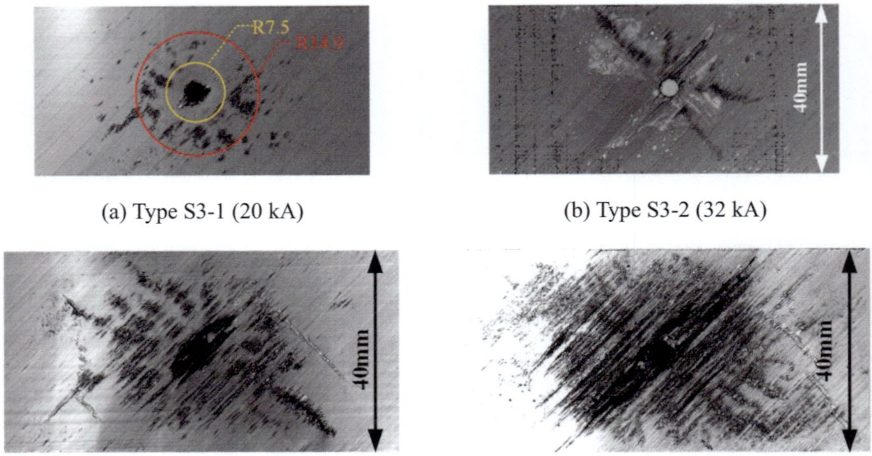

(a) Type S3-1 (20 kA)

(b) Type S3-2 (32 kA)

(c) Type S3-3 (48 kA)

(d) Type S3-4 (102 kA)

Fig. 3.44 Ultrasonic transmission scans of apparent damage area

(i) At 20 kA, minimal inter-ply current diversion produces small surface damage. Weak resin pyrolysis appears along conduction paths, with subtle damage at the fastener head and moderate gas-driven debonding in distal zones.

(ii) At 32 kA, damage areas enlarge, showing axial fiber fractures and ply uplift. Perpendicular pyrolysis cracks emerge, and bubble-induced ply protrusions form near the fastener hole.

(iii) At 48 kA, damage intensifies across fiber orientations. Ablation pits and fan-shaped symmetric zones appear, alongside significant resin pyrolysis in radial fiber regions.

(iv) At 102 kA, damage areas grow further, with larger ablation pits. Interrupted pyrolysis zones merge into continuous damage, slowing area expansion but increasing morphological complexity.

Ultrasonic scans (Fig. 3.45) correlate with surface observations, revealing cumulative damage through the thickness.

As the current peak rises, transmission damage areas increase. At 20 kA, the transmission damage zone is slightly larger and circular (Fig. 3.45a). At 32 kA, it extends toward edges. At 48 kA, damage spreads radially with 63 mm longitudinal reach. At 102 kA, the longitudinal zone reaches 74 mm after ply superposition.

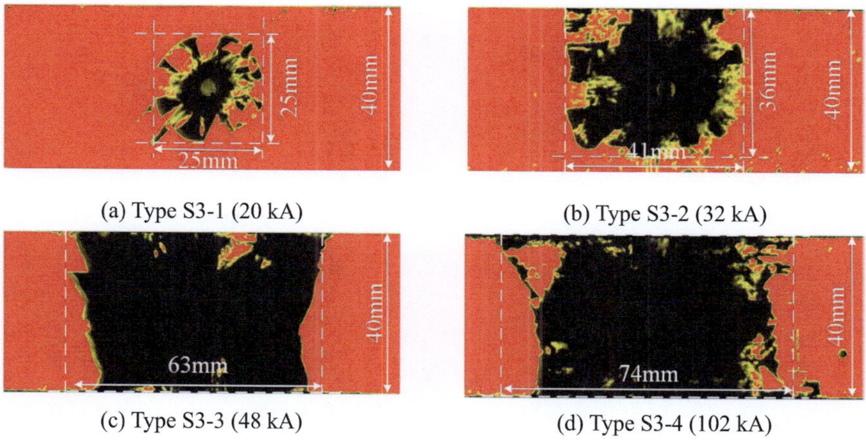

Fig. 3.45 Transmission scanning of fastener-CFRP composites

The Influence of Arc Injection Gap Distance on CFRP Lightning Damage.

To assess gap effects, experiments used 50 kA current with injection gaps of 26 mm (Type S3-5) and 50 mm (Type S3-6), as detailed in Table 3.6. The specimen structures were kept identical to Types S3-1 to S3-4 (refer to Table 3.4).

Surface and internal damage for Type S3-5 (26 mm gap) and Type S3-6 (50 mm gap) at 50 kA are shown in Fig. 3.46.

Observations reveal that:

(i) At a 50 mm gap, surface pyrolysis damage forms block-like protrusions, exceeding those at 26 mm.
(ii) Ultrasonic scans show the transmission damage area decreases as the gap widens.
(iii) Side delamination on fastener-CFRP interfaces is more pronounced at larger gaps, as illustrated in Fig. 3.47. Severe interlaminar delamination appears on the composite sides under lightning current component A. The extent and pattern of this delamination are strongly influenced by the fastener assembly and production process. Further research is required to fully elucidate the mechanisms driving lightning-induced delamination in fastener-CFRP structures.

Table 3.6 Experimental parameters for fastener-CFRP lightning tests

Test No	Experimental conditions	I (kA)	Rising rate (kA/μs)	Effect integral W (A^2s)
Type S3-5	Gap distance = 26 mm	53	2.21	152100
Type S3-6	Gap distance = 50 mm	52	2.17	142657

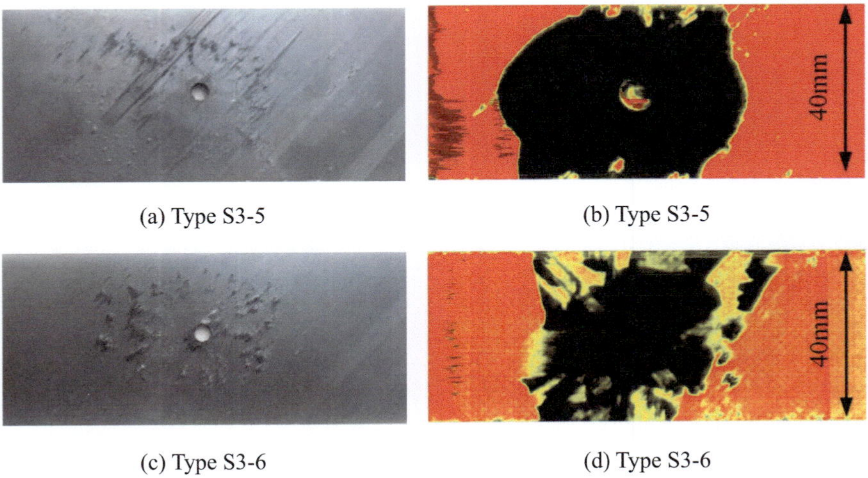

Fig. 3.46 Transmission scanning of fastener-CFRP composites under varying gaps

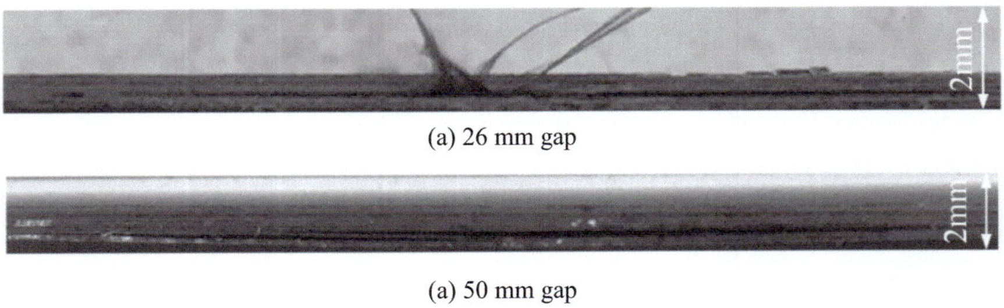

Fig. 3.47 Side delamination of fastener-CFRP composites under 50 kA

3.6 Summary

This chapter experimentally investigates lightning damage in unidirectional anisotropic laminates, two-dimensional woven architectures, and fastener-reinforced CFRP under lightning current component A. Using optical microscopy, ultrasonic scanning, X-ray tomography, and scanning electron microscopy, damage morphologies and modes are characterized, and underlying mechanisms are elucidated. Key findings include:

(1) Damage modes under component A encompass resin ablation/pyrolysis/carbonization, fiber-bundle ablation/sublimation/fracture, and ply delamination, driven by Joule heating thermal effects and explosive forces from accumulated pyrolysis gases.

(2) In unidirectional anisotropic laminates, total damage area combines brittle fiber fractures from gas shocks and resin-pyrolysis delamination. Damage area scales by a power law with current peak, while resin pyrolysis area correlates linearly with peak, and damage depth grows exponentially. Higher peaks enhance arc penetration, intensify interlayer pyrolysis, and trap gases that expand explosively, causing rapid damage propagation.
(3) For two-dimensional woven CFRP, fiber-bundle constraints suppress damage until peaks exceed the pyrolysis-burst threshold. As a result, fiber-fracture zones remain smaller and largely overlap with pyrolysis delamination. At low peaks, surface conductivity along radial/weft fibers limits penetration, yielding shallower damage. At high peaks, internal plies engage, but energy confinement concentrates damage, producing deeper penetration than in unidirectional laminates.
(4) Fastener-CFRP assemblies, which is critical in large aircraft, experience sparks, gas sputtering, and localized heating at the metal-composite interface. Increasing current peaks aggravates surface damage spread and internal delamination, whereas wider arc gaps reduce both damage area and delamination severity.

References

1. Feraboli P, Miller M. Damage resistance and tolerance of carbon/epoxy composite coupons subjected to simulated lightning strike[J]. Composites: Part A, 2009, 40(6–7): 954–967.
2. Hirano Y, Katsumata S, Iwahori Y, et al. Artificial lightning testing on graphite/epoxy composite laminate[J]. Composites Part A: Applied Science and Manufacturing, 2010, 41(10): 1461–1470.
3. Feraboli P, Kawakami H. Damage of carbon/epoxy composite plates subjected to mechanical impact and simulated lightning strike[J]. Journal of Aircraft, 2010, 47(3): 999–1012.
4. Hirano Y, Reurings C, Iwahori Y. Damage resistance of graphite/epoxy laminates with a fastener subjected to artificial lightning[C]. 18th International Conference on Composite Materials, Jeju Island, South Korea, 2011.
5. Li Y, Li R, Lu L, et al. Experimental study of damage characteristics of carbon woven fabric/epoxy laminates subjected to lightning strike[J]. Composites Part A: Applied Science and Manufacturing, 2015, 79: 164–175.
6. MIL-STD-464A. Electromagnetic environmental effects requirements for systems[S]. Washington DC, United States: Department of Defense Interface Standard, 2002.
7. SAE ARP 5412. Aircraft Lightning Environment and Related Test Waveforms[S]. Warrendale, PA, United States: SAE International Group, 2005.
8. Dong Q, Guo Y, Sun X, et al. Coupled electrical-thermal-pyrolytic analysis of carbon fiber/epoxy composites subjected to lightning strike[J]. Polymer, 2015, 56: 385–394.
9. Li S, Yin J, Yao X, et al. Damage analysis for carbon fiber/epoxy composite exposed to simulated lightning current, Journal of reinforced plastics and composites[J]. 35(15): 1201–1213.
10. Yao X, Guo C, Sun J, et al. Damage Simulation and Experiment of Carbon Fiber Composites Subjected to Lightning Current[J], High Voltage Engineering, 2017, 43(5): 1400–1408.
11. Yin J, Chang F, Li S, et al. Experimental and numerical simulation analysis of typical carbon woven fabric/epoxy laminates subjected to lightning strike[J]. Applied Composite Material. 2017, 24(6): 1353–1372.

12. Yi J, Li S, Chang F, et al. Tensile residual strength prediction of composite laminate with lightning strike thermal-mechanical coupling damage[J]. Acta Materiae Compositae Sinica, 2017, 34(1): 83–90.
13. Yi J, Li S, Yang Z, et al. Test on damage and mechanical property degradation of composite laminate with fastener subjected to lightning strike[J]. Acta Materiae Compositae Sinica, 2018, 35(5): 1131–1138.
14. Yi J, Li S, Yang Z, et al. Dynamic response and damage performance analysis for composite laminate under the act of lighting impact force[J]. Acta Materiae Compositae Sinica, 2018, 35(6): 1481–1489.
15. Yin J, Li S, Yao X, et al. Lightning strike ablation damage characteristic analysis for carbon fiber/epoxy composite laminate with fastener[J]. Applied Composite Material, 2016, 23(4): 821–837.
16. Yin J, Li S, Yao X, et al. Ablation damage characteristic analysis of composite laminate with fastener subjected to lighting strike[J]. Acta Materiae Compositae Sinica, 2017, 34(1): 104–111.
17. Sun J. Study on the damage mechanism of lightning pulse current on aircraft carbon fiber composite materials[D]. Xi'an. Xi'an Jiaotong University, 2019.
18. Tian X. Study on the damage mechanism of carbon fiber composites with fasteners under lightning current[D]. Xi'an. Xi'an Jiaotong University, 2022.
19. Tian X, Yao X, Sun J, et al. Damage characteristics of carbon fiber reinforced polymer composites with fastener subjected to lighting A current [J]. Acta Materiae Compositae Sinica, 2020, 37(2): 10.
20. Tian X, Yao X, Feng R,et al. Damage mechanism analysis for carbon fiber reinforced polymer composites with a fastener under lightning strike[J]. Composite structures, 2022(Jun.): 289.

Chapter 4
Multi-physical-Field Simulation of Lightning Damage in CFRP Composites

4.1 Simulation of CFRP Lightning Damage

4.1.1 Physical Process of Lightning Interaction with Aircraft

When an aircraft is struck, the lightning leader forms, attaches, and its current spreads and disperses over the CFRP surface, driving thermo-electric ablation, heat dissipation, and cooling. Damage arises from coupled effects: overvoltage, mechanical impact, electromagnetic forces, Joule heating, and resin vaporization due to rapid thermal expansion of the arc channel. Lightning is a gas-discharge phenomenon: once the field exceeds air breakdown (~30 kV/cm), a plasma channel conducts current. On aircraft, the arc adheres, propagates, and diffuses across CFRP panels. Impulse generators can simulate this process, replicating arc development, attachment, and dispersion, to evaluate CFRP damage. However, experimental methods face challenges:

(i) Plasma evolution complexity: with temperatures >30 000 K and electron speeds >5000 m/s, direct measurement of arc-plasma properties is infeasible.
(ii) Limited diagnostic accuracy: plasma diagnostics (laser, microwave) struggle against the many degrees of freedom in high-energy plasmas.

These factors hinder comprehensive experimental characterization of arc-material interactions, limiting insights into the microscopic damage mechanisms in CFRP.

Whether studying lightning attachment on large aircraft or assessing extensive structural damage, such research is typically conducted via simulated lightning strike experiments in laboratories. These high-voltage, high-current tests require specialized training and robust safety infrastructure, resulting in high costs and limited coverage of test conditions. In contrast, simulation offers a faster, safer, and more cost-effective approach for investigating lightning damage in CFRP composites.

With the rapid advancement of computational technology, numerical simulation has become a vital tool in materials science, plasma physics, and engineering. It

J. Sun et al., *Lightning Effects on Carbon Fiber-Reinforced Polymer Composites*,
https://doi.org/10.1007/978-981-95-3196-7_4

helps to visualize physical processes, interpret experimental observations, and facilitate new theoretical models. Moreover, it can optimize experimental design and reduce costs. For instance, magnetohydrodynamics (MHD) is widely used to simulate arc behavior during switchgear breaking, while the finite element (FE) method effectively resolves complex structural problems. These methods also apply to other engineering domains, such as heat conduction, mass diffusion, and thermoelectric coupling analysis.

4.1.2 Simulation Research Status of Lightning Damage in CFRP Composites

Tracing back to electrical-engineering studies, MHD-based arc models have long been used to simulate switchgear arcs. By solving coupled electromagnetic, thermal, and aerodynamic equations, these models relate arc shape and motion to injection current, applied fields, and electrode erosion, thereby explaining arc initiation, development, and quenching. However, switchgear electrodes are monolithic, with no interlaminar current shunting or multilayer damage, so MHD switch-arc models do not capture the surface ablation, current diversion, and intra-ply damage critical to CFRP composites under lightning.

In 2010, Ogasawara et al. [1] developed a thermo-electric coupling FEM model in ABAQUS to predict damage area and depth in CFRP under 2.6/10.5, 4/20, and 7/150 μs currents. They assumed through-thickness conductivity varies linearly with temperature, using the temperature field as a damage index, but did not account for temperature-dependent changes in thermal conductivity, specific heat, density, or mechanical properties. Li et al. [2, 3] extended this by incorporating temperature-dependent thermophysical and mechanical degradation, improving the model's ability to capture CFRP deterioration under lightning. Subsequent work [4] established a resistive-thermophysical model that analyzes how current waveform parameters, material properties, and laminate architecture govern lightning damage in CFRP composites.

In thermo-electric coupling simulations, a realistic conductivity model that reflects actual lightning strikes is essential for accurate results. CFRP conductivity is anisotropic and degrades with damage. Existing studies typically assume through-thickness conductivity varies linearly from 100 to 7.64×10^{-4} S/m over 600–3000 °C, but resin pyrolysis lags temperature rise, causing discrepancies between simulated and experimental damage modes and areas. To address this, researchers [5–7] integrated resin-pyrolysis kinetics into their models, creating a thermal-electric-pyrolysis coupling framework. Guo et al. [6] evaluated damage using both temperature-field and pyrolysis-extent criteria, finding that the temperature-based metric better predicts damage area, while the pyrolysis-based metric more accurately captures damage depth.

In addition to FEM-based thermo-electric coupling models, Katunin and Krukiewicz [8] treated the lightning arc as an external heat source, developing a thermal-ablation model for CFRP under strike. They parameterize arc heat flux by the current waveform and instantaneous channel radius, and include anisotropic, temperature-dependent thermal conductivity to emulate the spatiotemporal evolution of arc adhesion and movement on the surface. However, this model neglects the composite's electrical conductivity, Joule heating, and internal current scattering, leaving a gap between simulated and observed damage. The MHD arc model cannot capture delamination temperatures or interlaminar damage in semiconductive CFRP, while ABAQUS electro-thermal models reproduce current dispersion and damage initiation but omit the arc-attachment dynamics on CFRP.

This chapter introduces a space–time, fluid–solid coupling approach that integrates an arc-plasma evolution model with a thermo-electric coupling model for CFRP. By validating against experimental data, we establish critical damage criteria and evaluate how CFRP material parameters affect lightning-current interactions and damage distributions.

4.2 MHD Model and Simulation of Lightning Arcs

4.2.1 Basic Principle of Impulse Arcs

In a simulated lightning strike on CFRP, a faint glow, the step leader, forms between the electrode and composite before descending. Once the electric field exceeds the air breakdown threshold ($\approx$30 kV/cm), ionization occurs, and charges propagate downward. The CFRP surface blocks the leader, creating a conductive arc channel to the sample. Upon arc attachment, a bright white flash appears, and the material heats sharply due to direct arc radiation and Joule heating from ion neutralization. Here, arc heat flux represents thermal plasma effects, while Joule heat corresponds to resistive heating within the conductive path. Arc discharge concentrates energy into a high-temperature, high-luminosity plasma channel via coupled electromagnetic, thermal, and fluid-dynamic processes. Figure 4.1 divides this discharge into three regions: (i) Cathode region: electron emission and initial ionization; (ii) Arc column: sustained thermal ionization and conductivity; (iii) Anode region: charge collection and circuit completion.

Firstly, upon current injection into the metal electrode, rapid resistive heating occurs, radiating outward. Electrons are emitted from the cathode via field emission and thermionic emission, then accelerated toward the anode by the electric field. As they traverse the arc column, inelastic collisions with gas molecules ionize air, producing positive ions and free electrons. The thermal ionization process renders the plasma highly conductive. Free electrons are collected at the anode, while ions drift back to the cathode, sustaining the arc discharge. This discharge generates intense

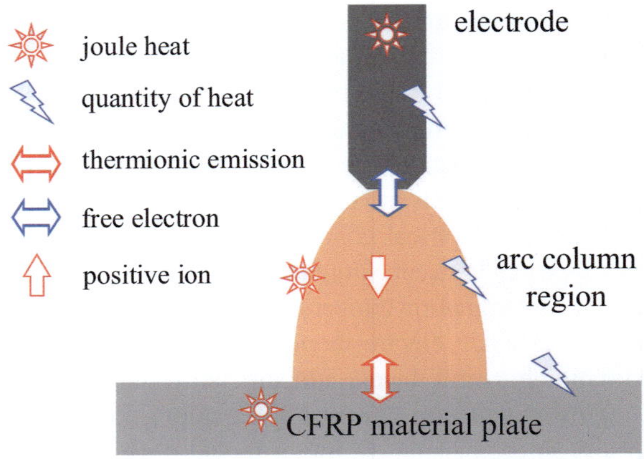

Fig. 4.1 Schematic of the three region in a lightning arc on CFRP composites

arc radiation and Joule heat, which dissipate through convection, conduction, and radiation, producing the characteristic bright white flash.

4.2.2 Basic Governing Equations of the Physical Field

To simulate fast-rise-time lightning arcs, the arc channel is approximated as a thermal plasma in local thermodynamic equilibrium (LTE), where frequent electron-particle collisions equilibrate electron and heavy-particle temperatures. The arc column is treated as a Newtonian, single-phase flow, and the surrounding air is modeled as a weakly compressible continuous fluid, with gravity neglected to reduce computational cost.

The MHD framework couples Maxwell's equations, the Navier-Stokes equations, and the heat-conduction equation to capture electromagnetic, fluid-dynamic, and thermal interactions in the plasma. In the solid electrodes (cathode and anode), only Maxwell's and the heat-conduction equations are solved. To simplify the geometry, a needle-plate discharge configuration is used to model impulse current injection. The model's governing equations include:

(i) Fluid flow equations;
(ii) Energy conservation equation;
(iii) Electromagnetic field equation;
(iv) Arc-material interface interaction.

These coupled equations resolve the dynamic evolution of the lightning arc and its thermal impact on CFRP composites.

Fluid Flow Equations

The laminar flow model describes the motion of the conductive plasma via classical mass and momentum conservation.

$$\frac{\partial \rho}{\partial t} + \nabla \cdot (\rho \overrightarrow{v}) = 0 \tag{4.1}$$

$$\rho \frac{\partial \overrightarrow{v}}{\partial t} + \rho(\overrightarrow{v} \cdot \nabla)\overrightarrow{v} = -\nabla p + \nabla \cdot \boldsymbol{\tau} + \overrightarrow{J} \times \overrightarrow{B} \tag{4.2}$$

Here, ρ, t, $\overrightarrow{v}$, p denote density (kg/m^3), time (s), velocity (m/s), and pressure (Pa), respectively. $\tau = 2\mu_D \boldsymbol{S} - \frac{2}{3}\mu_D(\nabla \cdot \overrightarrow{v})\boldsymbol{I}$ is the viscous stress tensor; μ_{D} is the dynamic viscosity coefficient; $\boldsymbol{S} = \frac{1}{2}(\nabla \overrightarrow{v} + (\nabla \overrightarrow{v})^T)$ is the strain rate tensor; $\overrightarrow{J} \times \overrightarrow{B}$ is the Lorentz force source term represents current-magnetic field interactions driving the flow.

Energy Conservation Equation

Heat transfer within the plasma is governed by the energy balance.

$$\rho C_p \left(\frac{\partial T}{\partial t} + \overrightarrow{v} \cdot \nabla T \right) = \nabla \cdot (k \nabla T) + \frac{5k_B}{2e} \overrightarrow{J} \cdot \nabla T + \overrightarrow{J} \cdot \overrightarrow{E} - 4\pi \varepsilon_N \tag{4.3}$$

where, C_{p} is the specific heat capacity, T is the absolute temperature, k is the thermal conductivity, k_{B} is the Boltzmann constant, e is the electronic charge, and $\overrightarrow{E}$ is the electric field strength. The electronic enthalpic flux $\frac{5k_B}{2e}\overrightarrow{J} \cdot \nabla T$, Joule heat $\overrightarrow{J} \cdot \overrightarrow{E}$ and radiation heat dissipation $4\pi \varepsilon_N$ together constitute the volume heat source, and ε_{N} is the net radiation coefficient of the plasma.

Electromagnetic Field Equations

The electromagnetic behavior model of arc plasma and electrode material is established by Maxwell equation. In the momentum and energy conservation equations, the Lorentz force and Joule heating source term depend on the distribution of current density $\overrightarrow{J}$, electric field strength $\overrightarrow{E}$ and magnetic flux density $\overrightarrow{B}$. According to Ohm's law (Eq. (4.4)), the current density can be calculated according to the electric field strength $\overrightarrow{E}$:

$$\overrightarrow{J} = \sigma \overrightarrow{E} \tag{4.4}$$

where, σ denotes the electrical conductivity of the corresponding regions, namely the cathode, the arc column, and the anode. The electric field $\overrightarrow{E}$ can be derived from the electromotive force φ(V) and the magnetic vector potential A(T/m):

$$\overrightarrow{E} = -\nabla\varphi - \frac{\partial \overrightarrow{A}}{\partial t} \tag{4.5}$$

where, φ is the electric potential and $\overrightarrow{A}$ is the magnetic potential vector. The electric field is mainly determined by the electromotive force gradient, but when the arc plasma moves in the discharge gap space, the time-varying magnetic field will generate a Neumann electric field. Although some researchers have proposed that when the current in the arc column is stable, the electromotive field $\partial \overrightarrow{A}/\partial t$ can be ignored compared with the electric field $\nabla\varphi$ [9–13]. However, in the context of this chapter, for a time-varying current characterized by the continuous action of multiple consecutive impulse currents, the potential field generated by the induction field cannot be neglected. Therefore, the current continuity equation ($\nabla \cdot \overrightarrow{J} = 0$) obtained by Gauss's law can be expressed in the form of electromotive force V and magnetic vector potential $\overrightarrow{A}$ as follows:

$$\nabla \cdot \left(\sigma\nabla V + \sigma\frac{\partial \overrightarrow{A}}{\partial t}\right) = 0 \tag{4.6}$$

In this chapter, the magnetic vector potential is used to calculate the magnetic induction intensity $\overrightarrow{B}$. According to the Maxwell-Ampere law, the magnetic induction intensity $\overrightarrow{B}$ is related to the gas dielectric constant μ and the current density $\overrightarrow{J}$, which can be expressed as:

$$\nabla \times \overrightarrow{B} = \mu\overrightarrow{J} \tag{4.7}$$

where μ is the permeability. Associative Eqs. (4.4) and (4.5) can be obtained.

$$\sigma\frac{\partial \overrightarrow{A}}{\partial t} + \nabla \times \left(\frac{1}{\mu}\nabla \times \overrightarrow{A}\right) + \sigma\nabla V = 0 \tag{4.8}$$

After calculating the magnetic potential vector $\overrightarrow{A}$ from Eq. (4.7), the magnetic induction intensity $\overrightarrow{B}$ can be derived based on the formula $\overrightarrow{B} = \nabla \times \overrightarrow{A}$.

Arc-Material Interface Interaction

Different from some studies that only simulate the plasma region, the electrode is assumed to be the boundary of the surface region, and the temperature or heat flux boundary conditions are simply set. Taking into account the current continuity and dynamic heat transfer between the electrode region and the plasma, along with the potential interaction effects between the arc plasma and the electrode regarding the spatial distribution of current injection and the attachment of the arc root, the cathode and anode materials are incorporated in the simulation domain, and the heat flux at the plasma-electrode interface is analyzed. Assuming that there is no ion current on the anode, the heat flux on the anode surface includes the heat transfer flux Q_{CA} generated by contact with the high-temperature plasma, the heat released by the electrons entering the anode-plasma interface, and the radiation cooling loss on the electrode surface, which can be written as [14].

$$\vec{q}_{anode} \cdot \vec{n} - \vec{q}_{plasma} \cdot \vec{n} = Q_{CA} + |\vec{J} \cdot \vec{n}|\varphi_A - \varepsilon_B \sigma_B T^4 \tag{4.9}$$

where, $|\vec{J} \cdot \vec{n}|$ is the normal current density of the interface, φ_A is the anode work function, ε_B and σ_B are the emissivity and Stefan-Boltzmann constant, respectively. Regarding the cathode, positive ions within the plasma are accelerated towards the cathode under the action of the electric field, and then are absorbed onto the cathode surface, generating heat. At the same time, the electron absorbs heat to overcome the work function of the cathode and leaves the cathode through thermal emission, resulting in the cooling of the cathode. Therefore, the heat flux at the boundary between the cathode and the plasma satisfies [15]:

$$\vec{q}_{cathode} \cdot \vec{n} - \vec{q}_{plasma} \cdot \vec{n} = Q_{CC} + j_{ion}\varphi_G - j_{electron}\varphi_C - \varepsilon_B \sigma_B T^4 \tag{4.10}$$

Among them, φ_G is the air ionization energy, φ_C is the cathode work function, j_{ion} and $j_{electron}$ are the ion current and the electron current, respectively. For the cathode thermal emission phenomenon, it is proposed that the electron current density j_e cannot exceed the Richardson-Dushman equation current density j_r. Therefore, the electron current and ion current can be expressed as [16]:

$$j_{electron} = \begin{cases} j_r & |\vec{J} \cdot \vec{n}| > j_r \\ |\vec{J} \cdot \vec{n}| & |\vec{J} \cdot \vec{n}| \le j_r \end{cases} \tag{4.11}$$

$$j_{ion} = |\vec{J} \cdot \vec{n}| - j_{electron} \tag{4.12}$$

Among them,

$$j_r = A_R T^2 \exp\left(-\frac{e\varphi_{\text{eff}}}{\sigma_B T}\right) \tag{4.13}$$

In Eq. (4.13), A_R is the Richardson constant, and φ_{eff} is the effective work function of the cathode considering the Schottky effect. Both of these constants are contingent upon the cathode material. For the tungsten cathode used in this paper, the values are 3.0 A/cm^2 K^2 and 2.63 V, respectively [17].

4.2.3 Numerical Model of MHD Arc

The coupled PDEs from Sect. 4.2.2 are solved via finite-element analysis under conditions replicating laboratory lightning strikes. Exploiting the axial symmetry of a free-burning arc, unchanged by external fields, a 2D axisymmetric model is implemented in the FEM environment to capture the spatiotemporal evolution of the arc column during current injection.

This model employs the Equilibrium Discharge Plasma module, integrating electric current (ec), magnetic field (mf), heat transfer (ht), and laminar flow (spf) physics. It solves current continuity in both plasma and electrodes, Maxwell's equations, heat-transfer equations, and the Navier-Stokes equations (in the plasma domain only). By incorporating multi-physics couplings, such as Joule heating, Lorentz forces, non-isothermal flow, and boundary discharge heat sources, it simulates the plasma-electrode heating/cooling at the interface and resolves the electrical, thermal, and fluid-dynamic coupling across all fields.

During modeling, cathode current-injection critically shapes the arc morphology in the gap. Consequently, prescribing the injection path or density at the cathode tip [10, 18–21] or at an artificial boundary [12, 13, 22] via numerical functions is suboptimal. Likewise, the anode's material properties govern current shunting and temperature rise by enforcing charge conservation between plasma and metal, thereby influencing arc attachment and spreading on the anode surface. Yet, many prior studies simplify the anode to a mere boundary [11, 18–20, 23, 24], overlooking these effects. Therefore, this work explicitly includes the cathode, arc plasma, and anode within the simulation domain and employs a time-dependent model to resolve heat transfer, electromagnetic, and fluid-dynamic interactions. The calculation domain of the model is shown in Fig. 4.2.

Taking a 10 mm arc gap as an example, the 25 $\times$ 35 mm simulation domain comprises: an air fluid region, a 20 mm tungsten cathode with a 60° tip, and an 8 mm thick, 25 mm radius steel anode. As the gap varies from 2 to 50 mm, the domain dimensions scale accordingly. Because the A-component peak occurs over tens of microseconds, we employ the Balanced Discharge Plasma module under an LTE assumption, in which electrons and heavy particles share temperature within the arc column. Near each electrode, however, heavy-particle temperatures (~2000–3500 K) match the electrode [14, 16, 25, 26], while hot electrons carry the current, creating NLTE sheaths. Following established practice [13, 22, 23, 25, 27–31], 0.1 mm NLTE layers at both cathode and anode aere modeled, as shown in Fig. 4.2. In these sheaths, the material acts as an ohmic conductor, using electrode conductivity, while retaining the gas's thermophysical properties. This approach ensures current continuity and captures the NLTE transition from plasma to metal conduction at the electrode interface.

The MHD-based thermal-plasma simulation relies on temperature-dependent air thermodynamic and transport properties. Under LTE, these parameters are tabulated as functions of temperature and pressure via the Chapman-Enskog method [32, 33]. Data for air plasma from 300 to 24,000 K are sourced from D'Angola et al. [32]

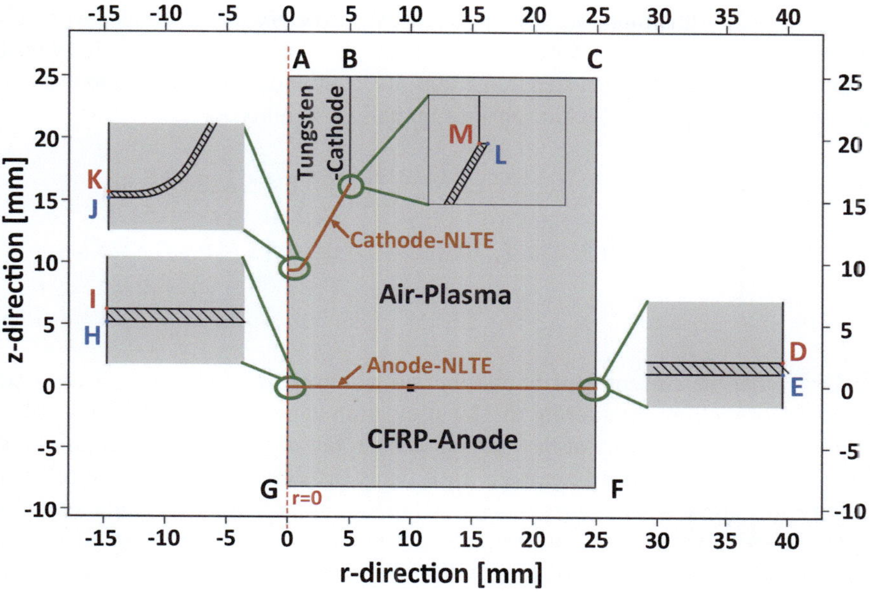

Fig. 4.2 Computational domain showing integrated cathode–plasma–anode geometry for MHD arc simulation

and Boulos et al. [33], while above 24 000 K, properties follow Capitelli et al. [34]. Radiative transfer uses the net radiation coefficient (ε_N) from Naghizadeh-Kashani et al. [35]. A lightning-arc simulation in an open atmosphere employs a large air domain with fixed-temperature boundaries (ambient) for heat transfer and open boundaries for fluid flow. Electrode surfaces are no-slip walls. The anode exchanges heat with 300 K air via convection ($h = 15$ W/m^2 K) and radiation ($\varepsilon = 0.4$). Surface discharge heat sources at both cathode and anode follow Eqs. (4.9) and (4.10).

To ensure consistency between the lightning current load applied in the MHD arc model and that in the experiment, data fitting is performed on the measured waveform curves of the continuous lightning current components A, B, and C. A piecewise function is employed to simulate the single or multiple continuous lightning current components injected into the air plasma region via the cathode. Specifically:

$$I(t) = \begin{cases} I_A(t) & 0 \le t \le 0.1\,\text{ms} \\ I_B(t) & 0.1\,\text{ms} \le t \le 2.5\,\text{ms} \\ I_C(t) & 2.5\,\text{ms} \le t \le 0.55\,\text{s} \end{cases} \tag{4.14}$$

The lightning current components A and B are described using a double exponential function.

$$\begin{aligned} i_A(t) &= 4I_A\left(e^{-\alpha t} - e^{-\beta t}\right), \quad \alpha = 3.6 \times 10^4, \quad \beta = 7.2 \times 10^4 \\ i_B(t) &= 2465I_A\left(e^{-\alpha t} - e^{-\beta t}\right), \quad \alpha = 1.5 \times 10^3, \quad \beta = 3.0 \times 10^4 \end{aligned} \tag{4.15}$$

According to this function, component B exhibits a peak time of approximately 100 μs, a duration of about 5 ms, and a peak amplitude of around 2 kA. To prevent solution non-convergence at step transitions, the continuous current component C is modeled using Eq. (4.16), which ensures numerical stability.

$$I_C(t) = \begin{cases} I_B(2.5\ \text{ms}) + \frac{(t-2.5\ \text{ms})\cdot[I_{av}-I_B(2.5\ \text{ms})]}{0.01}; & 2.5\ \text{ms} \le t \le 12.5\ \text{ms} \\ I_{av}; & 12.5\ \text{ms} \le t \le 0.5125\ \text{s} \\ I_{av} - (t-0.5125)\cdot I_{av}/0.01; & 0.5125\ \text{s} \le t \le 0.5225\ \text{s} \\ 0; & t \le 2.5\ \text{ms and } 0.5225\ \text{s} \le t \end{cases} \tag{4.16}$$

Given that the current density distribution within the cathode can influence arc plasma morphology by altering the local electromagnetic field, accurately determining the current density at the cathode-plasma interface requires calculating the current distribution within the cathode itself. Thus, in this model, the lightning current is uniformly applied to the top surface (AB) of the cathode, and the resultant current density distribution at the cathode interface is obtained by solving Maxwell's equations and the heat transfer equation within the cathode domain. It is important to ensure that the cathode length is sufficient, such that the influence of the current injection boundary condition on the arc plasma behavior becomes negligible. Table 4.1 lists the boundary conditions defined in the two-dimensional axisymmetric arc model.

The grid division of the MHD model is illustrated in Fig. 4.3. The computational domain consists of a total of 59,391 elements, primarily triangular meshes (57,955 elements). The quadrilateral elements (1,436) are concentrated in the boundary layers near the electrode-plasma interfaces to better accommodate non-slip wall boundary conditions and resolve the non-local thermodynamic equilibrium (NLTE) regions adjacent to the electrodes. To enhance convergence in regions exhibiting high gradients in physical fields, mesh refinement is applied near the axis of symmetry and plasma boundaries. The maximum element size on the surfaces of the cathode and anode is 0.15 mm, while in the NLTE regions, the maximum element sizes are refined to 0.04 mm (cathode) and 0.06 mm (anode). For the bulk fluid domain, the maximum mesh size is 0.4 mm.

To address numerical instability and convergence challenges caused by the near-zero electrical conductivity of air at room temperature, several studies have proposed strategies for setting the initial conditions of MHD arc models. Based on thermal field emission theory, Tanaka et al. [36, 37] suggested applying a high-temperature boundary condition (3500 K) at the cathode surface to facilitate arc initiation in air. The elevated temperature preheats the surrounding air, increasing its conductivity due to Joule heating, thereby promoting arc formation and expansion. This fixed high-temperature boundary approach has been widely adopted [11–13, 18–22, 38, 39]. Alternatively, some works [10, 13, 40–43] define a preheated cylindrical region (5000–8000 K) in the fluid domain to simulate the initial plasma channel created by voltage breakdown between electrodes. In Ref. [16], instead of imposing a fixed cathode temperature, a 1 mm thick high-conductivity layer was introduced in the

Table 4.1 Boundary conditions of 2D symmetric model of pulsed arc

Entry	AB	BC	CD	DE	EF	FA	DG	BH
T, q	$T = 300$ K	$T = 300$ K	$T = 300$ K	$-\boldsymbol{n} \cdot \boldsymbol{q} = h(T_{ext} - T)$	$-\boldsymbol{n} \cdot \boldsymbol{q} = h(T_{ext} - T)$	Axial symmetry	Equation (4.4)	–
$\vec{v}, p$	–	Open boundary $(p\boldsymbol{I} + \boldsymbol{K}) \cdot \boldsymbol{n} = 0$	Open boundary $(p\boldsymbol{I} + \boldsymbol{K}) \cdot \boldsymbol{n} = 0$	–	–	Axial symmetry	$\boldsymbol{v} = 0$	$\boldsymbol{v} = 0$
$\vec{J}, \varphi$	–	$\boldsymbol{n} \cdot \boldsymbol{J} = 0$	$\boldsymbol{n} \cdot \boldsymbol{J} = 0$	$\varphi = 0$	$\varphi = 0$	Axial symmetry	Equations (4.5)–(4.13) and (4.14)	–
$\vec{A}$	$\boldsymbol{n} \times \boldsymbol{A} = 0$	$\boldsymbol{n} \times \boldsymbol{A} = 0$	$\boldsymbol{n} \times \boldsymbol{A} = 0$	$\boldsymbol{n} \times \boldsymbol{A} = 0$	$\boldsymbol{n} \times \boldsymbol{A} = 0$	Axial symmetry	–	–

q (W/m^2): heat flow; h (W/m^2 K): heat transfer coefficient; T_{ext} (K): outside temperature; K (Pa): viscous stress tensor

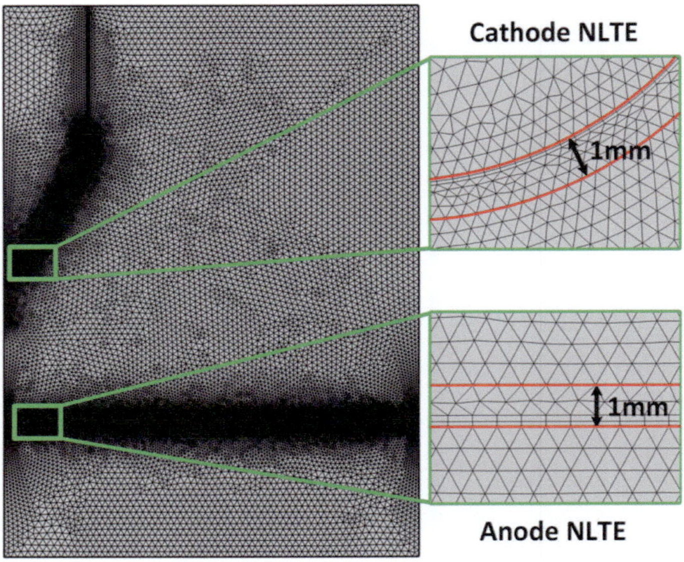

Fig. 4.3 Mesh setting of the MHD simulation model for air arc

plasma region to support arc initiation. Abdelal et al. employed a 1D electron transport model to compute field-induced electron emission and gas ionization in the electrode gap, thereby estimating the initial plasma conductivity required for the arc model.

In the present study, a small direct current (2 A) is used to obtain a steady-state solution of the arc model, which serves as the initial condition for the time-domain simulation of lightning current injection. During transient simulation, the arc rapidly reaches a state largely independent of its initial conditions; hence, the influence of the small pre-applied current on the subsequent arc evolution is minimal. A fully coupled direct solver is employed in the transient analysis. Given the sharp rise rate of the lightning current in its early stage, the initial time step is set to 1×10^{-10} s to ensure computational accuracy.

4.2.4 Simulation and Characterization of Lightning Arcs on CFRP

Using an arc-plasma MHD model, we compute the discharge-gap and arc-anode interface fields under repeated lightning strikes [44, 45]. On a 64-core (2.3 GHz) node with 192 GB RAM, each multi-component arc simulation requires $\approx$156 h. Figure 4.4 shows the transient temperature of the arc column during continuous discharge of components A (20 kA), B (500 A), and C (335 A).

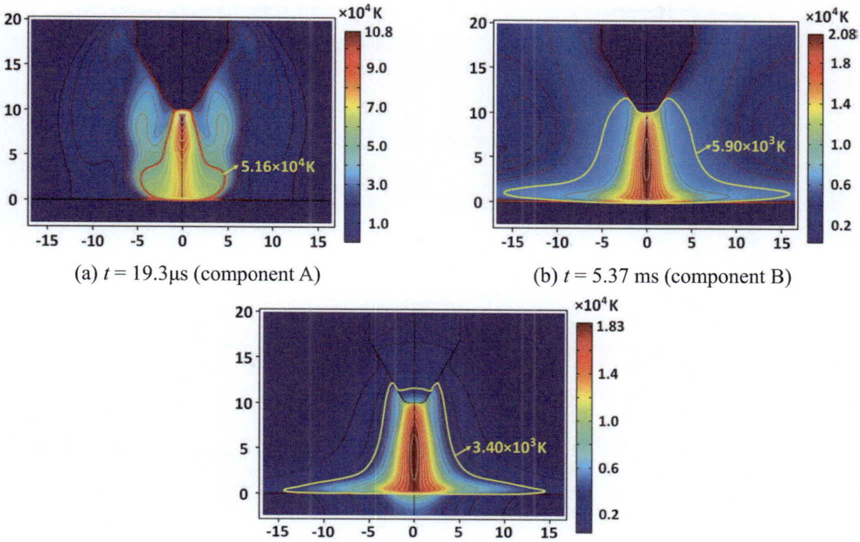

(a) $t = 19.3\mu s$ (component A)

(b) $t = 5.37$ ms (component B)

(c) $t = 0.518$ s (component C)

Fig. 4.4 Temperature distribution during multiple lightning current components (A + B + C)

The breakdown begins with corona at the cathode tip, followed by Joule heating, which raises local temperature and enhances air conductivity. Initially, the hot core is confined near the cathode. Driven by thermal conduction, gas convection, and the Lorentz force, the plasma channel rapidly extends toward the anode, forming a pinpoint, high-temperature arc. During the transient temperature of the arc column during continuous discharge of components A (20 kA), B (500 A), and C (335 A), it can be seen that:

(i) Component A (20 kA, ~20 μs): its ultrashort duration yields an approximately adiabatic, slender arc from cathode to anode with minimal heat exchange (Fig. 4.4a).
(ii) Component B (500 A, ~5 ms): with lower current, the steady-state core reaches ≈20800 K. Over milliseconds, heat spreads radially, expanding the arc attachment area on the anode (Fig. 4.4b).
(iii) Component C (335 A, 0.58 s): the extended duration allows energy to transfer fully to the anode and surrounding gas, producing a bell-shaped temperature profile (Fig. 4.4c). Although the core temperature drops, the anode-surface air remains at ≈3400 K, sufficient to sublimate carbon fibers (3316 K), ablate fibers, and pyrolyze resin in the CFEP interlayer.

As a conductive fluid, the arc plasma behaves hydrodynamically: its high temperature causes the surrounding gas to expand rapidly, forming a supersonic jet that impacts the anode surface. With a 10 mm discharge gap, the surface pressure distribution under Components A, B, and C is shown in Fig. 4.5 [20, 21].

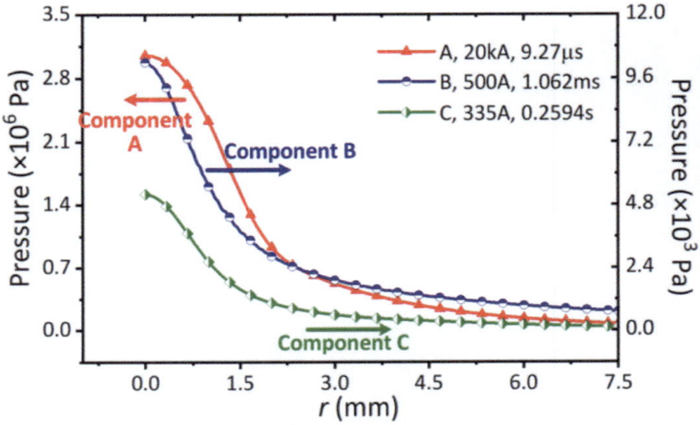

Fig. 4.5 Pressure distribution on the anode surface during multi-component (A–B–C) lightning currents

From Fig. 4.5, the peak pressure occurs at the arc center. During repeated strikes, the highest overpressure on the anode appears at 9.27 μs (rising edge of Component A), reaching 3.05 MPa. In the later Component B and C phases, the maximum pressures fall to 1.02×10^4 Pa and 5.21×10^3 Pa, respectively. Thus, for the majority of the strike (Component C duration), the arc pressure is well below atmospheric pressure. At ambient temperature, such arc-induced pressures are insufficient to produce mechanical deformation in CFRP composites.

According to the mechanisms of Joule heating and arc-thermal ablation, lightning damage initiation and progression in CFRP composites depend on the local current and heat-flux distributions at the arc-anode interface. Using MHD simulations for a 10 mm discharge gap, Figs. 4.6 and 4.7 show these physical fields on the sample surface during Components A–C.

For Component A, the first arc, the anode current density is initially low but rises sharply after ≈12.5 μs, peaking at 1.37×10^9 A/m^2 at 12.9 μs (Fig. 4.6a), indicating the arc attachment occurs around 12.5 μs. The heat-flux (Fig. 4.7a) shows a lag, which is due to finite thermal-conduction and convection times, and peaks at 19.0 μs. Sustained heating of the surrounding air increases its conductivity, causing current to spread radially over a wider anode area, and the plasma jet to expand outward. Consequently, after the peak time, both the current–density and heat-flux footprints broaden, while their peak intensities decrease, as seen in Figs. 4.5, 4.6a, and 4.7a.

At the onset of Component B, the high-temperature channel from Component A remains, with the core region still above 10^4 K. Consequently, Fig. 4.7b shows no new arc attachment; the peak current and heat flux remain at the original arc center, and both decrease with radial distance. For Components B and C, their longer steady-state phases allow energy injection and losses (via convection and radiation) to balance, so the surface fields maintain consistent distributions and magnitudes over time. Although Component A's peak current and heat-flux densities are 1–2 times higher

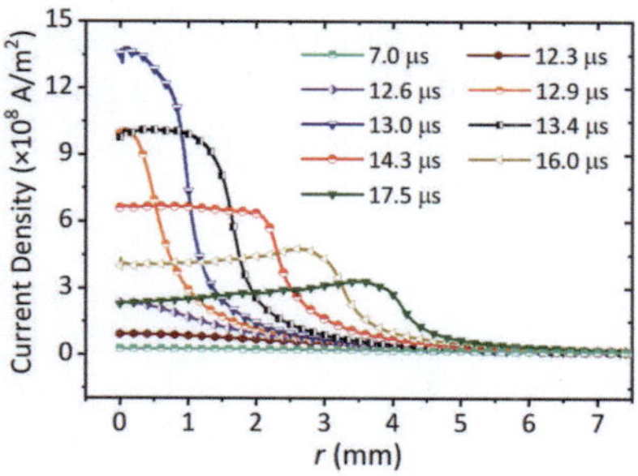

(a) Component A ($I_A = 20$ kA)

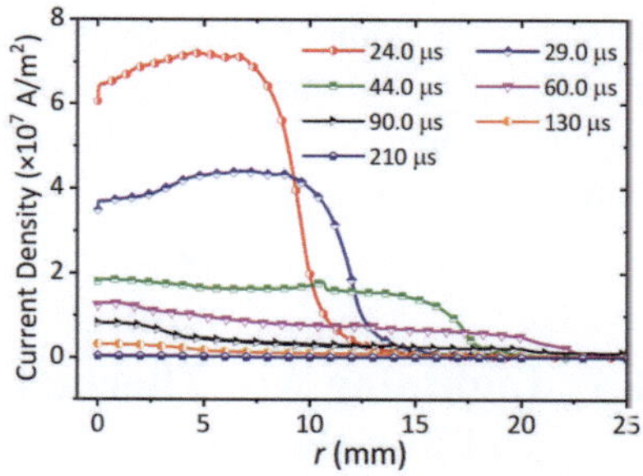

(b) Component B

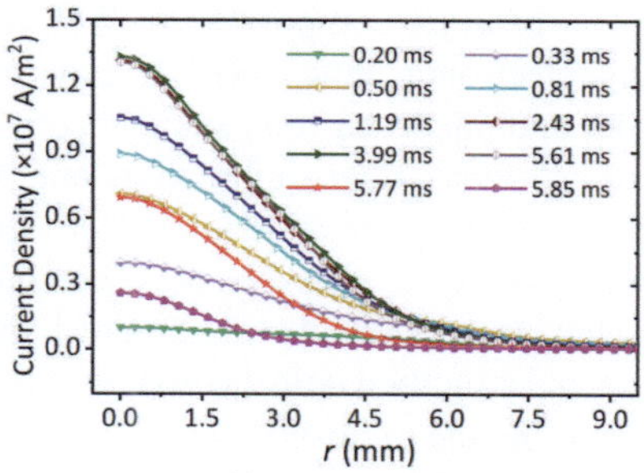

(c) Component C

Fig. 4.6 Radial distribution of anode-surface current density (A + B + C)

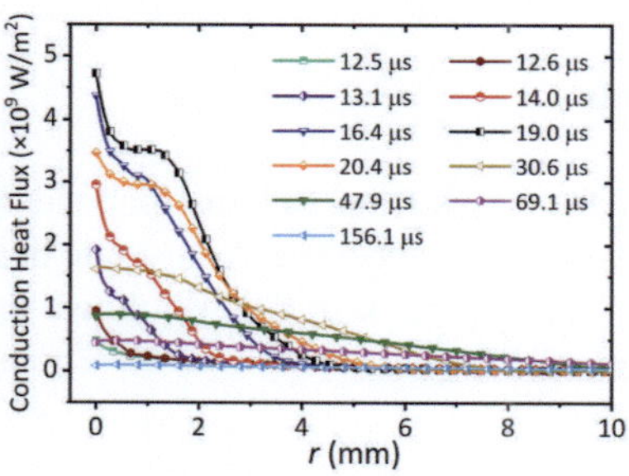

(a) Component A ($I_A = 20$ kA);

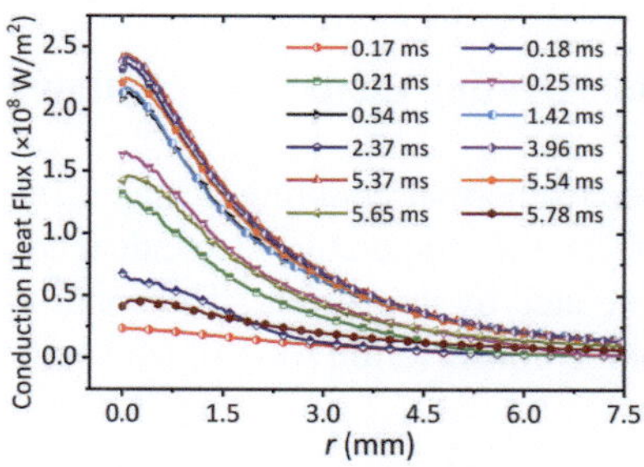

(b) Component B

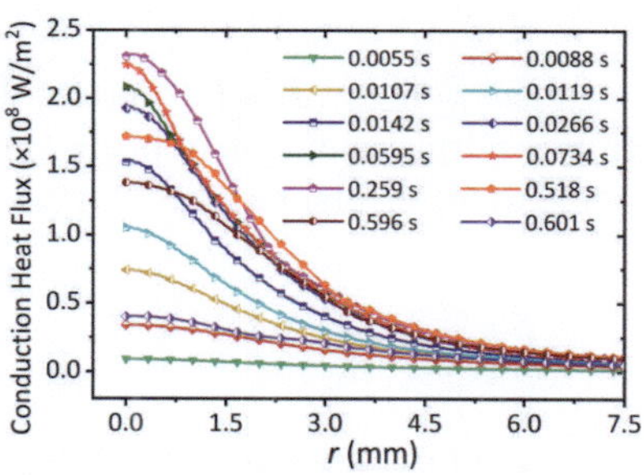

(c) Component C

Fig. 4.7 Radial distribution of heat-flux on the anode surface (A + B + C) than those of Component B, its short duration limits the total thermoelectric energy transferred. In contrast, Component C, with its prolonged action, delivers a greater cumulative thermal load, which will be quantitatively compared in the following damage-analysis section.

4.3 Lightning Arc-Thermal Coupling Model and Simulation

The lightning-strike response of CFRP composites involves simultaneous coupling of plasma arcs, Joule heating, electromagnetic forces, and thermal-mechanical effects. When a pulsed arc attaches and injects current, Joule heating raises the composite's temperature, causing resin pyrolysis, fiber sublimation, and internal thermal expansion. This thermal damage also alters the material's electrical conductivity, thermal diffusivity, and mechanical integrity. Conversely, these conductivity changes modify the Joule heating rate and the electric-field distribution inside the composite. Finally, evolving thermal transport and structural deformation reshape the current-flow paths and heat dissipation, completing the multi-physics coupling loop.

4.3.1 Lightning Electrothermal Coupling Control Equations

Joule Heating Control for Lightning Currents

Electro-thermal coupling during a lightning strike arises from two interdependent effects: the temperature-dependent conductivity of the composite and the Joule heat generated by the current. The conduction of lightning current in CFRP follows the charge-conservation form of Maxwell's equations:

$$\int_S J \cdot \boldsymbol{n} dS = \int_V r_c dV \tag{4.17}$$

where, V represents the element volume; $\boldsymbol{J}$ is the current density vector; S is the unit surface area; $\boldsymbol{n}$ is the outer normal direction of plane S; r_c is the charge volume density in the unit. According to the divergence theorem, we can get:

$$\int_V \left[\frac{\partial}{\partial X} \cdot J - r_c \right] dV = 0 \tag{4.18}$$

Since dV is an arbitrary volume, the above equation can be simplified as a differential equation.

$$\frac{\partial}{\partial X} \cdot J - r_c = 0 \tag{4.19}$$

By introducing an arbitrarily changing potential field $\delta\varphi$, and integrating it in volume, we can obtain:

$$\int_V \delta\varphi\left[\frac{\partial}{\partial X} \cdot J - r_c\right]dV = 0 \tag{4.20}$$

Using the divergence theorem, the above equation can be written as:

$$-\int_V \frac{\partial \delta\varphi}{\partial X} \cdot JdV = \int_S \delta\varphi JdS + \int_S \delta\varphi r_c dV \tag{4.21}$$

According to the differential form of Ohm's law:

$$J = \sigma^E \cdot E = -\sigma^E \cdot \frac{\partial \varphi}{\partial X} \tag{4.22}$$

where, $\boldsymbol{E}$ represents the electric field intensity vector; $\varphi(x, y, z, t)$ is the potential value at t time inside the material (x, y, z); σ^{E} is the three-dimensional conductivity matrix. After introducing the Ohm's law, the charge conservation control equation can be written as:

$$\int_V \frac{\partial \delta\varphi}{\partial X} \cdot \sigma^E \cdot \frac{\partial \varphi}{\partial X} dV = \int_S \delta\varphi JdS + \int_S \delta\varphi r_c dV \tag{4.23}$$

The Joule heat power P_{ec} due to the current flow can be calculated based on the electric field strength and the current density in the material:

$$P_{ec} = E \cdot J = E \cdot \sigma^E \cdot E \tag{4.24}$$

In transient analysis, the average Joule heat power P_{ec} can be expressed as:

$$\overline{P}_{ec} \approx E \cdot \sigma^E \cdot E - E \cdot \sigma^E \cdot \Delta E + \frac{1}{3}\Delta E \cdot \sigma^E \cdot \Delta E \tag{4.25}$$

Considering the energy conversion efficiency, the thermal energy converted by electric energy during lightning strike is:

$$Q_V = \eta_V \cdot P_{ec} \tag{4.26}$$

Lightning-Induced Joule Heat Balance Equation of CFRP Composites

When a lightning pulse flows through a CFRP composite, Joule heating raises the local temperature, triggering resin pyrolysis and thermal degradation. From a heat-transfer perspective, this damage process can be modeled as nonlinear heat conduction with an internal heat source. Applying Fourier's law and energy conservation [17, 46] yields the transient heat-conduction equations for an anisotropic medium:

$$k_{xx}\frac{\partial^2 T}{\partial^2 x} + k_{yy}\frac{\partial^2 T}{\partial^2 y} + k_{zz}\frac{\partial^2 T}{\partial^2 z} + \dot{q} = \rho c_\rho \frac{\partial T}{\partial t} \tag{4.27}$$

where, k_{xx}, k_{yy}, k_{zz} represent the thermal conductivity of CFRP composites in the principal axis direction, respectively; T is the material temperature; $\dot{q}$ is the heat generation rate, that is, the heat generated by the Joule heat effect of the lightning current per unit time; ρ is the material density; c_p is the material specific heat; t is the calculation time.

CFRP composites is an orthotropic material, and its heat transfer performance is obviously directional. The thermal conductivity matrix $\left[\bar{K}\right]$ in the overall coordinate of the material can be obtained by the dominant thermal coefficient matrix $[K] = \begin{bmatrix} k_{xx} & & \\ & k_{yy} & \\ & & k_{zz} \end{bmatrix}$, which can be written as [47]:

$$\left[\bar{K}\right] = [n]^T[K][n] \tag{4.28}$$

where, $[n] = \begin{bmatrix} l_1\ m_1\ n_1 \\ l_2\ m_2\ n_2 \\ l_3\ m_3\ n_3 \end{bmatrix}$. l_i, m_i and n_i represent the corresponding direction cosine, respectively. $[n]$ is the transformation matrix. Under the Joule-heating induced by the lightning pulse, the temperature of the CFRP rises sharply. Subsequently, heat is transferred from the high-temperature regions to colder areas within the material and to the surrounding environment.

Boundary Conditions

The surface response in the CFRP lightning-damage model includes both conduction and radiation. At temperatures below 400 °C, thermal radiation is minimal, so convection dominates heat dissipation. The convective heat flux density q_c between the CFRP surface and ambient air is given by:

$$q_c = k_g(\theta_B - \theta) \tag{4.29}$$

where k_g is the convective heat-transfer coefficient between the CFRP surface and air, θ_B is the ambient temperature, and θ is the surface temperature. As the CFRP temperature rises, thermal radiation becomes significant. At high temperatures, radiation dominates heat loss. The radiative heat flux density q_r from the CFRP surface to the environment is given by:

$$q_r = \tau\sigma\left[(\theta_B - \theta^z)^4 - (\theta - \theta^z)^4\right] \tag{4.30}$$

In the formula, τ is the radiation coefficient; σ is the Stefan-Boltzmann constant; θ^z is the Absolute zero.

Heat Balance Equations

The basic heat balance equation is

$$\int_V \rho \dot{U} dV = \int_S q dS + \int_V r dV \tag{4.31}$$

where, V represents the surface area is a certain volume of S; ρ *is* the density of the material; $\dot{U}$ is the rate of change of internal energy; q is the heat flux per unit area of inflow volume V; r is the Heat generated per unit volume. Using the divergence theorem and the Fourier heat conduction law, the heat balance equation can be written as a variational form:

$$\int_V \rho \dot{U} \delta\theta dV + \int_V \frac{\partial \delta\theta}{\partial X} \cdot k \cdot \frac{\partial \theta}{\partial X} dV = \int_V \delta\theta r dV + \int_S \delta\theta q dS \tag{4.32}$$

where, **k** is the heat conduction coefficient matrix; θ is temperature. By introducing surface thermal interaction and Joule heat as heat release, the heat balance equation is rewritten as [48].

$$\begin{aligned} \int_V \rho \dot{U} \delta\theta dV + \int_V \frac{\partial \delta\theta}{\partial X} \cdot k \cdot \frac{\partial \theta}{\partial X} dV = \int_S \delta\theta \eta_v P_{ec} dV \\ + \int_S \delta\theta q dS + \int_S \delta\theta (q_c + q_r) dS \end{aligned} \tag{4.33}$$

Thermal Expansion

Carbon fibers remain dimensionally stable below 400 °C, so CFRP's thermal deformation is governed by the epoxy matrix. Microscopically, rising temperatures induce nonlinear lattice vibrations in the resin, causing thermal expansion. Since the resin's coefficient of thermal expansion exceeds that of the carbon fibers, thermal strains concentrate at the fiber-matrix interfaces, generating interfacial stresses that constrain resin expansion. When the material experiences a temperature difference ΔT, the thermal-expansion stress in an anisotropic CFRP laminate is [49]:

$$\begin{cases} \varepsilon_{xx} = \left[\sigma_{xx} - \mu\left(\sigma_{yy} + \sigma_{zz}\right)\right]/E_{xx} + \alpha_{xx}^{T}\Delta T \\ \varepsilon_{yy} = \left[\sigma_{yy} - \mu(\sigma_{xx} + \sigma_{zz})\right]/E_{yy} + \alpha_{yy}^{T}\Delta T \\ \varepsilon_{zz} = \left[\sigma_{zz} - \mu\left(\sigma_{xx} + \sigma_{yy}\right)\right]/E_{zz} + \alpha_{zz}^{T}\Delta T \end{cases} \tag{4.34}$$

where, ε is the element normal stress; σ is the normal strain; τ is the shear stress; γ is the shear strain; μ is the Poisson's ratio; E is the elastic modulus; and there is

$$\begin{cases} \gamma_{xy} = \tau_{xy}/G \\ \gamma_{yz} = \tau_{yz}/G \\ \gamma_{zx} = \tau_{zx}/G \end{cases} \quad G = \frac{E}{2(1+\mu)} \tag{4.35}$$

where, G is the shear modulus. According to Eq. (4.34), the distribution of thermal expansion stress under a certain temperature distribution field can be calculated.

4.3.2 Construction of the CFRP Lightning Damage Model

Finite Element Model Setup

Using finite-element software, we built the CFRP lightning-damage model shown in Fig. 4.8. The full model dimensions for a unidirectional layup ($[-45/45/0/90]_{3S}$) are 300 × 45 × 3.6 mm, comprising 24 plies of ≈0.15 mm each.

Boundary conditions mimic the laboratory test: the impulse current is injected at the model center; both ends are clamped and tied to ground (0 V). Heat exchange with the environment occurs via convection ($h = 1$ W/m^2 K), radiation ($\varepsilon = 0.9$), and ambient temperature (25 °C). Based on experiments showing damage confined

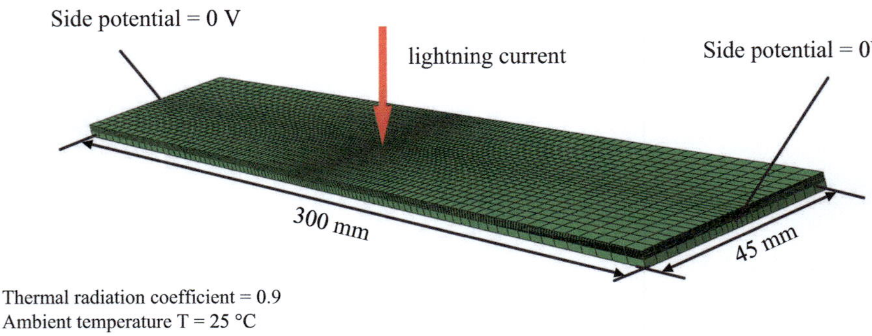

Fig. 4.8 Lightning-damage finite-element model of CFRP composites

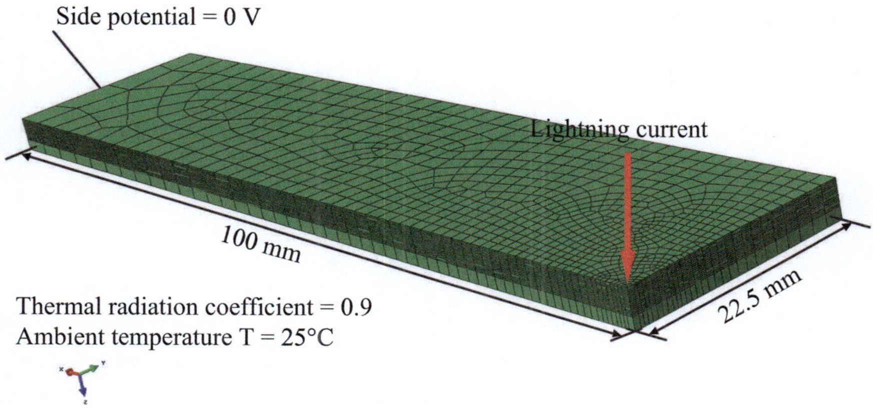

Fig. 4.9 Quarter-model of CFRP for lightning-damage simulation

to 60 cm^2 and $\leq$1.6 mm depth (10–11 plies) under a 100 kA Component A strike, we reduced the model to 200 × 45 × 3.6 mm. The upper 16 plies retain their actual stacking; the lower 8 are condensed into a single quasi-isotropic ply. Exploiting symmetry, we simulate using a 1/4 model (Fig. 4.9). A finer mesh is applied around the injection zone, with coarser elements at the edges. We perform thermal-electrical-structural coupling using Q3D8R elements.

Simulated Impulse Lightning Current

For the simulation, the measured component A waveform, with a peak of ≈40 kA, was digitized and applied to the CFRP model, as shown in Fig. 4.10. Using the MHD-derived anode-surface current densities and heat-flux profiles from Figs. 4.6 and 4.7, we impose these boundary sources to model the Joule-heating effects of the lightning impulse on the CFRP composite.

Thermal Degradation of Resin Matrix of CFRP Composites

Under the thermal loading of the simulated component A impulse, the CFRP heats rapidly, inducing resin pyrolysis and potentially carbon-fiber sublimation. In order to evaluate the lightning ablation damage of CFRP composites, it is necessary to describe the pyrolysis behavior of the resin. Based on the thermal weight loss experiment, the definition of pyrolysis degree from thermogravimetric data:

$$C = \frac{m_0 - m}{m_0 - m_t}, 0 \leq C \leq 1 \tag{4.36}$$

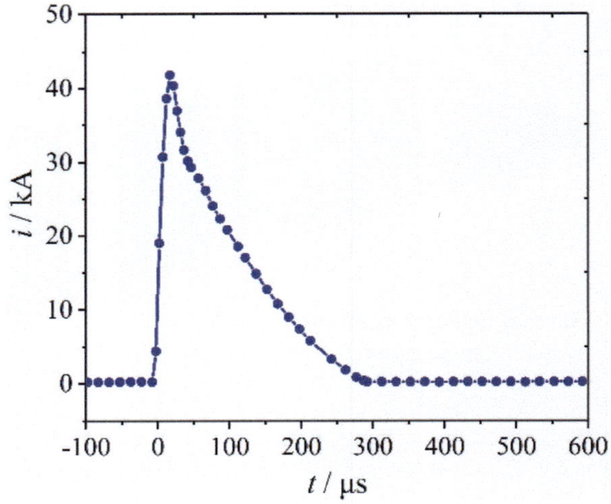

Fig. 4.10 Simulation of component A of impulse lightning current

where, m is the sample mass at time t. m_0 and m_t are the initial mass and residual mass after heating, respectively. However, because thermogravimetric tests heat at ≤100 K/min, they cannot capture the extreme heating rates (to 10^3–10^4 °C within 10^2 μs) seen in lightning events, making direct use of C(T) curves impractical for ablation prediction.

Resin degradation involves depolymerization, chain scission, and side-group elimination, all of which follow chemical-kinetic behavior. Therefore, the thermal-degradation of CFRP can be modeled by an n-th order reaction rate equation, coupling the temperature history from the arc-thermal simulation with kinetic parameters to predict the local degree of pyrolysis and resulting material loss [50].

$$\frac{dC}{dt} = K(T)(1 - C)^n \tag{4.37}$$

where, C is the pyrolysis degree; n is the pyrolysis reaction order; $K(T)$ is the kinetic reaction rate constant, related to activation energy Q, determined by Arrhenius equation:

$$K(T) = A\exp\left(-\frac{Q}{RT}\right) \tag{4.38}$$

where, A is the pre-exponential factor. $R = 8.314$ J/mol/K is the gas constant. Equation (4.37) can be written as:

$$-\frac{d(1 - C)}{dt} = K(T)(1 - C)^n \tag{4.39}$$

Bring $K(T)$ into the above equation and integrate it to obtain:

$$(1 - C)^{1-n} = (n - 1)A\exp\left(-\frac{Q}{RT}\right)t + C_0 \tag{4.40}$$

The pyrolysis degree depends on both time and temperature, but a lightning strike induces a non-isothermal, spatially varying temperature field. Each point in the material experiences a unique temperature profile and heating rate, so C cannot be computed directly from Eq. (4.40). Instead, applying the Scheil summation principle, the continuous temperature change process is discretized into small time intervals Δt, treating each as an isothermal step. The overall pyrolysis degree is then the sum of the degrees C_i at time t_i. Thus the final thermal degradation degree can be obtained by superimposing the thermal degradation degree of each isothermal process. According to Eq. (4.40), the degree of pyrolysis at the initial time can be written as:

$$(1 - C_1)^{1-n} = (n - 1)A\exp\left(-\frac{Q}{RT_1}\right)\Delta t + C_0 \tag{4.41}$$

The introduction of virtual time t_i^* represents the time required for the degree of pyrolysis to reach C_{i-1} under the condition of constant temperature T_i, which can be obtained from the Eq. (4.40):

$$(1 - C_{i-1})^{1-n} = (n - 1)A\exp\left(-\frac{Q}{RT_i}\right)t_i^* + C_0,\ i \geq 2 \tag{4.42}$$

Calculation predicts

$$t_i^* = \frac{(1 - C_{i-1})^{1-n} - C_0}{(n - 1)A\exp\left(-\frac{Q}{RT_i}\right)} \tag{4.43}$$

Assuming that the pyrolysis degree reaches C_i after Δt, there is

$$(1 - C_i)^{1-n} = (n - 1)A\exp\left(-\frac{Q}{RT_i}\right)(t_i^* + \Delta t) + C_0,\ i \geq 2 \tag{4.44}$$

Combining Eqs. (4.42) and (4.44), we can get

$$(1 - C_i)^{1-n} = (1 - C_{i-1})^{1-n} + (n - 1)A\exp\left(-\frac{Q}{RT_i}\right)\Delta t,\ i \geq 2 \tag{4.45}$$

Therefore, the numerical calculation expression of pyrolysis degree is:

$$\begin{cases} C_1 = 1 - \left[(n - 1)A\exp\left(-\frac{Q}{RT_1}\right)\Delta t + C_0\right]^{1/(1-n)} & \\ C_i = 1 - \left[(1 - C_{i-1})^{1-n} + (n - 1)A\exp\left(-\frac{Q}{RT_i}\right)\Delta t\right]^{1/(1-n)} & i \geq 2 \end{cases} \tag{4.46}$$

Property Parameters

(1) Normal electrical conductivity model for CFRP composites

Based on the dynamic conductivity tests presented in Chap. 2, the variation of the equivalent dynamic resistance of CFRP composites with impulse current amplitude was obtained for various fiber orientations under non-destructive lightning current conditions [51–53]. When Component A of the lightning current is applied, the dynamic electrical conductivity of the composites can be estimated using test data under a 20/60 μs impulse current, which has a similar waveform. From the experimental results of $[0]_{24}$-ply unidirectional CFRP subjected to non-destructive impulses, a curve-fitting analysis shows that the electrical conductivity along the 0° ply direction (i.e., along the fiber direction) increases with current amplitude according to a power-law relationship. This fitted curve $\sigma_{xx} = 0.58198 \times I_{peak}^{0.33507}$ is illustrated in Fig. 4.11 [51].

Using the same approach, power-law fitting relationships were also established for the conductivity versus current amplitude in the 90° ply direction (perpendicular to the fibers) and in the thickness (Z) direction, which are $\sigma_{yy} = 0.37959 \times I_{peak}^{0.36098}$ and $\sigma_{zz} = 0.0323 \times I_{peak}^{0.65571}$, respectively. Accordingly, the equivalent conductivities in the fiber direction (X), transverse in-plane direction (Y), and thickness direction (Z) under different amplitudes of Component A can be extrapolated, as listed in Table 4.2 [51]. The data in Table 4.2 highlight the pronounced anisotropy of CFRP composites with a single fiber orientation. Within the same plane (X and Y directions), the conductivity differences are within one order of magnitude. However, the conductivity in the through-thickness direction (Z) is approximately three orders of magnitude lower than that in the in-plane directions, underscoring the strong directional dependence of CFRP electrical behavior under impulse loading.

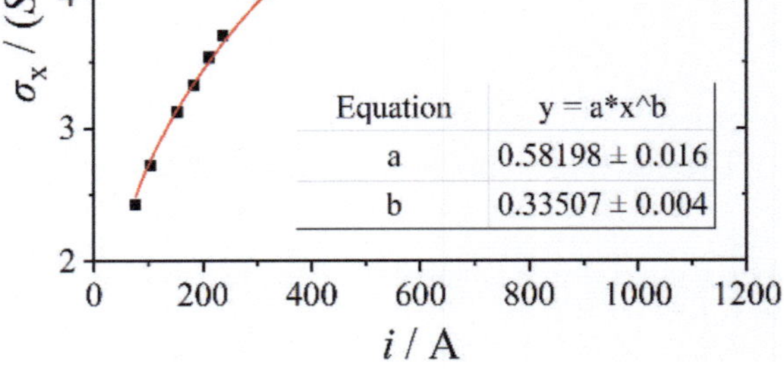

Fig. 4.11 Variation of equivalent electrical conductivity of $[0]_{24}$-ply CFRP composites with current amplitude under a 20/60 μs impulse current

Table 4.2 Equivalent electrical conductivity of CFRP composites under different impulse current amplitudes

Impulse current amplitudei_{peak} (kA)	σ_{xx} (S/m)	σ_{yy} (S/m)	σ_{zz} (S/m)
20	16070	13550	21.35
40	20270	17400	33.64
60	23220	20140	43.88
100	27559	24222	61.34

(2) **CFRP material properties as functions of pyrolysis degree**

CFRP composites exhibit anisotropy not only in electrical conductivity (Table 4.2) but also in their thermal and mechanical properties, which vary with ply orientation. To capture this in the simulation, we define a 3D coordinate system (Fig. 4.12) aligned with the length (*X*), width (*Y*), and thickness (*Z*) of the composite. Each ply's fiber angle relative to the *X*-axis is then set according to the actual layup sequence (Fig. 4.13).

Once the multiphysics lightning-damage geometry is built, each material region is assigned orthotropic properties. Based on our dynamic conductance measurements and literature values, the initial thermal parameters (thermal conductivity k and thermal expansion coefficient γ) and mechanical parameters (Young's modulus E, Poisson's ratio μ and shear modulus G) along the fiber direction, transverse in-plane, and through-thickness directions are listed in Table 4.3 [50] and Table 4.4 [54].

The ply rotation alters the composite's anisotropic properties. For example, when a ply is rotated by an angle α relative to the reference frame, the electrical-conductivity tensor in that frame becomes:

$$\sigma^E = \begin{bmatrix} \sigma_{xx}\cos^2\alpha + \sigma_{yy}\sin^2\alpha & (\sigma_{yy} - \sigma_{xx})\sin\alpha\cos\alpha & 0 \\ (\sigma_{yy} - \sigma_{xx})\sin\alpha\cos\alpha & \sigma_{xx}\sin^2\alpha + \sigma_{yy}\cos^2\alpha & 0 \\ 0 & 0 & \sigma_{zz} \end{bmatrix} \tag{4.47}$$

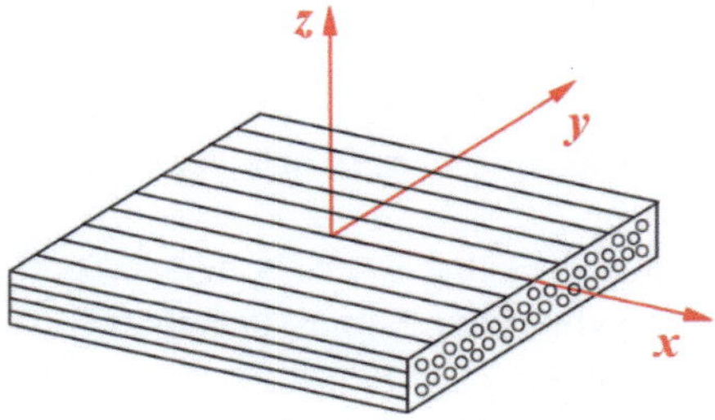

Fig. 4.12 Three-dimensional coordinate system of the simulation model

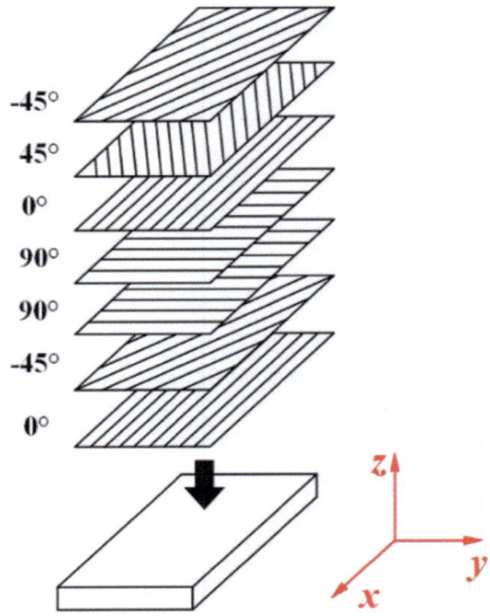

Fig. 4.13 Schematic of the CFRP ply orientations

Table 4.3 Density and thermal parameters of CFRP composites

ρ (kg/m^3)	c_p (J/kg·K)	k_{xx} (W/m K)	k_{yy} (W/m K)	k_{zz} (W/m K)	γ_{xx} (/K)	γ_{yy} (/K)	γ_{zz} (/K)
1520	1065	11.8	0.609	0.609	1.8×10^{-8}	2.2×10^{-5}	2.2×10^{-5}

Table 4.4 Mechanical parameters of CFRP composites

E_x (GPa)	E_y (Gpa)	E_z (Gpa)	μ_{xy}	μ_{yz}	μ_{xz}	G_{12} (GPa)	G_{13} (GPa)	G_{23} (GPa)
137	8.2	8.2	0.30	0.30	0.45	4.36	4.36	3.00

where $\sigma(\alpha)$ is the rotation matrix and σ is the conductivity tensor in the ply's principal directions. Material parameters in the lightning-damage model depend not only on the laminate architecture but also on the evolving material state under high current. Using kinetic parameters $n = 3.5$, $A = 3 \times 10^{15}$ s^{-1}, $Q = 180$ kJ/mol/K, the resin pyrolysis degree at various heating rates is obtained via Eq. (4.47) and plotted in Fig. 4.14.

It can be seen from Fig. 4.14 that when the temperature exceeds 200 °C, the epoxy resin matrix begins pyrolysis. At approximately 600 °C, the pyrolysis degree approaches 1, indicating that the resin is nearly fully decomposed. The pyrolysis behavior of resin matrix CFRP composites is strongly dependent on the heating rate. Specifically, higher heating rates result in a delayed onset of matrix pyrolysis. Under lightning current injection, the temperature rise rate in CFRP composites can reach 10^7–10^8 °C/s. Therefore, the initial onset of lightning-induced ablation in the resin matrix is significantly delayed.

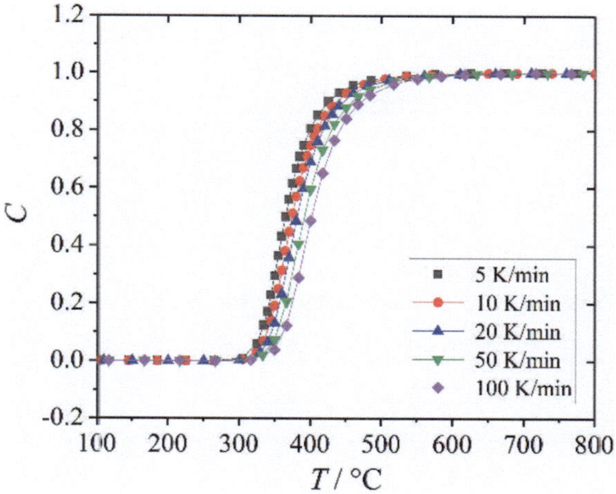

Fig. 4.14 Resin pyrolysis degree versus temperature for different heating rates

The pyrolysis of the resin matrix leads to variations in key thermophysical properties, including electrical conductivity σ^{E}, specific heat capacity c_{p}, thermal conductivity k^{E}, and density ρ. According to the thermophysical mixing rule, the relationship between these properties and the pyrolysis degree $C(T, t)$ is expressed as:

$$\begin{cases} \sigma^E = \sigma_0^E[1 - C(T,t)] + \sigma_1^E C(T,t) \\ c_p = c_{\mathrm{p}0}[1 - C(T,t)] + c_{\mathrm{p}1} C(T,t) \\ k^E = k_0^E[1 - C(T,t)] + k_1^E C(T,t) \\ \rho = \rho_0[1 - C(T,t)] + \rho_1 C(T,t) \end{cases} \tag{4.48}$$

where, σ_0^E, $c_{\mathrm{p}0}$, k_0^E and ρ_0 represent the material parameters when the pyrolysis degree $C = 0$; σ_1^E, $c_{\mathrm{p}1}$, k_1^E and ρ_1 represent the material parameters when the resin is completely pyrolyzed ($C = 1$).

In resin pyrolysis, water, low-molecular-weight volatiles, and gases first evolve from the polymer matrix. As temperature rises, the polymer backbone cleaves, and upon complete ablation, residues carbonize. Between 1000 and 1500 °C, carbon fibers and carbonized products undergo graphitization, with crystalline rearrangement and growth. Above 3316 °C, the carbonized residue sublimates. The pyrolysis degree $C(T, t)$ reflects these temperature-time-chemical transformations, which alter density, thermal, mechanical, and electrical properties. In our model, all material parameters vary linearly with C, as listed in Tables 4.5 and 4.6.

Ultimately, post-ablation, the CFRP surface becomes a carbon-rich, highly conductive residue. When $C \approx 1$, the surface conductivity approaches that of pristine carbon fiber, enabling rapid current transfer to deeper plies. The electrical parameters used in this model are given in Table 4.7 [53].

Table 4.5 Density and thermal properties versus pyrolysis degree

C	ρ (kg/m^3)	c_p (J/kg·K)	k_{xx} (W/m K)	k_{yy} (W/m K)	k_{zz} (W/m K)	γ_{xx} (/K)	γ_{yy} (/K)	γ_{zz} (/K)
0	1520	1065	11.8	0.609	0.609	1.8×10^{-8}	2.2×10^{-5}	2.2×10^{-5}
1	1100	2509	1.736	0.1	0.1	0	0	0

Table 4.6 Mechanical propertoes versus pyrolysis degree

C	E_x (GPa)	E_y (Gpa)	E_z (Gpa)	μ_{xy}	μ_{yz}	μ_{xz}	G_{12} (GPa)	G_{13} (GPa)	G_{23} (GPa)
0	137	8.2	8.2	0.30	0.30	0.45	4.36	4.36	3.00
1	137	0.0041	0.0041	0.30	0.30	0.45	0.00174	0.00174	0.0012

Table 4.7 Conductive properties versus pyrolysis degree ($I_{peak} = 40$ kA)

C	σ_{xx} (S/m)	σ_{yy} (S/m)	σ_{zz} (S/m)
0	20270	17400	33.64
1	29300	29300	2000

4.3.3 Computational Modeling of Arc-Thermoelectric Coupling Effects in Lightning-Strike Damage of CFRP Composites

Coupling of Impulse Lightning Arc and Thermoelectricm Model

When CFRP composites are struck, lightning current rapidly injects into the material at the arc attachment points, generating Joule heat during internal conduction. Simultaneously, thermal conduction and enthalpy transfer occur at the high-temperature arc-CFRP interface, where the injected heat flux induces localized temperature spikes. Fiber sublimation begins once the temperature exceeds the CFRP sublimation threshold, while interlaminar resin pyrolysis and gas-driven delamination proceed from rapid heating and gas expansion. This chapter develops a 3D FEM thermoelectric-coupling model, employing the Electric Currents and Heat Transfer modules with electromagnetic-thermal coupling interfaces. By including material phase transitions and convective boundary conditions, the model simulates temperature evolution and progressive damage mechanisms, including fiber sublimation, resin pyrolysis, and delamination, driven by coupled arc-thermoelectric interactions.

The multiphysics model applies loads, including current density J(r,t), conductive heat flux Q(r,t), and enthalpy flux, extracted from the arc-anode interface fields of the MHD simulation. Because these loads vary in space and time, direct parameterization is impractical. Instead, we implement spatial discretization of the physical-field outputs and temporal continuity, mapping them onto the CFRP model's surface nodes, as shown in Fig. 4.15 [55].

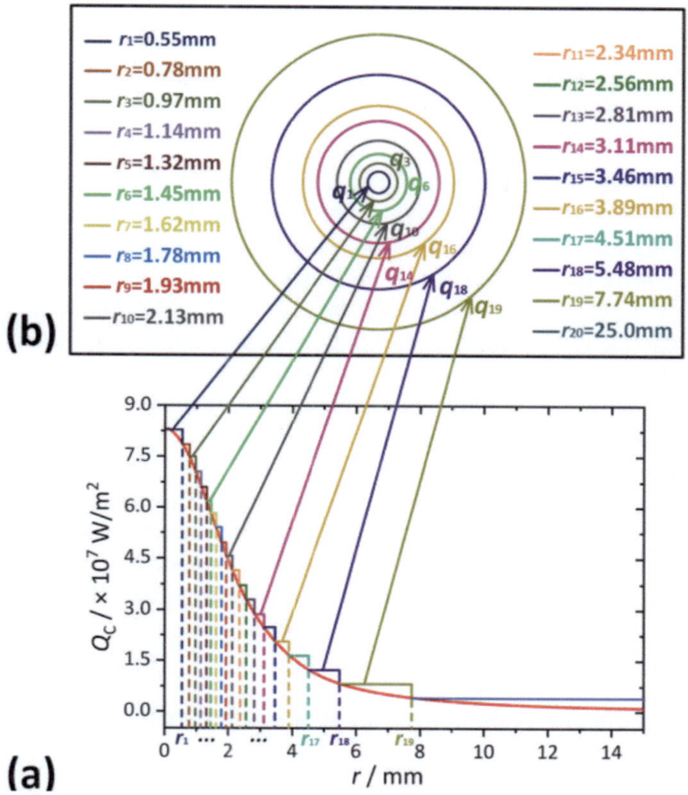

Fig. 4.15 Load application methodology in the coupling model: **a** Mirrored distribution and spatial discretization of conductive heat flux on the planar anode surface; **b** definition of $q(\mathrm{t})$ and segmented loading zones

Based on the MHD results, the radial profile of anode-surface heat flux at the peak current is shown in Fig. 4.15a. The heat flux density peaks at the center and decays with radius. To implement this in the thermoelectric model, we discretize the profile into 21 concentric zones $(0, r_1)$, (r_1, r_2), (r_2, r_3), …, (r_{19}, r_{20}), (r_{20}, r_{21}), each assigned a time-dependent flux $\mathrm{q}_i(t)$ $(i = 1, \ldots, 21)$, as depicted in Fig. 4.15b. Within each annulus, $\mathrm{q}_i(t)$ is uniform spatially but varies in time, capturing the transient conduction during the lightning pulse. We employ the Foster et al. field-mapping method. Abdelal and Murphy [56] to apply these axisymmetric circular/annular loads on a structured mesh via a binary 2D indicator F_{2D}, enabling precise application of axisymmetric circular/annular loads on structured meshes. The governing procedure is expressed as:

$$F_{2D}=\begin{cases}\left[\left\lfloor\frac{r_i}{\sqrt{x^2+y^2}}\right\rfloor\Big/\left(\left\lfloor\frac{r_i}{\sqrt{x^2+y^2}}\right\rfloor+1\right)\right] & i=1\\ \left[\left\lfloor\frac{r_i}{\sqrt{x^2+y^2}}\right\rfloor\Big/\left(\left\lfloor\frac{r_i}{\sqrt{x^2+y^2}}\right\rfloor+1\right)\right]-\left[\left\lfloor\frac{r_{i-1}}{\sqrt{x^2+y^2}}\right\rfloor\Big/\left(\left\lfloor\frac{r_{i-1}}{\sqrt{x^2+y^2}}\right\rfloor+1\right)\right] & i=2,\ \ldots,\ 20\end{cases} \tag{4.49}$$

$F_{2D} = 1$ activates the load in zone *i*, and 0 deactivates it elsewhere. To map from the 2D arc model to the 3D CFRP model, we project the flux onto the CFRP's upper surface, partitioned by node (*x*, *y*) via Eq. (4.49). Because the CFRP surface lies at a different *z*-level than the 2D anode, each zone's flux is scaled by an adjustment factor $\sqrt{x^2+y^2}/\sqrt{x^2+y^2+z^2}$ to match the arc-model values $q_i(t)$. Current density loads $j_i(t)$ are processed similarly, ensuring the spatial heterogeneity and temporal dynamics of both thermal and electrical loads are accurately represented.

A lightning damage model for CFRP composites is implemented in finite-element software by developing custom subroutines within the electro-thermal-structural analysis framework. During the simulation, the material's electrical, thermal, and mechanical properties are dynamically updated based on the resin pyrolysis degree. A fully coupled electro-thermal-mechanical analysis interlinks the electric field, temperature field, and pyrolysis field. The resulting temperature and degradation distributions post-strike are then extracted to characterize the material's lightning damage. The complete workflow is outlined in Fig. 4.16.

Analysis of Simulated Lightning-Strike Damage in CFRP Composites

A numerical model examined a CFRP specimen (300 × 45 × 3.6 mm) with layup $[-45/45/0/90]_{3S}$ under a 40 kA lightning component A waveform. During the strike, temperature, resin pyrolysis, electric potential, and stress fields evolve, tightly coupling with ablation damage initiation and propagation.

(1) **Temperature field distribution**

The electro-thermal-mechanical model resolves transient temperatures in each ply. Figure 4.17 ($t = 0.5$ μs) and Fig. 4.18 ($t = 16.9$ μs) show these fields, with regions exceeding 600 °C shaded gray.

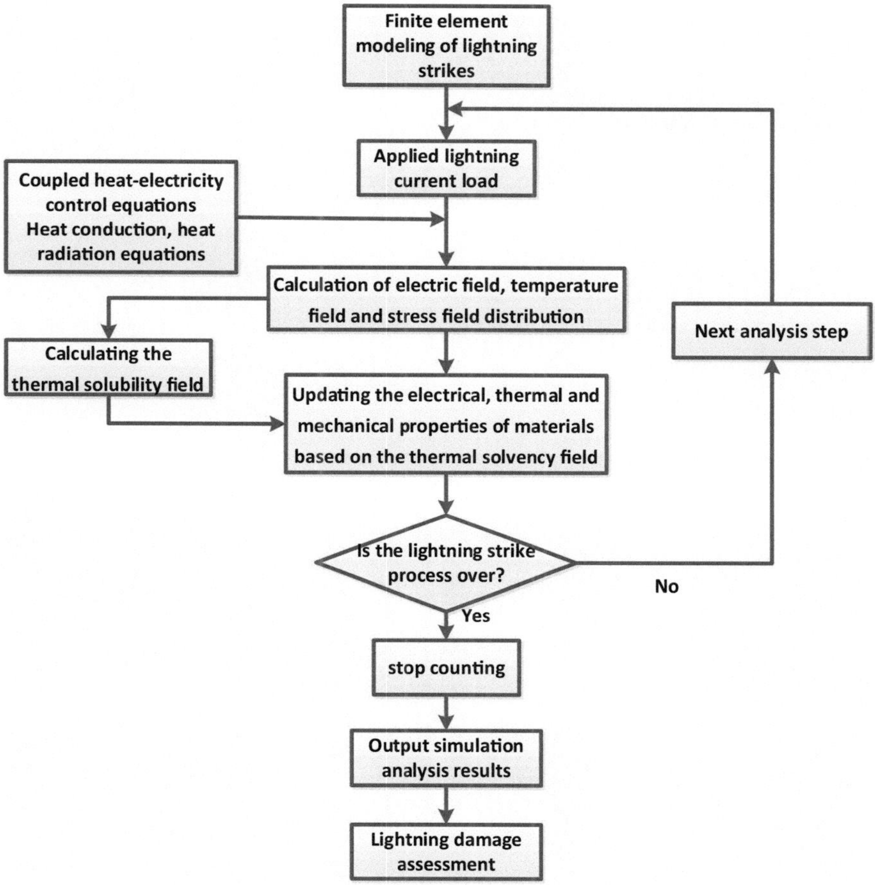

Fig. 4.16 Flowchart of the CFRP lightning-damage simulation and analysis

The specimen is grounded at both ends with electrically isolated lateral edges, and current is injected centrally, producing an electric field mainly along x, from the injection point toward the left boundary. This yields a high-temperature elliptical zone ($T > 600$ °C) aligned with the x-axis. Regression of dynamic conductive behavior under 40 kA gives in-plane conductivities of 20270 S/m (fiber direction) and 17400 S/m (transverse), resulting in minimal anisotropy of the damage footprint. From Figs. 4.17 and 4.18, the high-T zone shrinks with depth. At $t = 0.5$ μs, heat penetrates to the third ply; successive ply currents and downward conduction homogenize superficial ply temperatures, while deeper plies remain cooler. By $t = 16.9$ μs, the pyrolysis front reaches the seventh ply (see Fig. 4.19 [53]).

Analysis of the temperature curves in Fig. 4.19 shows that surface plies (1–2) reach ~3,300 °C, near the carbon fiber sublimation point (3,316 °C), confirming sublimation under a 40 kA component A strike. Joule heating drives this rapid rise, with the

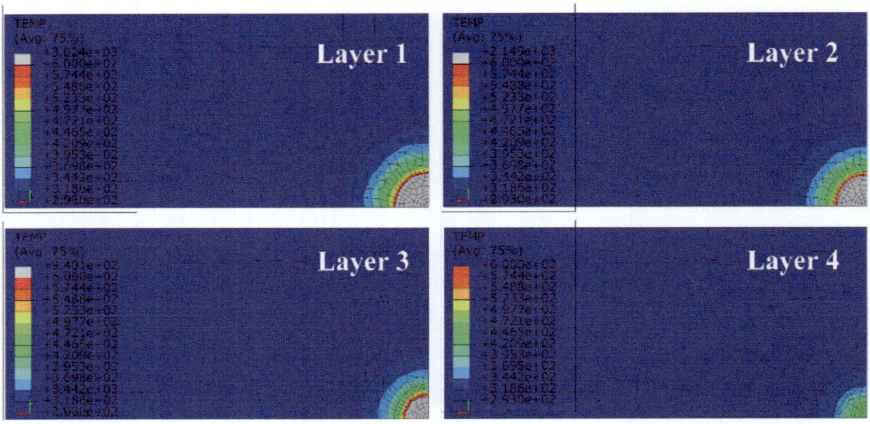

Fig. 4.17 Temperature field distribution in each ply at $t = 0.5$ μs

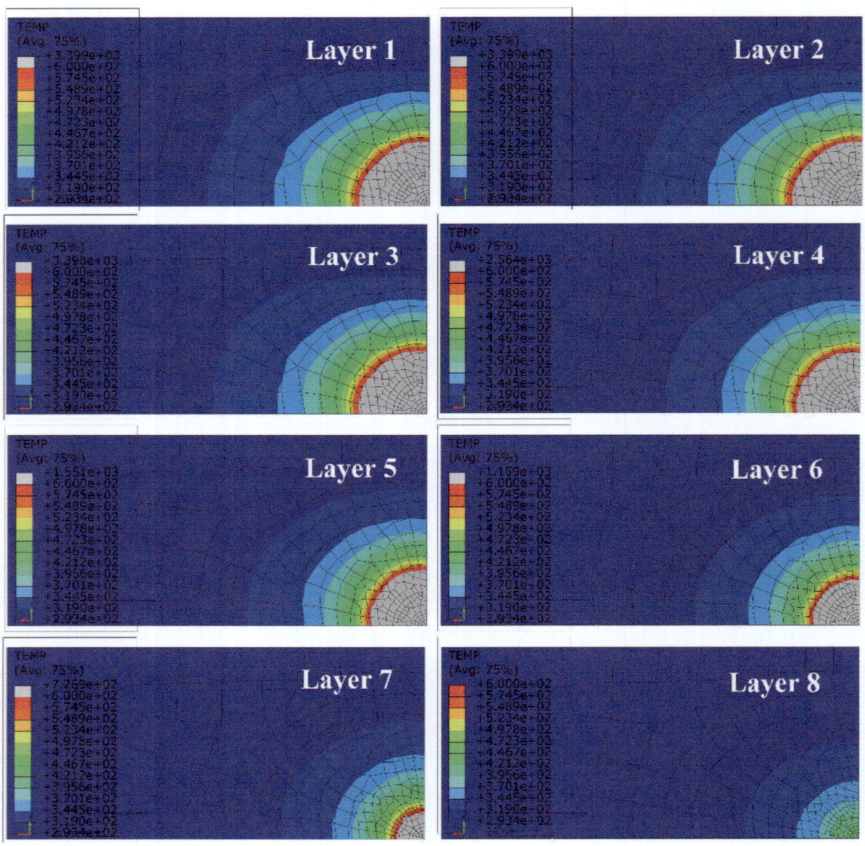

Fig. 4.18 Temperature field distribution in each ply at $t = 16.9$ μs

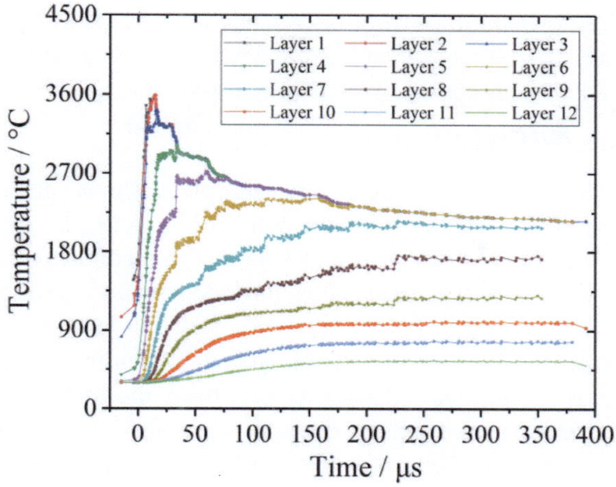

Fig. 4.19 Temporal evolution of temperature at ply centerpoints in CFRP under lightning current

first six plies exceeding 600 °C at peak current ($t = 16.9$ μs). The epoxy matrix (pyrolysis onset ~400 °C) carbonizes, forming amorphous carbon with higher conductivity, which produces two effects: (i) Reduced Joule heating in pyrolyzed zones (lower resistivity); (ii) Faster heat spread to neighboring regions and deeper current penetration. As a result, surface temperatures decline after the peak, while subsurface plies continue heating, albeit with slower ramp rates at greater depths. Beyond nodal analysis, spatial temperature profiles reveal through-thickness thermal penetration. Figure 4.20 plots the depth (D) where $T > 600$ °C at several time instants [53].

As illustrated in Fig. 4.20, before peak current, surface heating drives the 600 °C isotherm deeper than 0.8 mm. During the current decay, thermal energy from the

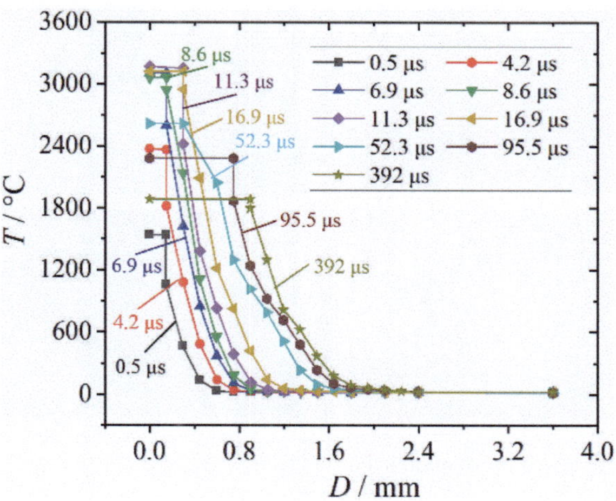

Fig. 4.20 Temperature distribution along the thickness at the laminate over time

hot surface passes downward, causing surface cooling and growth of the isothermal zone from 0.3 mm to ~0.9 mm. After current cessation and subsequent conduction/dissipation, the CFRP will approach a uniform, thermally equilibrated state.

(2) **Pyrolysis degree field distribution**

At t = 16.9 μs, the pyrolysis degree across each ply is shown in Fig. 4.21. While its spatial pattern mirrors the temperature field in Fig. 4.18, the pyrolysis zone is notably smaller than the region exceeding 600 °C. This confirms that temperature alone is necessary but not sufficient-time at elevated temperature also governs resin decomposition.

Figure 4.22 plots the time history of pyrolysis degree at the center of each ply's attachment zone. It shows that under the 40 kA strike, the first six plies reach $C > 0.8$ by t = 16.9 μs whereas the seventh ply peaks around 0.3 and the eighth ply only 2.33×10^{-5}. Deeper plies undergo delayed pyrolysis as thermal energy accumulates during the subsequent current decay. Defining normalized temperature $T^* = T/T_{\max}$, we compare each ply's peak pyrolysis degree with T^* in Fig. 4.23.

As shown in Fig. 4.23, pyrolysis within each ply lags behind its thermal peak. For example, when the ninth ply center reaches 450 °C, its pyrolysis degree is only 0.04. This reflects the time-dependent kinetics of polymer decomposition: brief temperature spikes contribute little to matrix breakdown, as rapid heating does not allow

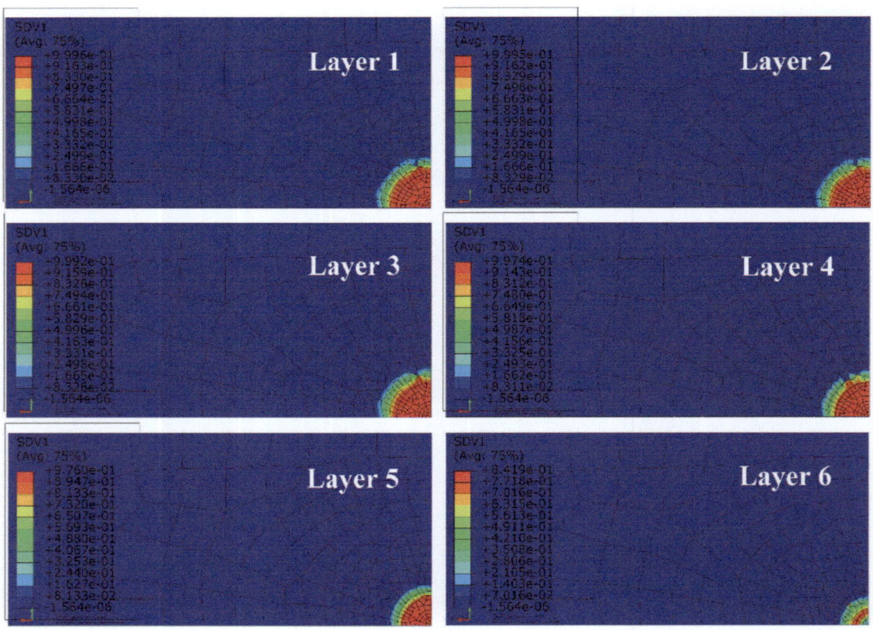

Fig. 4.21 Pyrolysis degree distribution in each ply at t = 16.9 μs

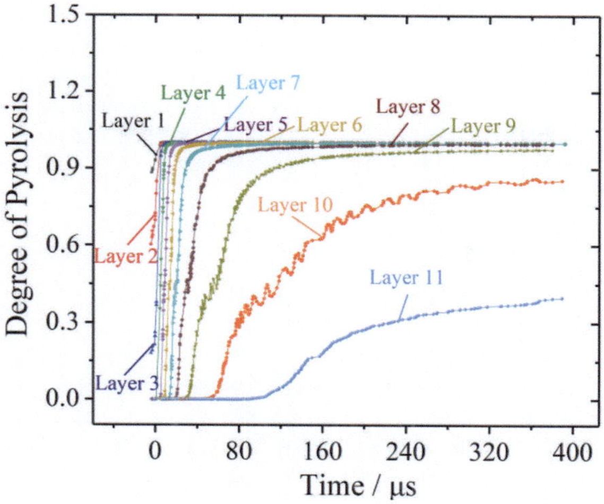

Fig. 4.22 Temporal evolution of resin pyrolysis degree at the center of each ply under a 40 kA lightning strike

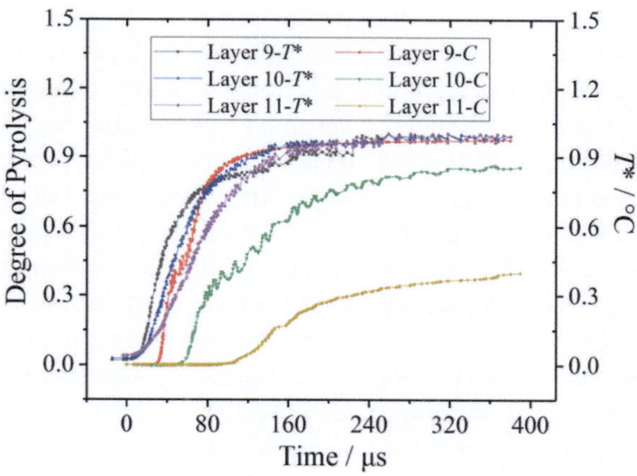

Fig. 4.23 Comparison of pyrolysis degree and normalized temperature $T*$ at ply center

sufficient time for cumulative bond cleavage, shifting effective pyrolysis to higher temperatures. Figure 4.24 plots the pyrolysis-degree profile C versus depth D at the strike center over time. By $t = 16.9$ μs, the top ~0.8 mm (≈first five plies) approaches complete decomposition. The evolution of these curves captures the through-thickness progression of damage. Ultimately, in regions deeper than 1 mm, the resin is almost fully degraded, indicating near-total thermal ablation across the laminate thickness.

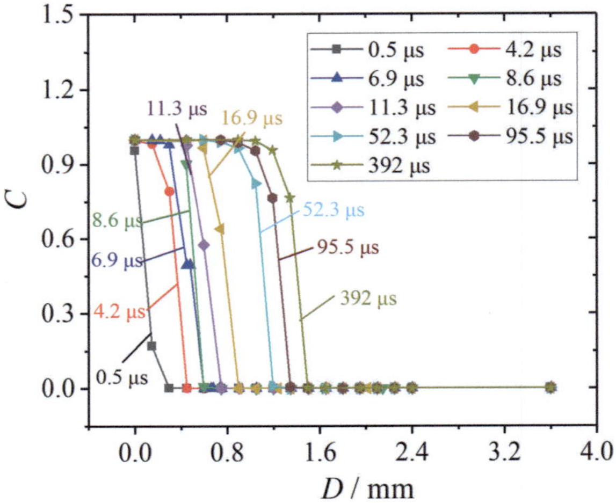

Fig. 4.24 Distribution of resin pyrolysis degree versus depth at the laminate center over time

(3) **Electric field distribution**

Figure 4.25 shows the planar electric potential across each ply at various times.

From Fig. 4.25, the potential gradient lies along the x-axis, reflecting the grounded short edges and isolated long edges. Over time, equipotential regions expand radially from the strike point. Pyrolyzed zones, rich in conductive amorphous carbon, display minimal potential drop, forming approximate equipotential islands that grow in concert with ablation damage. This equipotential spread also penetrates through the thickness, as seen in Fig. 4.26. Prolonged application of component A current drives equipotential zones deeper, mirroring the increasing damage depth. Quantitative traces of electric potential at each ply's center are plotted in Fig. 4.27, showing the transient potential decay from the strike centroid through successive plies.

As shown in Fig. 4.27, the peak node potential during a 40 kA strike exceeds 1000 V, driven by current amplitude and material conductivity. During the early injection phase, current spreads across surface plies before resin pyrolysis, causing a rapid rise in potential. As pyrolysis advances, resin carbonization sharply increases conductivity, so even as current continues rising, the node potential ($I \cdot R$) decreases due to the lowered resistance. At $t = 16.9$ μs, which is the current peak, the first six plies all reach their maximum potentials, indicating complete pyrolysis at their centers, consistent with the temperature peaks in Fig. 4.19.

The electric-field intensity $\boldsymbol{E}$, which is the spatial gradient of potential, defines current pathways during the strike and is derived from the potential distribution:

$$E = \frac{\partial \varphi}{\partial X} \tag{4.50}$$

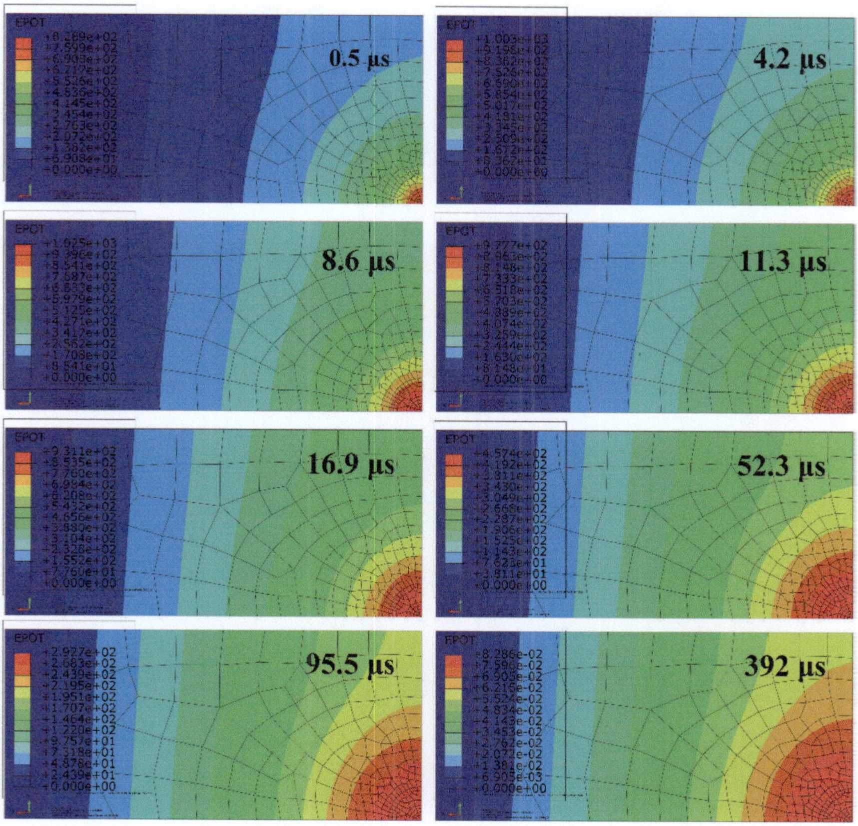

Fig. 4.25 Temporal evolution of planar electric potential distributions in CFRP composites

The magnitude and direction of the electric-field intensity $\boldsymbol{E}$ at each node map the current-dispersion pathways during a lightning strike. Its temporal evolution at the strike centroid is plotted in Fig. 4.28, while Fig. 4.29 shows the lateral distributions of E_z across the laminate at key time points.

During the initial current-injection phase, current flows only through the surface plies, leaving deeper plies at near-zero potential (Fig. 4.26). As a result, the through-thickness field E_z dominates, exceeding 2.5 kV/mm, while E_x and E_y remain comparatively low. With current confined to the first two plies (total thickness $\approx$ 0.3 mm) and a peak potential of ~1 000 V (Fig. 4.27), one estimates $E_z \approx$ 1000 V/0.3 mm $\approx$ 3 kV/mm. The 300 mm specimen length renders Ex negligible by comparison. After resin pyrolysis, the ablated surface becomes equipotential ($E_z = 0$), and internal fields vanish within the damaged zone. High field gradients now concentrate at the interface between damaged and undamaged regions through the thickness. As shown in Fig. 4.29, the upper (damaged) section, which is now rich in conductive

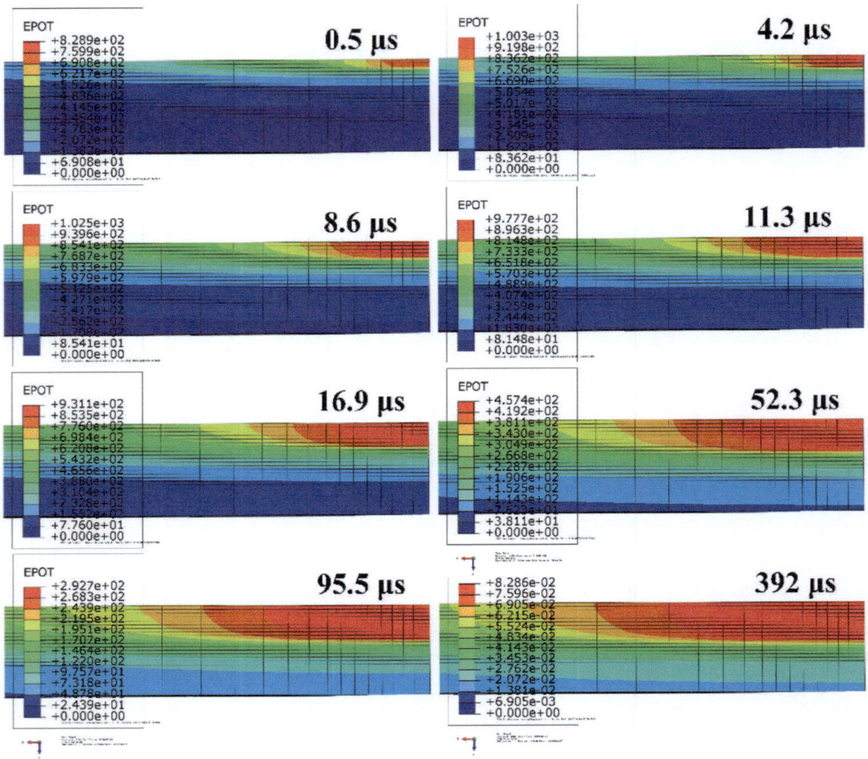

Fig. 4.26 Temporal evolution of lateral electric potential distributions in CFRP composites

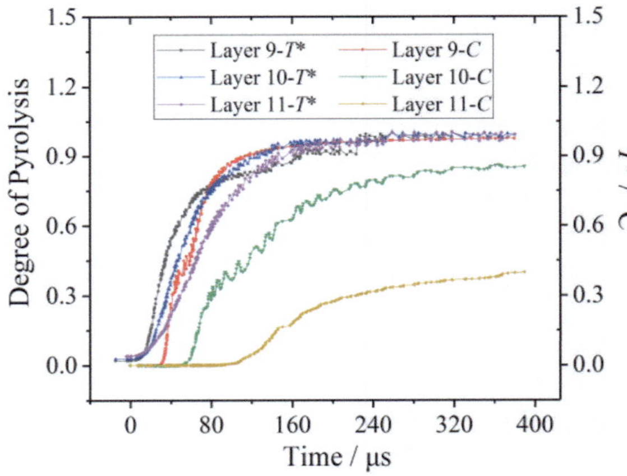

Fig. 4.27 Temporal evolution of electric potential at ply center positions in CFRP composites

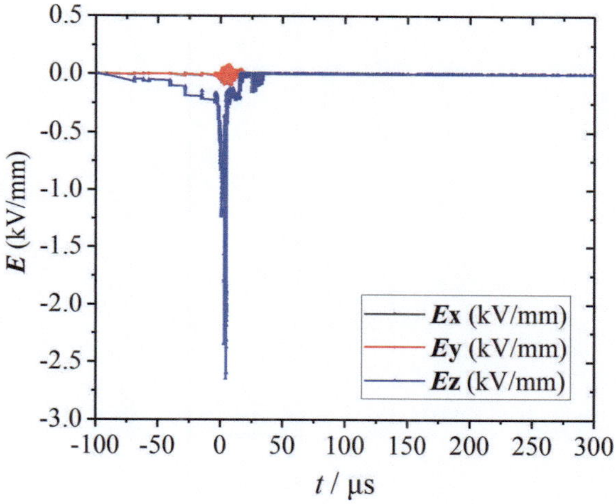

Fig. 4.28 Temporal evolution of electric-field intensity at the lightning-strike centroid

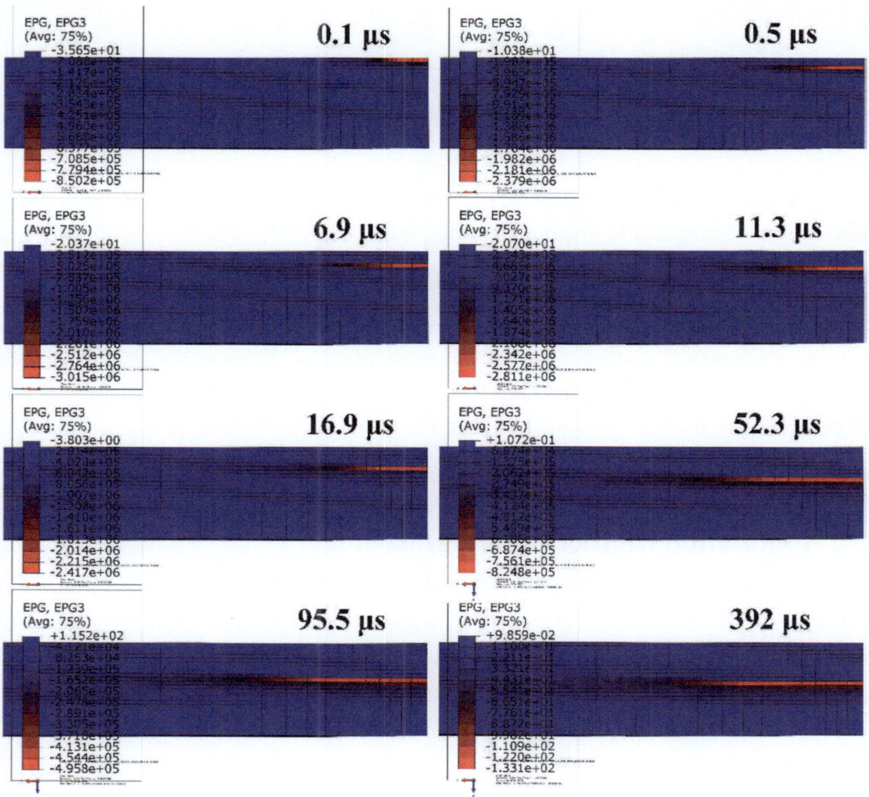

Fig. 4.29 Lateral distributions of electric-field intensity Ex in CFRP composites at selected times

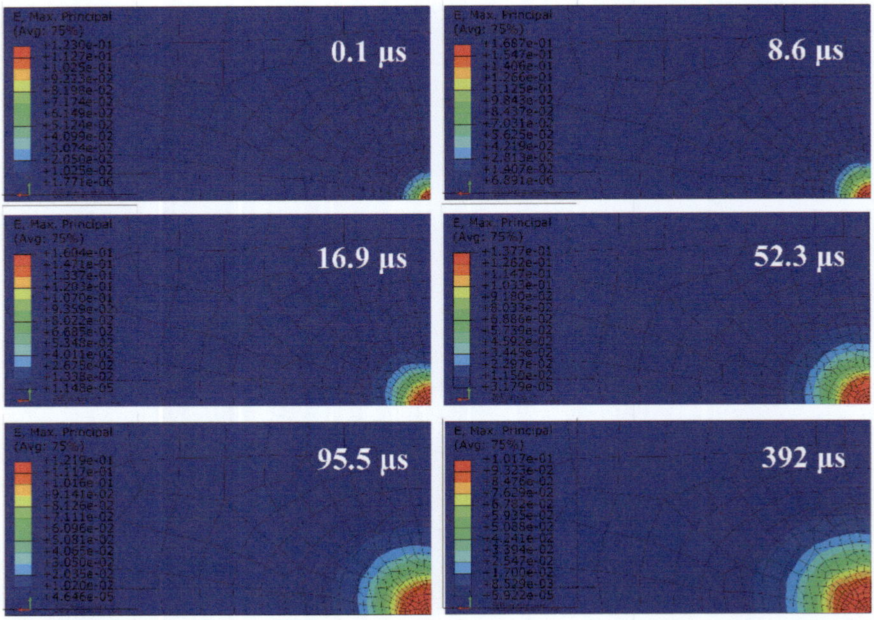

Fig. 4.30 Planar displacement and strain distributions in CFRP composites at selected times

carbon, remains equipotential, whereas the lower, intact plies exhibit conductivity three orders of magnitude lower, sustaining large E_z at their boundary.

(4) **Stress field distribution**

Figure 4.30 shows the surface displacement and strain evolution in the electro-thermal-mechanical model under the lightning impulse. The strains in CFRP arise primarily from thermal expansion, and their spatial patterns closely follow the temperature gradients established during the strike.

4.4 Evaluation Methods and Validation for Lightning-Strike Damage in CFRP Composites

Lightning-induced ablation in CFRP results from resin thermal decomposition under arc heating. Therefore, temperature and pyrolysis-degree fields are the primary diagnostics for assessing damage under impulse currents.

4.4.1 Evaluation Methods for Lightning-Strike Damage in CFRP Composites

Damage Criteria for CFRP Composites

(1) **Temperature-field criteria**

The pyrolysis onset temperature is key to evaluating ablation via the temperature field. Because resin degradation kinetics accelerate with heating rate, the pyrolysis initiation temperature rises at higher rates, as shown in Fig. 4.31.

During a lightning strike, the non-uniform, time-varying heating of CFRP plies can be approximated as a superposition of constant-rate segments. Although higher rates elevate the initial pyrolysis temperature, oxygen ingress at the strike site enhances resin breakdown, whereas carbon fibers impede ablation. To capture these competing effects, we adjust the degradation kinetics and adopt a 450 °C threshold for ablation damage.

(2) **Pyrolysis-degree field criteria**

Resin pyrolysis is a time- and temperature-dependent chemical process: the pyrolysis degree C reflects both the current temperature and its history. For example, at 450 °C, resin reaches $C = 0.98$ under 100 K/min, but only $C = 0.26$ under 10^6 K/min. Thus, a temperature-only criterion overestimates damage when heating is too brief. To remedy this, we integrate the pyrolysis-degree field into our electrothermal-mechanical model. Zones where $C > 0.9$ are classified as ablation-damage regions, capturing both spatial and temporal aspects of thermal degradation more accurately than temperature thresholds alone.

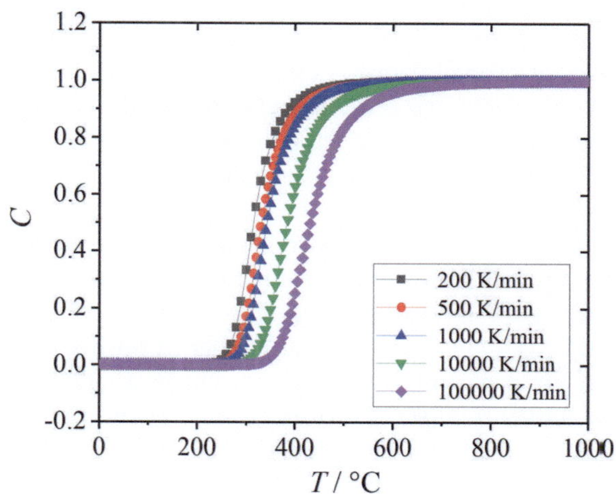

Fig. 4.31 Pyrolysis-degree profiles of resin at varying heating rates

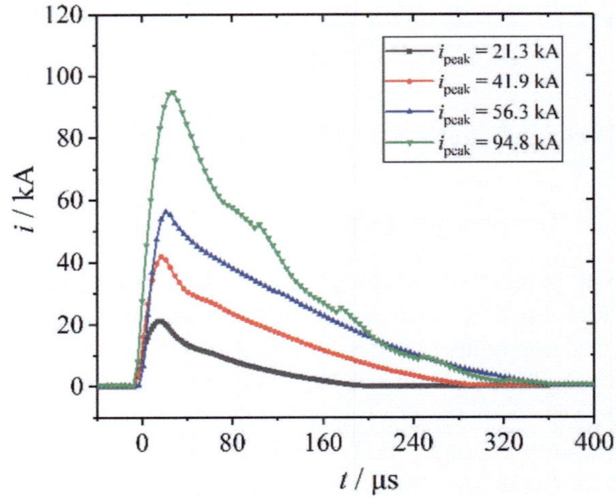

Fig. 4.32 Lightning current impulse component A waveform used in simulation

Experimental Validation of Lightning-Damage Criteria

Two CFRP composite types, which are differing in material system and ply sequence, were subjected to component A currents of 20, 40, 60, and 100 kA. Damage area and depth from simulations were compared against experimental measurements. The simulated waveform of component A is shown in Fig. 4.32.

Damage-Area Evaluation

A unidirectional laminate (150 × 100 × 3.6 mm, $[-45/45/0/90]_{3S}$, 24 plies at ~ 0.15 mm each) was modeled. Total damage area under strike combines pyrolytic delamination and fiber-fracture regions, with higher currents reducing the pyrolysis fraction of total damage. Delamination stems from interlaminar cracks driven by thermal stress and resin degradation. We use the 450 °C isotherm as the ablation boundary, marking damage zones in gray (see Fig. 4.33).

Pixel-based quantification of simulated damage areas for 20–100 kA strikes is shown in Fig. 4.34.

In Fig. 4.34, the simulated delamination area (from the 450 °C isotherm) aligns closely with the experimental pyrolytic-delamination measurements but underestimates the total damage area. This discrepancy arises because the electro-thermal model captures only thermal effects (resin pyrolysis, expansion), whereas fiber-fracture damage that driven by mechanical failure mechanisms is not represented. To further validate, a woven laminate (150 × 100 × 4 mm, $[0/90]_F$, 16 plies) was tested at 20, 40, 60, and 100 kA. Here, fiber-bundle constraints limit fracture, so damage is dominated by ablation and thus well captured by thermal criteria. Figure 4.35 compares simulated and experimental transmission-scan areas, and Table 4.8 quantifies their deviations.

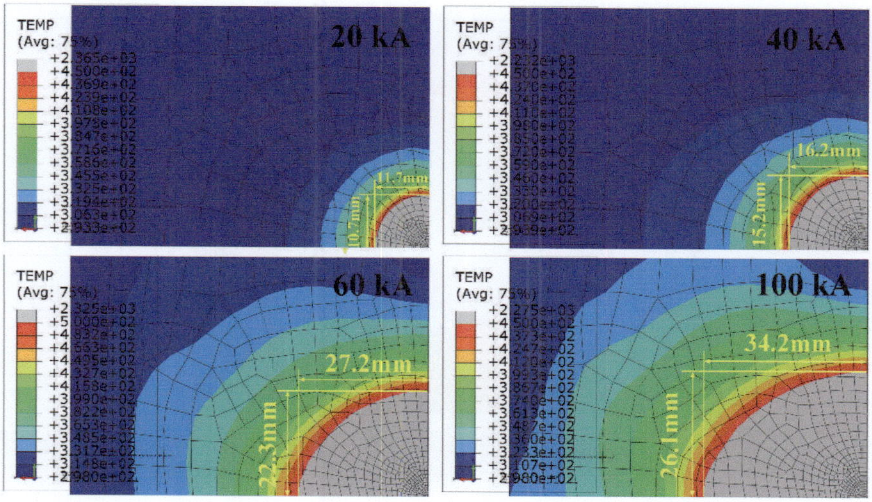

Fig. 4.33 Evaluation of specimen damage area under varying-amplitude lightning pulses using the temperature field

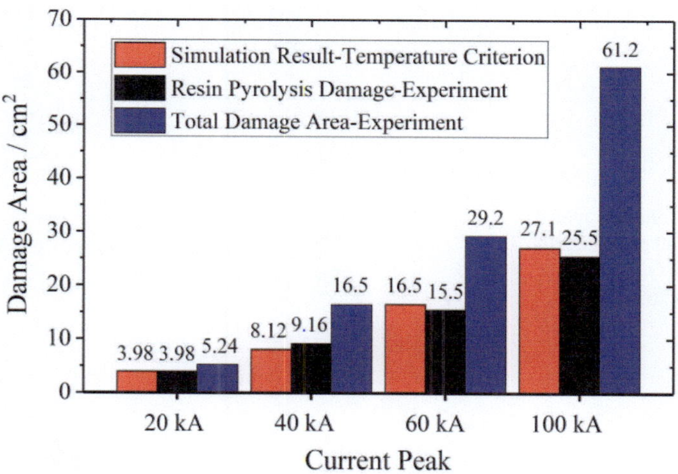

Fig. 4.34 Comparison of simulated and experimental damage areas using the temperature-field criterion

As shown in Figs. 4.34 and 4.35, our electro-thermal-mechanical coupling model accurately predicts ablation damage areas via temperature-field criteria, with simulation–experiment deviations $\leq 8.15\%$. Under the same conditions, using pyrolysis degree $C = 0.5$ ($\approx$12.5 wt% resin loss) as the damage threshold produces the results in Fig. 4.36.

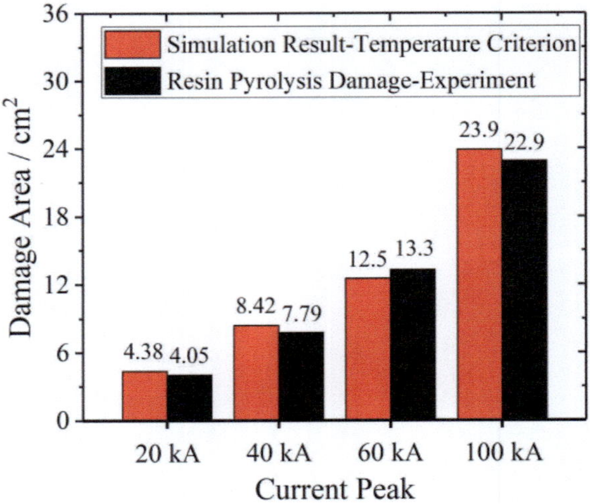

Fig. 4.35 Comparison of simulated and experimental damage areas in woven composites

Table 4.8 Simulated versus experimental lightning-damage areas

No.	Experimental damage area/cm^2	Simulated damage area/cm^2	Deviation (%)
Type 1–1	4.05	4.38	8.15
Type 1–2	7.79	8.42	8.08
Type 1–3	13.3	12.5	6.02
Type 1–4	22.9	23.9	4.37

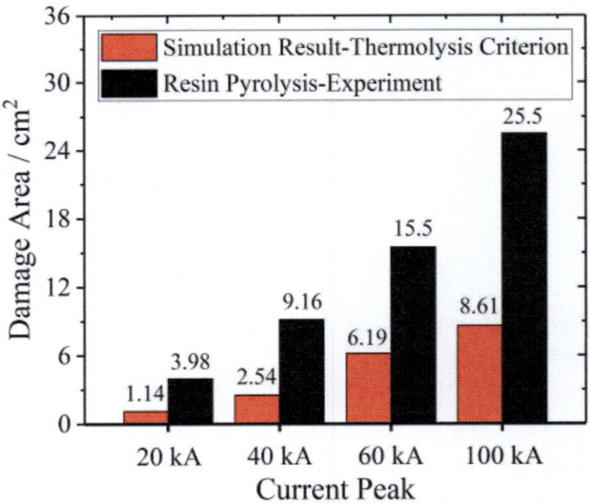

Fig. 4.36 Comparison of simulated and experimental damage areas using the pyrolysis-degree criterion

These comparisons reveal that $C = 0.5$ yields smaller simulated damage areas than observed experimentally. Resin decomposition follows dynamic kinetic laws, requiring both high temperature and sufficient exposure time. While pyrolysis continues during the strike and cooling phases, our current model only simulates the strike interval, leading to an underestimation of total damage area relative to experiment.

Damage-Depth Evaluation

Beyond area, penetration depth critically gauges lightning severity in CFRP. For unidirectional laminates, simulated depths from the post-strike temperature field are compared to experimental values in Fig. 4.37.

In Fig. 4.37, simulated depths overestimate experiments. Transient exceedance of the pyrolysis threshold does not always produce permanent damage, and temperature fluctuations (Fig. 4.19) from heat loss and internal conduction complicate the identification of the deepest isotherm. By contrast, resin pyrolysis in CFRP is irreversible, so the pyrolysis-degree field offers a more robust depth metric. Applying $C > 0.9$ to define through-thickness ablation yields excellent agreement with experimental depths in Fig. 4.38, with deviations detailed in Table 4.9.

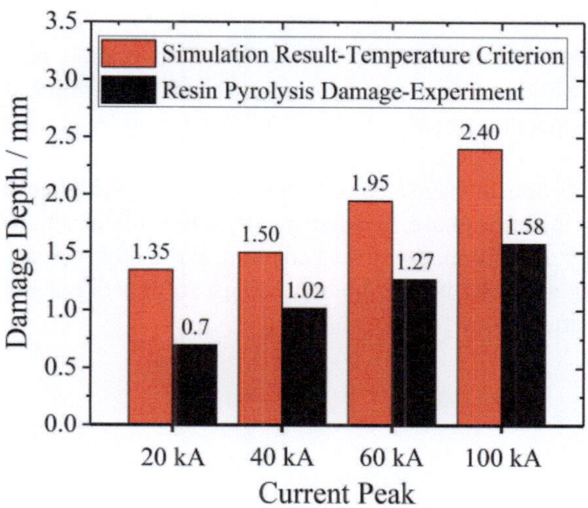

Fig. 4.37 Evaluation of simulated damage depth via the temperature-field criterion

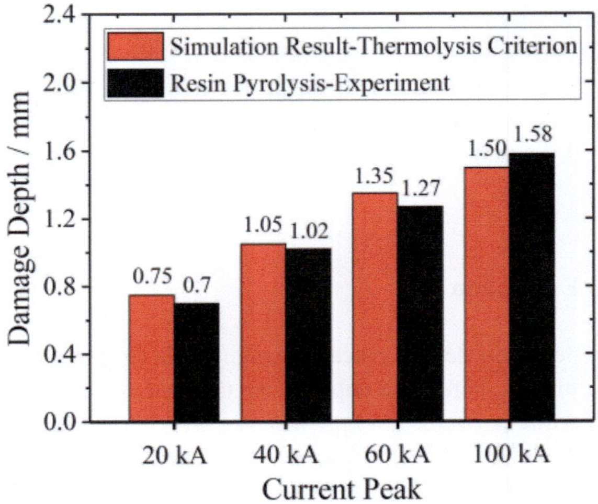

Fig. 4.38 Evaluation of simulated damage depth via the pyrolysis-degree criterion

Table 4.9 Simulated versus experimental lightning-damage depths

No.	Experimental results/mm	Simulated results/mm	Deviation (%)
Type 1–1	0.70	0.75	7.14
Type 1–2	1.02	1.05	2.94
Type 1–3	1.27	1.35	6.30
Type 1–4	1.58	1.50	5.06

4.4.2 Lightning-Damage Evolution in CFRP Composites

Based on the evaluation methods, temperature-field criteria accurately predict ablation areas, while pyrolysis-degree criteria reliably determine damage depths. Figure 4.39 presents the temporal evolution of damage under varying component A peak currents.

From Fig. 4.39, both damage area and depth increase with higher current intensity. During the early injection phase, they rise rapidly. The thermal action integral, which quantifies energy deposition, dominates ablation. For a 60 kA peak, 72.55% of the total action integral occurs within the first 100 μs, so growth rates of both area and depth diminish after 100 μs.

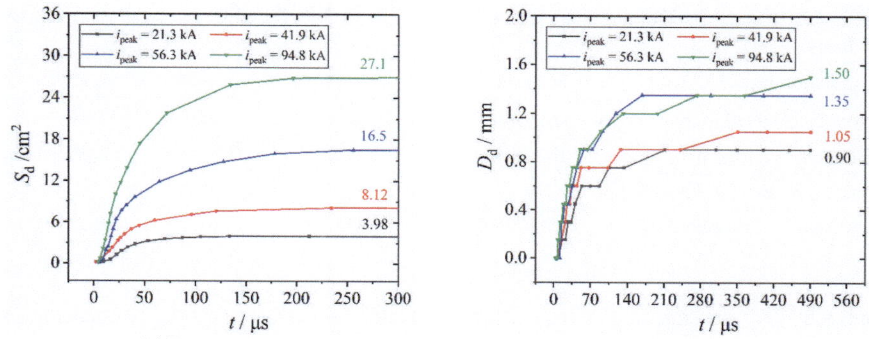

(a) Damage area (b) Damage depth

Fig. 4.39 Temporal evolution of damage area and depth in CFRP composites

4.5 Impact of Electrical Conductivity on Lightning Ablation Damage in CFRP Composites

The interaction between lightning current components and CFRP composites encompasses complex physical processes, including lightning leader development, arc attachment phenomena, streamer propagation/current dispersion at material surfaces, thermoelectric ablation, and thermal dissipation. During lightning current injection, the electrical conductivity characteristics of CFRP composites not only govern interlayer current distribution but also critically determine Joule heating generation, which exhibits strong spatial correlation with local conductivity. Consequently, the electrical conductivity of CFRP composites significantly influences their lightning ablation damage effects.

4.5.1 Impact of Electrical Conductivity on Lightning Ablation Damage in CFRP Composites

In existing literature, many researchers rely on the static electrical conductivity of CFRP composites, which is measured under low-voltage DC, for simulation purposes. However, these static values deviate markedly from the dynamic conductivity properties observed under impulse current conditions in this study, as summarized in Table 4.10.

To assess how conductivity choice affects simulated lightning damage, we developed two models using: static conductivities from prior studies ($\sigma_{xx} = 29300$ S/m; $\sigma_{yy} = 77.8$ S/m; $\sigma_{zz} = 0.000794$ S/m) and dynamic conductivities measured here ($\sigma_{xx} = 20270$ S/m; $\sigma_{yy} = 17400$ S/m; $\sigma_{zz} = 33.64$ S/m). Figure 4.40 compares the resulting damage areas and depths. As shown in Fig. 4.40, the dynamic conductivity model aligns closely with experimental data, with −4.9% deviation in damage area

Table 4.10 Comparison of nominal electrical conductivity values for CFRP composites in various studies

Simulation model	Material system	σ_{xx} (S/m)	σ_{yy} (S/m)	σ_{zz} (S/m)
Model employed in this study	IM600/133	20270	17400	33.64
Guo et al.'s model [6]	IM600/133	29300	77.8	0.000794
Ogasawara et al.'s model [1]				
Abdelal et al.'s model [56]	IM600/133	35970	1.145	0.003876

and +2.9% in damage depth. In contrast, the static conductivity model overestimates damage area by 304% and underestimates depth by 39.2%.

During impulse lightning current injection into CFRP composites, the current–density distribution directly controls Joule heating generation. To evaluate this, a cross-section taken 5.6 mm from the strike center along the X-axis is defined Fig. 4.41). Figure 4.42 compares the lightning current densities in the static and dynamic conductivity models.

In the static model, current concentrates within $Y = 1.0$–1.5 mm (surface plies 1–3), reaching a peak density of 3.5×10^9 A/m^2. In contrast, the dynamic model shows a uniform distribution across $Y = 0.6$–1.5 mm (surface plies 1–6), with a lower maximum of 1.1×10^9 A/m^2. Comparing ply-wise densities (Fig. 4.43), the static model's through-thickness conductivity is ~10^5 times lower than the dynamic value, so far less current shunts into deeper plies, focusing it on the top four layers. Therefore, the static model predicts shallower damage and excessive surface ablation, whereas the dynamic model distributes current, and thus heating, more evenly into deeper layers, increasing damage depth.

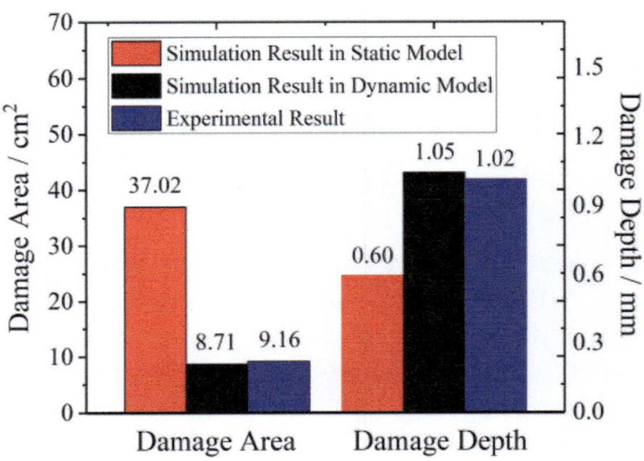

Fig. 4.40 Comparison of damage results between static and dynamic conductivity models for CFRP composites

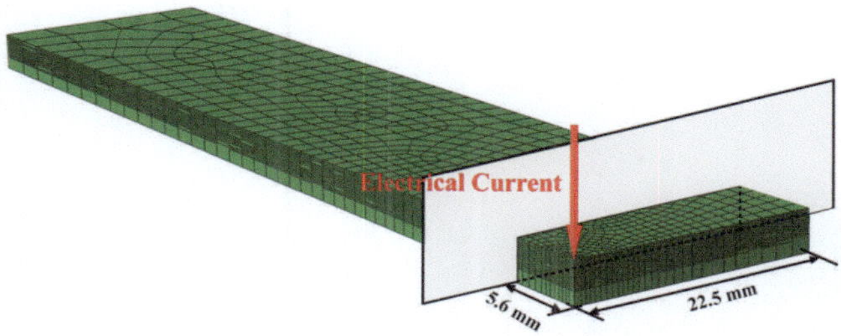

(a) Cross-section setup

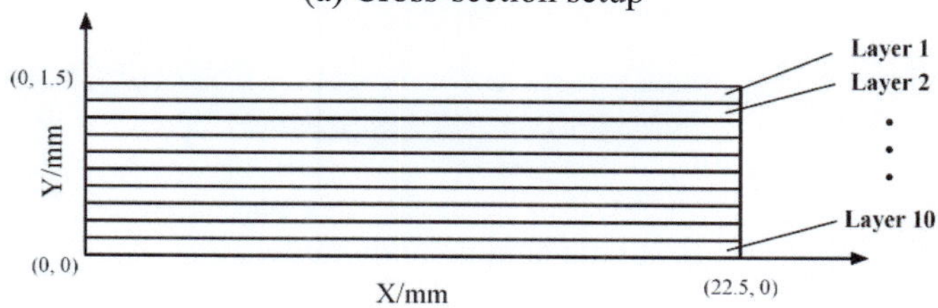

(b) Cross-sectional layup schematic

Fig. 4.41 Schematic diagram of the cross-section setting in CFRP composites

4.5.2 Influence of Conductivity on Lightning Damage Area

Given the significant discrepancies between simulation results from dynamic and static conductivity models demonstrated earlier, this study further investigates how CFRP conductivity affects ablation damage area and depth. A series of interpolated conductivity values between the static and dynamic parameters are selected as the ambient-temperature initial conductivities for CFRP composites (Fig. 4.44).

Directionality of Lightning Damage Areas

Two sets of in-plane anisotropic conductivities are compared: (i) $\sigma_{xx} = 20270$ S/m, $\sigma_{yy} = 17400$ S/m; (ii) $\sigma_{xx} = 27494$ S/m, $\sigma_{yy} = 3559$ S/m. A comparison of simulated lightning ablation zones across plies under these two conductivity configurations is presented in Fig. 4.45.

Analysis shows that a larger contrast between fiber-aligned conductivity (σ_{xx}) and transverse conductivity (σ_{yy}) enhances damage anisotropy. In Fig. 4.45a, with minimal divergence ($\sigma_{xx} \approx \sigma_{yy}$), all plies exhibit nearly circular damage zones. Conversely, Fig. 4.45b, with pronounced contrast ($\sigma_{xx} \gg \sigma_{yy}$), displays distinctly elliptical damage patterns:

(i) Plies 1–2 (−45°/45° orientation) show smaller surface damage, being grounded at both ends.
(ii) Ply 3 (0° orientation) extends most along the x-axis, reflecting high σ_{xx}.

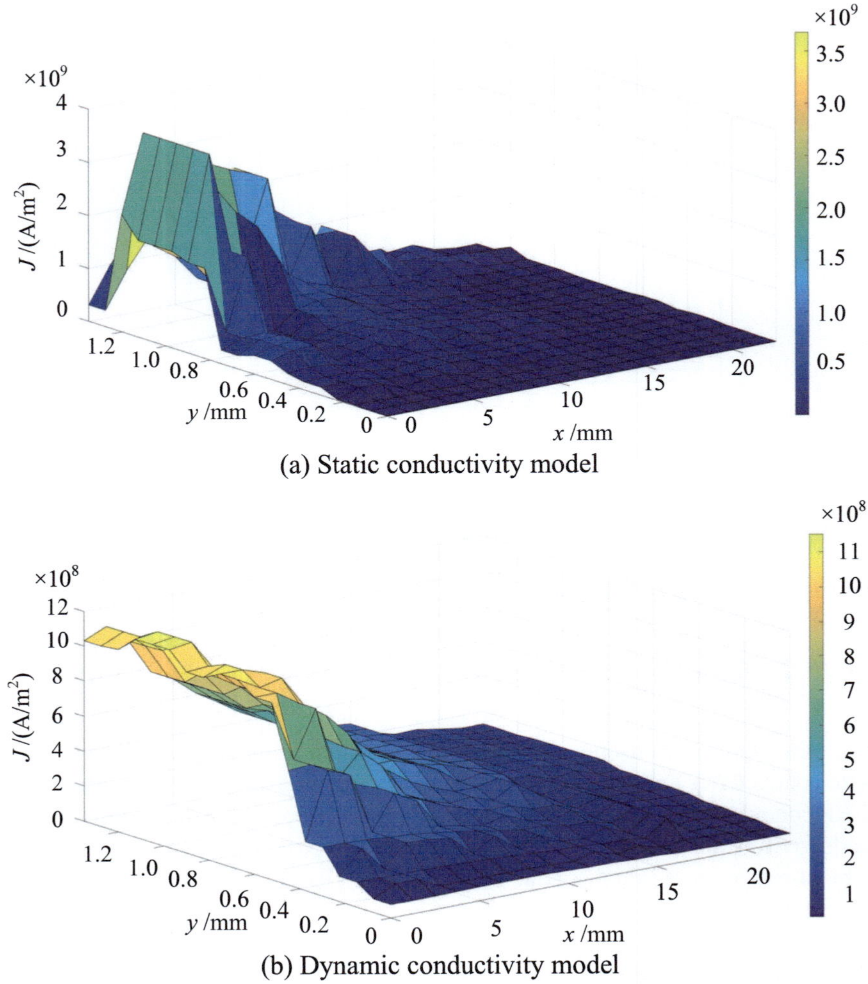

Fig. 4.42 Comparison of current–density distributions between static and dynamic conductivity models

(iii) Plies 4–5 (90° orientation) exhibit limited x-extension but greater spread along the y-axis, due to lower σ_{yy} and the underlying heat from Ply 3.

Influence of Through-Thickness Conductivity σ_{zz} on Damage Area

Reducing σ_{zz} from 33.64 to 4.5 S/m increases the simulated ablation area from 7.97 to 39.3 cm^2, an approximately fivefold expansion (Fig. 4.46). Because damage anisotropy varies by ply orientation, the total damage area is computed as the union of each ply's projected damage. As σ_{zz} decreases, less current shunts through the

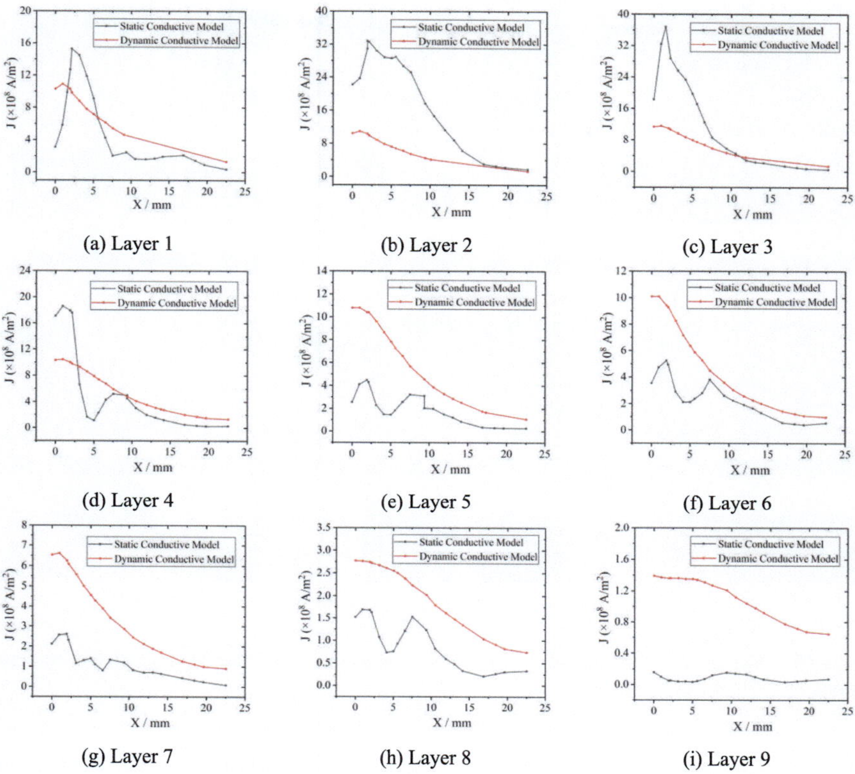

Fig. 4.43 Comparison of current–density distributions by ply between static and dynamic conductivity models

Fig. 4.44 Conductivity sampling point selection

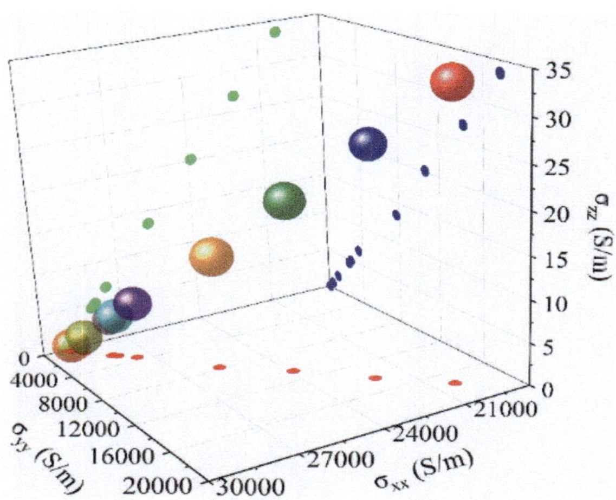

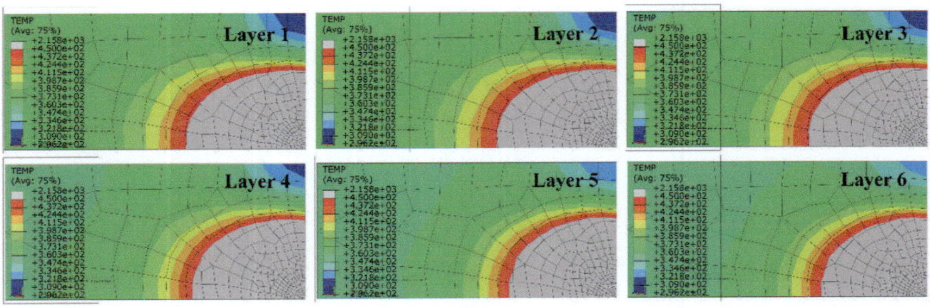

(a) $\sigma_{xx} = 20270$ S/m, $\sigma_{yy} = 17400$ S/m

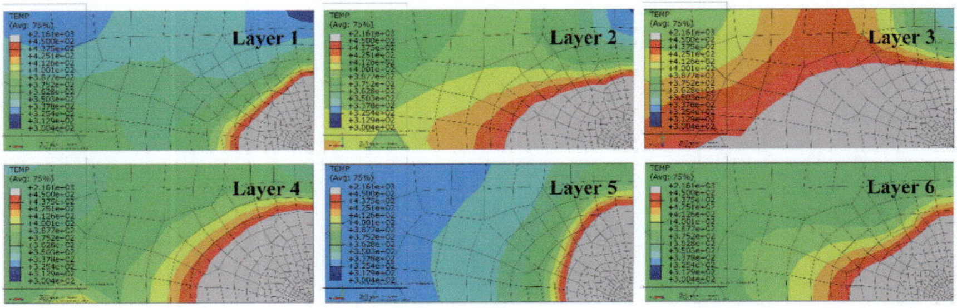

(b) $\sigma_{xx} = 27494$ S/m, $\sigma_{yy} = 3559$ S/m

Fig. 4.45 Comparison of simulated ablation zones across plies for different conductivity configurations

thickness, forcing more current in-plane, which amplifies Joule heating in the surface plies and thus enlarges the ablation footprint.

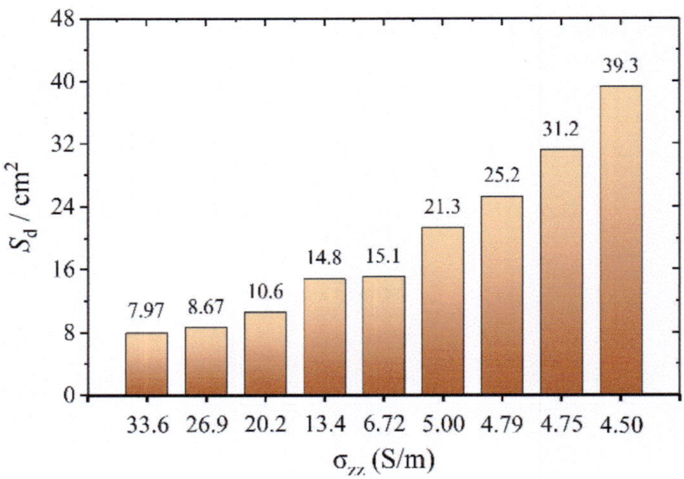

Fig. 4.46 Lightning ablation damage area for models with different σ_{zz}

Influence of In-Plane Conductivity σ_{xx} and σ_{yy} on Damage Area

Using the dynamic conductivity baseline, we proportionally reduce σ_{xx}, σ_{yy}, and σ_{zz} and evaluate ablation areas (Figs. 4.47 and 4.48). When all conductivities drop by the same factor, the relative current-shunt ratios among directions stay constant. However, lower in-plane conductivities (σ_{xx} and σ_{yy}) increase I^2R losses for a given current, producing more Joule heat and thus larger damage areas.

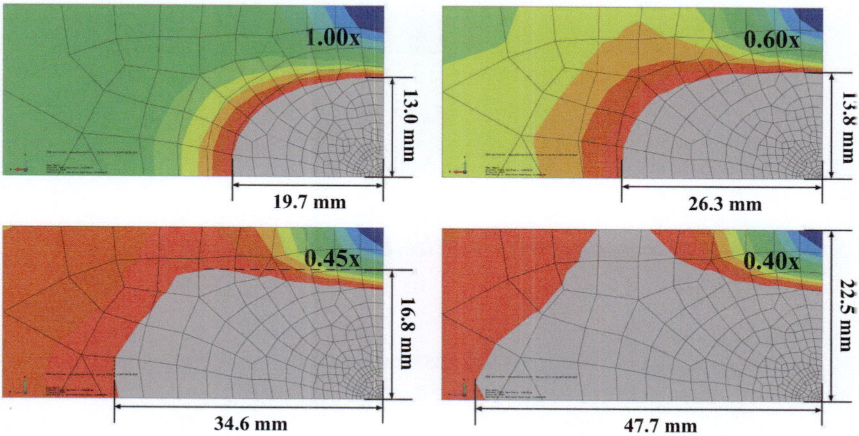

Fig. 4.47 Ablation damage area for models with varying conductivity reduction ratios

Fig. 4.48 Lightning ablation damage area versus conductivity reduction ratio

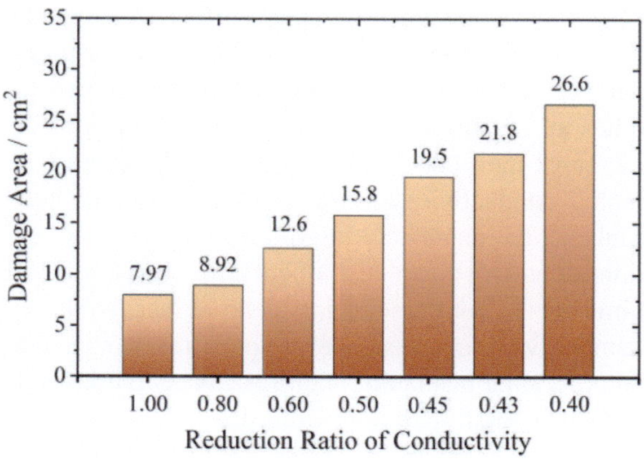

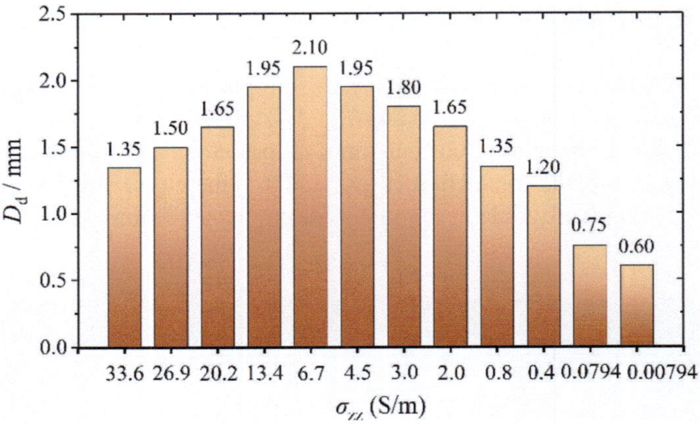

Fig. 4.49 Simulated lightning ablation damage depth versus σ_{zz}

4.5.3 Effect of Conductivity on the Depth of Lightning Damage

The impact of CFRP's through-thickness conductivity σ_{zz} on damage depth is nonlinear, as shown in Fig. 4.49. Reducing σ_{zz} from 33.6 to 6.7 S/m increases damage depth from 1.35 to 2.10 mm, reflecting enhanced thermal penetration at moderate conductivity. Further reducing σ_{zz} to 0.00794 S/m reverses this trend, decreasing depth to 0.60 mm.

Joule heating (I^2R) depends on both current density and material conductivity. Lower σ_{zz} reduces interlaminar current shunting, limiting deep-layer heating, yet increases resistance, amplifying Joule heat for a given current and thus deepening damage. When σ_{zz} becomes very low, current penetration stalls and energy concentrates at the surface, shallowing damage again. Consequently, damage depth first increases then decreases as σ_{zz} declines. From a lightning-resistance standpoint, higher conductivity lowers resistance to current flow, reducing temperature rise. Thus, CFRP manufacturing often enhances conductivity via fiber/matrix modifications or adding conductive fillers (e.g., carbon nanotubes, metal foams). Conversely, ultra-low σ_{zz} confines heating to the surface: while surface area damage grows, depth damage, which more severely impacts mechanical integrity, is minimized. Therefore, incorporating insulating, high-resistance interlayers may be an effective strategy to mitigate overall lightning damage.

4.6 Summary

A 2D axisymmetric MHD arc-plasma model with Joule heating coupling was developed alongside a 3D lightning-damage model for CFRP composites integrating arc, thermoelectric, and mechanical interactions with material properties varying by pyrolysis degree. Through analysis of temperature evolution, pyrolysis-degree distributions, and electric-field dynamics during component A injection, the processes of arc attachment, current dispersion, and intra-/inter-laminar damage evolution were elucidated. Experimental and simulation comparisons established damage-identification criteria, and the role of electrical conductivity in damage distribution was quantified. The main conclusions are as follows:

(1) A non-constant heating-rate thermal-degradation model for the resin matrix was combined with chemical-kinetics analysis to yield dynamic material parameters linked to pyrolysis degree. The resulting pyrolysis model defines an ablation onset temperature, while spatial-temporal pyrolysis-degree fields enhance prediction accuracy.
(2) The arc-thermoelectric coupling simulation-solving Maxwell's, heat-transfer, and mechanical equations, successfully reproduced lightning ablation behaviors. Damage boundaries defined by the 450 °C isotherm exhibited 8.15% area error, and depths identified by pyrolysis degree >0.9 showed 7.14% deviation.
(3) Significant deviations observed when using static DC conductivity prompted the integration of dynamic conductivity, determined via numerical fitting of impulse-current measurements, into the multiphysics lightning-damage model for the first time. Deviations in simulated damage area and depth were reduced from +304% and −39.2% (static model) to −4.9% and +2.9%, respectively. This enhancement validates the model's accuracy and demonstrates that dynamic conductivity parameters more faithfully represent the true electrical state of CFRP under lightning, whereas static parameters lead to overestimated areas and underestimated depths.
(4) Within the arc-thermoelectric coupling framework, electrical conductivity emerges as the key parameter governing lightning damage. Conductivity not only controls Joule heating but also steers current shunting, thereby shaping damage patterns:

    (i) In-plane anisotropy: damage directionality intensifies as the contrast between σ_{xx} (fiber direction) and σ_{yy} (transverse) increases.
    (ii) Damage area: expands as in-plane conductivities (σ_{xx}, σ_{yy} and through-thickness conductivity (σ_{zz}) decrease.
    (iii) Damage depth: exhibits a non-monotonic dependence on σ_{zz}, increasing with moderate reductions in σ_{zz} but decreasing once σ_{zz} falls below a critical threshold, due to restricted current penetration and surface-localized heating.

References

1. Ogasawara T, Hirano Y, Yoshimura A. Coupled thermal-electrical analysis for carbon fiber/epoxy composites exposed to simulated lightning current[J]. Composites: Part A, 2010, 41(8): 973–981.
2. Li S, Yin J, Yao X, et al. Damage analysis for carbon fiber/epoxy composite exposed to simulated lightning current, Journal of reinforced plastics and composites[J]. 2017, 35(15): 1201–1213.
3. Yin J, Chang F, Li S, X.L. Yao, et al. Lightning strike ablation damage influence factors analysis of carbon fiber/epoxy composite based on coupled electrical-thermal simulation[J]. Applied Composite Material, 2017, 24(5): 1089–1106.
4. Yin J, Li S, Yao X, et al. Lightning strike ablation damage characteristic analysis for carbon fiber/epoxy composite laminate with fastener[J]. Applied Composite Material, 2016, 23(4): 821–837.
5. Dong Q, Guo Y, Chen J, et al. Influencing factor analysis based on electrical-thermal-pyrolytic simulation of carbon fiber composites lightning damage[J], Composite Structures, 2016, 140: 1–10.
6. Guo Y, Dong Q, Chen J, et al. Comparison between temperature and pyrolysis dependent models to evaluate the lightning strike damage of carbon fiber composite laminates[J]. Composites: Part A, 2017, 97: 10–18.
7. Yin J, Li S, Yao X, et al. Numerical simulation of ablation damage and propagation behavior analysis for composite laminates subjected to lightning strike[J]. Transactions of Materials and Heat Treatment, 2016, 37(3): 203–209.
8. Katunin A, Krukiewicz K. Preliminary analysis of thermal response of dielectric and conducting composite structures during lightning strike[J]. Composites Theory and Practice, 2016, 16(1): 8–14.
9. Chemartin L, Lalande P, Peyrou B, et al. Direct effects of lightning on aircraft structure: analysis of the thermal, electrical and mechanical constraints[J]. AerospaceLab, 2012 (5): p. 1–15.
10. Chemartin L, Lalande P, Montreuil E, et al. Three dimensional simulation of a DC free burning arc. Application to lightning physics[J]. Atmospheric Research, 2009, 91(2–4): 371–380.
11. Xiangteng M A, Fusheng W, Han C, et al. Thermal damage analysis of aircraft composite laminate suffered from lightning swept stroke and arc propagation[J]. Chinese Journal of Aeronautics, 2020, 33(4): 1242–1251.
12. Aider Y, Wang Y, Abdelal G F, et al. Modeling of the Electric Arc Plasma Discharge Produced by a Lightning Strike Continuing Current[C]//2019 International Conference on Lightning and Static Electricity (ICOLSE), Wichita, Kansas, USA. 2019.
13. Chemartin L, Lalande P, Delalondre C, et al. Modelling and simulation of unsteady dc electric arcs and their interactions with electrodes[J]. Journal of Physics D: Applied Physics, 2011, 44(19): 194003.
14. M. Tanaka et al., "Time-dependent calculations of molten pool formation and thermal plasma with metal vapour in gas tungsten arc welding," Journal of Physics D: Applied Physics, vol. 43, no. 43, p. 434009, 2010.
15. Solana P, Kapadia P, Dowden J. A mathematical analysis of heating effects and electrode erosion in conical electrical arc cathodes[J]. Journal of Physics D: Applied Physics, 1998, 31(24): 3446.
16. Traidia A, Roger F. Numerical and experimental study of arc and weld pool behaviour for pulsed current GTA welding[J]. International journal of heat and mass transfer, 2011, 54(9–10): 2163–2179.
17. Sillero J A, Ortega D, Munoz-Serrano E, et al. An experimental study of thoriated tungsten cathodes operating at different current intensities in an atmospheric-pressure plasma torch[J]. Journal of Physics D: Applied Physics, 2010, 43(18): 185204.
18. Fusheng W, Xiangteng M A, Han C, et al. Evolution simulation of lightning discharge based on a magnetohydrodynamics method[J]. Plasma Science and Technology, 2018, 20(7): 075301.

19. Chen H, Wang F S, Ma X T, et al. The coupling mechanism and damage prediction of carbon fiber/epoxy composites exposed to lightning current[J]. Composite Structures, 2018, 203: 436–445.
20. Gharghabi P. Experimental and Numerical Studies of Lightning Strike Induced Damage to Carbon Fiber Epoxy Composites[D]. Mississippi State University, 2020.
21. Gonzalez J J, Lago F, Freton P, et al. Numerical modelling of an electric arc and its interaction with the anode: part II. The three-dimensional model—influence of external forces on the arc column[J]. Journal of Physics D: Applied Physics, 2005, 38(2): 306.
22. F. Lago, J.J. Gonzalez, P. Freton, A. Gleizes, A numerical modelling of an electric arc and its interaction with the anode: Part I. The two-dimensional model, J. Phys. D: Appl. Phys. 37 (2004) 883–897.
23. Liu Y, Wang Y. Modeling the lightning continuing current electric arc discharge and material thermal damage: Effects of combinations of amplitude and duration[J]. International Journal of Thermal Sciences, 2021, 162: 106786.
24. Bigand A. Damage assessment on aircraft composite structure due to lightning constraints[D]. Toulouse, ISAE, 2020.
25. Lowke JJ, Tanaka M. 'LTE-diffusion approximation' for arc calculations. Journal of Physics D: Applied Physics. 2006; 39(16): 3634.
26. M. Lisnyak, "Theoretical, numerical and experimental study of DC and AC electric arcs," Université Orléans, 2018.
27. Aider Y. A unified plasma-materials finite element model of lightning strike interaction with carbon fiber composite materials[D]. Mississippi State University, 2019.
28. Wu C S, Gao J Q. Analysis of the heat flux distribution at the anode of a TIG welding arc[J]. Computational Materials Science, 2002, 24(3): 323–327.
29. S.L.J. Millen, A. Murphy, G. Abdelal, G. Catalanotti, Sequential finite element modelling of lightning arc plasma and composite specimen thermal-electric damage, Comput. Struct. 222 (2019) 48–62.
30. A. Traidia, F. Roger, E. Guyot, Optimal parameters for pulsed gas tungsten arc welding in partially and fully penetrated weld pools, Int. J. Therm. Sci. 49 (2010) 1197–1208.
31. Menart J and Lin L 1999 Numerical study of a free burning argon arc with copper contamination from the anode Plasma Chem. Plasma Process. 19 153–70.
32. D'Angola A, Colonna G, Gorse C, et al. Thermodynamic and transport properties in equilibrium air plasmas in a wide pressure and temperature range. Eur Phys J D 2008; 46(1): 129–50.
33. Murphy AB. Transport coefficients of air, argon-air, nitrogen-air and oxygen-air plasmas. Plasma Chem Plasma Process 1995; 15: 279–307.
34. Capitelli M, Colonna G, Gorse C, et al. Transport properties of high temperature air in local thermodynamic equilibrium[J]. The European Physical Journal D-Atomic, Molecular, Optical and Plasma Physics, 2000, 11(2): 279–289.
35. Naghizadeh-Kashani Y, Cressault Y, Gleizes A. Net emission coefficient of air thermal plasmas. Journal of Physics D: Applied Physics. 2002; 35(22): 2925.
36. M. Tanaka, H. Terasaki, M. Ushio, and J. J. Lowke, "A unified numerical modeling of stationary tungsten-inert-gas welding process," *Metall. Mater. Trans. A*, vol. 33, no. 7, pp. 2043–2052, 2002.
37. M. Tanaka and J. J. Lowke, "Predictions of weld pool profiles using plasma physics," *J. Phys. D, Appl. Phys.*, vol. 40, no. 1, pp. R1–R23, 2007.
38. Hsu KC, Etemadi K, Pfender E. Study of the free-burning high-intensity argon arc. Journal of applied physics. 1983; 54(3): 1293–301.
39. Lago F, Gonzalez J J, Freton P, et al. A numerical modelling of an electric arc and its interaction with the anode: part III. Application to the interaction of a lightning strike and an aircraft in flight[J]. Journal of Physics D: Applied Physics, 2006, 39(10): 2294.
40. Tholin F, Chemartin L, Lalande P. Simulation of the Lightning Arc Root Interaction with Anisotropic Materials[C]//ICOLSE 2017. 2017.
41. Millmore S, Nikiforakis N. Multi-physics simulations of lightning strike on elastoplastic substrates[J]. Journal of Computational Physics, 2020, 405: 109142.

42. A. Larsson, P. Lalande, A. Bondiou-Clergerie, P. Lalande, A. Delannoy, The lightning swept stroke along an aircraft in flight. Part I: thermodynamic and electric properties of lightning arc channels, J. Phys. D, Appl. Phys. 33(15) (2000) 1866–1875.
43. F. Tholin, L. Chemartin, P. Lalande, Numerical investigation of the interaction of a lightning and an aeronautic skin during the pulsed arc phase, in: IET Conference Proceedings, 2015.
44. Jinru Sun, Xueling Yao, Jiaxin Ren, et al. Analytical Investigation of Lightning Strike-Induced Damage of OPGWs Based on a Coupled Arc-Electrical-Thermal Simulation[J]. IEEE Transactions on Power Delivery, 2022, 37(6).
45. Sun, S. Li, Yao. X, et al. Damage Characteristics of OPGWs Under Consecutive Lightning Strikes Based on a Coupled Arc-Thermal-Electric Simulation, 2023, IEEE Transactions on Power Delivery, 38(3):1927–1937.
46. Zhang, J.K., Guan, Z.D., & Li, Z.N. (2006). Three-dimensional finite element analysis of the temperature field during curing of thermosetting composites. Journal of Composite Materials, 23(2), 175–179.
47. Yan, X.Q., & Wu, H.P. (2003). Finite-element formulation for three-dimensional heat conduction in orthotropic anisotropic materials. Journal of Harbin Institute of Technology, 35(4), 405–409.
48. Weber M, Kamal MR. Microstructure and volume resistivity of composites of isotactic polypropylene reinforced with electrically conductive fibers[J]. Polymer Composites, 2010, 18(6): 726–740.
49. Ding, N., Zhao, B., Liu, Z.Q., et al. (2013). Simulation of lightning-strike ablation damage in composite laminates. Acta Aeronautica et Astronautica Sinica, 34(2), 301–308.
50. Dong Q, Guo Y, Sun X, et al. Coupled electrical-thermal-pyrolytic analysis of carbon fiber/epoxy composites subjected to lightning strike[J]. Polymer, 2015, 56: 385–394.
51. J. Sun, X. Tian, Y. Yao*, et al. Lightning strike-induced dynamic conduction characteristics and damage behavior of carbon fiber-reinforced polymer composites. Composite Structures, 2021, 275: 114391.
52. Cao, C.L., Sun, J.R., Xu, W.J., Chen, J.L., & Yao, X.L. (2018). Electrical conductivity characteristics of carbon-fiber composites under impulse current. Journal of Xi'an Jiaotong University, 52(2), 103–109.
53. Sun, J.R. (2019). Investigation of damage mechanisms in aircraft carbon-fiber composites under lightning impulse current (Doctoral dissertation). Xi'an Jiaotong University, Xi'an, China.
54. Yin, J., Li, S.L., Yao, X.L., Chang, F., & Zhang, X.H. (2017). Prediction of residual tensile strength of composite laminates with thermoelectric-coupled lightning damage. Journal of Composite Materials, 34(1), 83–90.
55. P. Foster, G. Abdelal and A. Murphy, "Understanding how arc attachment behaviour influences the prediction of composite specimen thermal loading during an artificial lightning strike test," Compos Structure, 2018, 192: 671–683.
56. Abdelal GF, Murphy A. A multiphysics simulation approach for efficient modeling of lightning strike tests on aircraft structures. IEEE Transactions on Plasma Science. 2017;45(4): 725–35.

Chapter 5
Lightning Damage Experiment and Evaluation of CFRP Composites Under Multiple Lightning Effects

The lightning strike process is a complex physical phenomenon involving multiple continuous, time-sequenced lightning current components. Due to variations in the electrical characteristics of each component, their damage effects on CFRP composites differ. The combined effect of sequential lightning components, acting continuously without interruption, forms a cumulative process in which each subsequent component acts on the already-damaged material state created by the preceding ones. As a result, the resulting damage differs significantly from that caused by isolated or time-separated components.

To address the limitation that single-component tests cannot replicate the coupled effects of transient mechanical forces, electromagnetic fields, and Joule heating in real lightning scenarios, and therefore fail to accurately represent actual lightning-induced damage to CFRP composites. This chapter conducts an experimental study on direct damage induced by multiple sequential time-series lightning components. Tests were conducted on unidirectional CFRP laminates, both with and without a copper mesh protective layer, using a dedicated experimental platform for simulating sequential lightning effects. Damage was characterized through morphological observation, infrared thermography, and ultrasonic scanning, allowing for the analysis of damage modes and characteristics across different CFRP structures.

The relationship between damage severity and lightning parameters, including peak current, rise rate, charge, and action integral, was established, and the individual contributions of each component to the overall damage were quantitatively evaluated. A regression analysis was then performed to explore the correlation between damage and multiple influencing factors. These results provide experimental validation and a theoretical foundation for further investigation of direct lightning effects on CFRP composites and structural elements, and offer guidance for the optimized design of lightning protection systems.

J. Sun et al., *Lightning Effects on Carbon Fiber-Reinforced Polymer Composites*,
https://doi.org/10.1007/978-981-95-3196-7_5

5.1 Experimental Conditions for Multiple Continuous Time Sequence Lightning Damage to CFRP Composites

5.1.1 Experimental Samples of CFRP Composites

In this chapter, an experimental study on direct damage induced by multiple continuous time-sequence lightning current components was conducted using CFRP composites with typical unidirectional lamination and copper mesh protective structures. The detailed specifications are as follows:

(i) Unidirectional laminated CFRP composites (Type S4) adopted the T300/3021 material system, with a carbon fiber volume fraction of 60%. The specimens measured 250 × 45 × 3.6 mm, with a single-ply thickness of 0.15 mm, totaling 24 layers, and a layup sequence of $[-45/45/0/90]_{3S}$.
(ii) Unidirectional laminated CFRP composites with copper mesh protection (Type S5) used the same material system and specimen size (250 × 45 × 3.6 mm), with 24 layers and the same layup configuration $[-45/45/0/90]_{3S}$. A metallic copper mesh protective layer with a thickness of 0.25 mm and mesh count of 1100 was applied to the surface of the composite.

5.1.2 Experimental Conditions for Multiple Continuous Time Sequence Lightning Damage

The experimental fixture configuration was consistent with that in Fig. 3.3 of Chap. 4. The sample was clamped at both ends, and the lightning impulse arc injection gap was adjustable within 50 ± 10 mm. The CFRP composites were divided into five test groups, and the lightning current combination modes for multiple continuous, uninterrupted time-sequenced components are shown in Table 5.1.

Figure 5.1 illustrates the waveforms of each lightning current component and their mutual time-sequence relationships under the ABCD quadruple continuous time sequence lightning current action mode, in which the B component is represented by a square wave current with a duration of 5 ms.

During the experiment, the B component of the lightning current was applied within 500 μs following the initiation of the A component, the C component was applied within 5 ms after the termination of the B component, and the D component was applied within 250 ms to 1 s after the start of the C component. The electrical parameter conditions for each lightning current component acting on the unidirectional lamination structure CFRP composite samples are listed in Table 5.2.

In Table 5.2, the test configurations and corresponding parameters for each CFRP sample are as follows:

(i) Type S4-1: Single application of lightning current component A, peak 41.67 kA.

Table 5.1 Multiple continuous uninterrupted time-sequence lightning current component test modes

Experimental modes	Multiple continuous, time sequence lightning current component combinations	Type S4 sample number
A mode	Applying a single lightning current component A	Type S4-1
AB mode	Continuously applying the lightning current components A and B	Type S4-2
DBC mode	Continuously applying lightning current components D, B, and C	Type S4-3
ABC mode	Continuously applying lightning current components A, B, and C	Type S4-4
ABCD mode	Continuously applying lightning current components A, B, C and C	Type S4-5

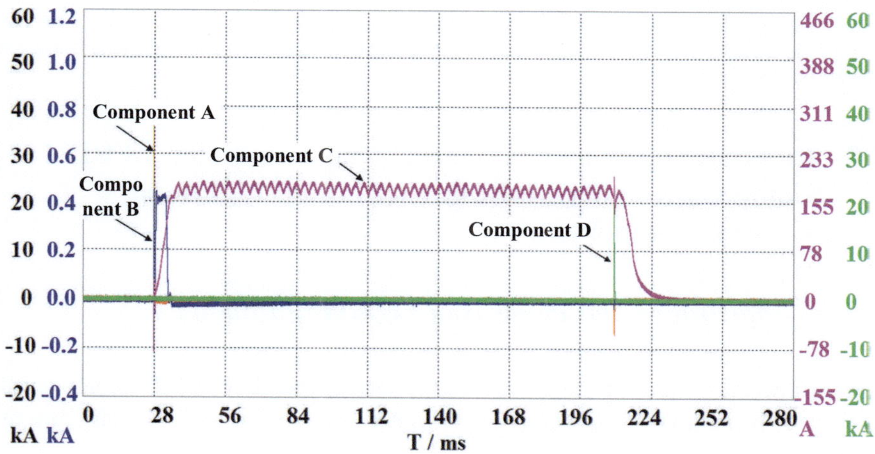

Fig. 5.1 Waveforms and time-sequence relationships of each lightning current component in ABCD mode

(ii) Type S4-2: Sequential application of component A followed by component B, with A peaking at 41.28 kA and B averaging 344 A.

(iii) Type S4-3: Component D → component B → component C sequence, in which D peaks at 35.21 kA, B averages 362.9 A, and C has an amplitude of 163.0 A.

(iv) Type S4-4: Component A → component B → component C time sequence, featuring A at 41.11 kA, B at 364.3 A, and C at 158.4 A.

(v) Type S4-5: Full ABCD sequence, which is A at 40.81 kA, B at 375.4 A, C at 170.3 A, and D at 19.89 kA.

Table 5.2 Electrical parameters of multiple lightning current components (unprotected unidirectional lamination structure CFRP composites)

Sample number	Component A			Component B			Component C			Component D		
	$I_{p\text{-}A}$ / kA	T_f / μs	W_A/ A^2s	$I_{ave\text{-}B}$/ A	T_L/ ms	Q_B/C	$I_{ave\text{-}C}$/ A	T_L/ ms	Q_C/C	$I_{p\text{-}C}$/ kA	T_f /μs	W_D/ A^2s
Type S4-1	41.67	26.76	79225	–	–	–	–	–	–	–	–	–
Type S4-2	41.28	25.75	81659	344.0	5.036	1.883	–	–	–	–	–	–
Type S4-3	–	–	–	362.9	5.070	1.840	163.0	187.8	30.61	35.21	14.00	47968
Type S4-4	41.11	25.50	78359	364.3	5.060	1.843	158.4	189.5	30.00	–	–	–
Type S4-5	40.81	25.50	76353	375.4	5.054	1.897	170.3	191.3	32.57	19.89	14.50	18804

Note $I_{p\text{-}A}$ and $I_{p\text{-}D}$ represent the peak current values of the lightning current A and D components, respectively; T_f denotes the wavefront time; T_L refers to the current duration; $I_{ave\text{-}B}$ and $I_{ave\text{-}C}$ are the average current values of lightning current components B and Cs; W_A and W_D denote the action integrals of components A and D; Q_B and Q_C are the transferred charges of components B and C

5.2 Lightning Damage to Unidirectional-Laminated CFRP Composites Under Multiple Continuous Lightning Current Components

5.2.1 Appearance and Patterns of Lightning Damage

The surface lightning damage of unidirectional-laminated CFRP composites subjected to multiple continuous, uninterrupted, time-sequence lightning current components is shown in Fig. 5.2.

Analysis of Fig. 5.2 reveals:

(1) Under component A (peak ~40 kA, transferred charge ~3 C), Type S4-1 exhibits fiber bundle swelling and fracture along the 45° fiber direction at the injection point. Delamination appears along both the 45° and −45° ply directions, symmetrically outward from the fracture zone. Scattered local delamination and resin pyrolysis spots are also observed (red circle in Fig. 5.2a). Type S4-2 shows similar damage modes but with greater severity (Fig. 5.2b).
(2) With addition of component C (transferred charge ~30 C), Types S4-3, S4-4, and S4-5 exhibit markedly increased damage, in ascending order of severity (Fig. 5.2c–e). A subcircular ablation crater forms at the attachment point, surrounded by extensive resin pyrolysis, multi-layer fiber breakage, and severe delamination extending to the composite edges. Flaky carbon deposits and sintered residues (region A in Fig. 5.2e) result from fiber sublimation and resin carbonization at lightning temperatures.

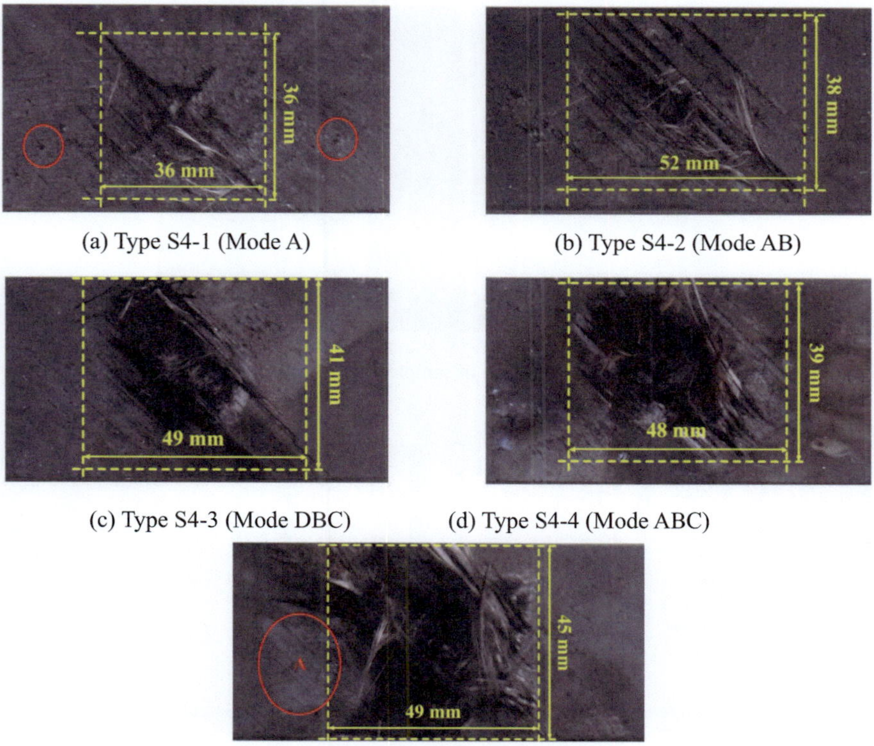

(a) Type S4-1 (Mode A)

(b) Type S4-2 (Mode AB)

(c) Type S4-3 (Mode DBC)

(d) Type S4-4 (Mode ABC)

(e) Type S4-5 (Mode ABCD)

Fig. 5.2 Lightning damage appearance of CFRP composite material samples

(3) Post-strike, fiber bundles are fractured and warped along the laminate surface, with fan-shaped damage zones symmetrically flanking the fracture line. Following the damage delineation approach from Sect. 3.2, Type S4-4 damage is partitioned into three zones: Zone I: attachment ablation, Zone II: carbon-fiber fracture, and Zone III: resin-pyrolysis delamination, as depicted in Fig. 5.3.

5.2.2 *Temperature Characteristics of CFRP Composites After Multiple Lightning Strikes*

During the application of multiple continuous, uninterrupted, time-sequence lightning current components, the temperature at the attachment point and surrounding region of the CFRP rose sharply. Given the thermophysical properties of the resin and carbon fibers, resin pyrolysis initiates at 180–200 °C (in oxygen), and carbon-fiber sublimation occurs above 3000 °C. Figure 5.4 presents the infrared temperature maps of sample Type S4-4 at successive times following components A, B, and C.

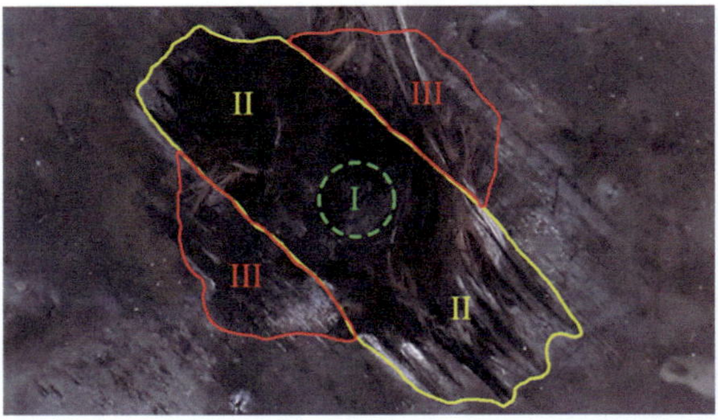

Fig. 5.3 Lightning damage area delineation of sample Type S4-4

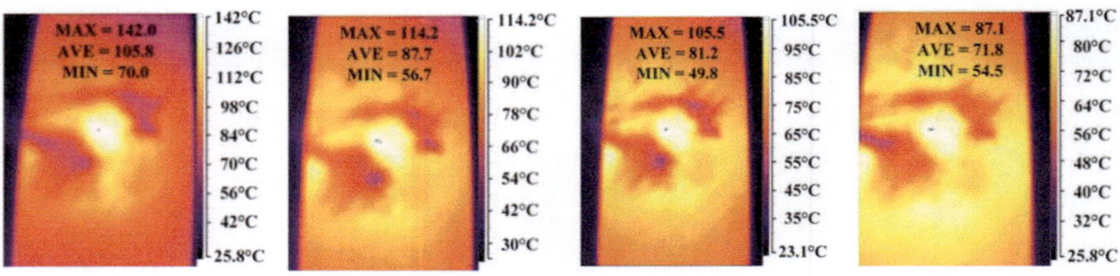

(a) $t = 13$ s (b) $t = 36$ s (c) $t = 50$ s (d) $t = 87$ s

Fig. 5.4 Infrared temperature distribution of sample Type S4-4 after successive application of components A, B, and C

Inspection of Fig. 5.4 shows that the high-temperature zone is localized at the attachment point, matching the circular ablation crater observed post-strike. At $t =$ 13 s, the peak temperature of Type S4-4 reached 142 °C. As cooling proceeds via heat exchange with the environment, the surface temperature decays rapidly. Figure 5.5 plots the time-history of surface temperature under different component sequences.

Comparing the lightning damage appearance of CFRP samples in Fig. 5.2 with the measurement results of sample temperatures after lightning strikes, it can be seen that as the number of lightning current components increased in the applied multiple-strike mode, the surface temperature of CFRP samples rose correspondingly, and the degree of lightning-induced damage became more pronounced. In addition, the surface temperatures of samples Type S4-3 (ABC mode) and Type S4-4 (ABCD mode) were significantly higher than those of samples Type S4-1 (A mode) and Type S4-2 (AB mode) after lightning strikes, indicating that the inclusion of lightning current component C led to a marked increase in the post-strike surface temperature of the samples. According to the measured surface temperature data of the CFRP composites after lightning strike tests, and taking into account the thermophysical properties of resin and carbon fiber as well as the characteristics of arc temperature (typically on the order of 10^3–10^4 K), the temporal evolution pattern of the CFRP composite surface temperature following the lightning strike on sample Type S4-4 can be obtained through curve fitting, as shown in Fig. 5.6:

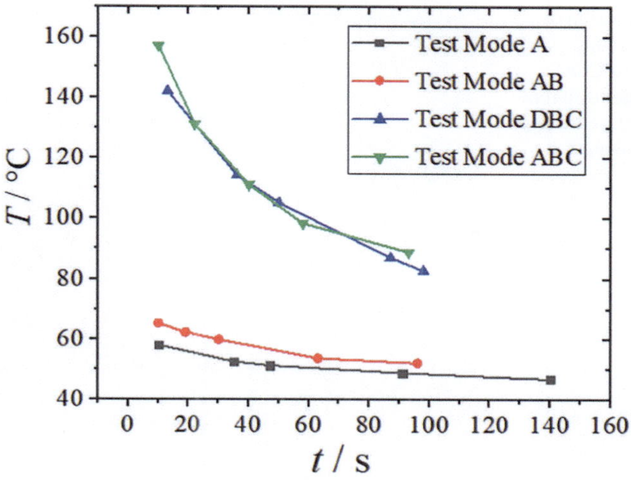

Fig. 5.5 Temporal evolution of CFRP surface temperature under multiple continuous lightning strikes

$$T = 277.2 \times t^{-0.2555} \tag{5.1}$$

From Eq. (5.1), it is indicated that the instantaneous surface temperature of the CFRP composites exceeds 1600 °C at $t = 1$ ms following the lightning strike. Since the duration of the lightning current component C is approximately 200 ms, and the heat exchange between the composite sample and the surrounding air medium occurs continuously during the action of the lightning current component, the transient maximum surface temperature of the CFRP composites under multiple continuous uninterrupted lightning strikes may exceed this value. The thermal energy associated

Fig. 5.6 Fitted curve of temperature variation over time for sample Type S4-4 after multiple continuous uninterrupted lightning strikes

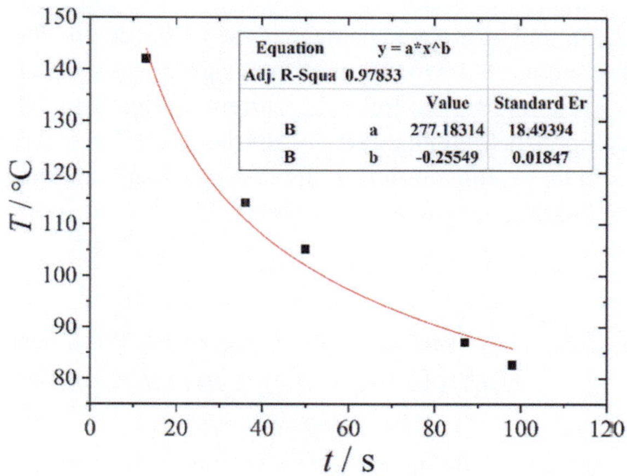

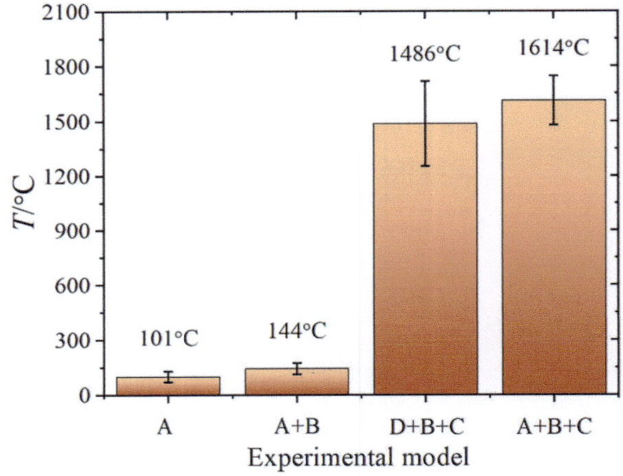

Fig. 5.7 Comparison of transient temperature characteristics of CFRP composites under different combinations of lightning current components

with the lightning current component causes a substantial increase in the temperature of the CFRP composite, leading to thermal sublimation of the carbon fibers and pyrolysis of the resin matrix. In addition, the sample temperature remains at approximately 270 °C around $t = 1$ s after the lightning strike, indicating that the resin decomposition process persists for a certain period following the termination of the lightning current pulse. If the temperature at $t = 0.1$ s is adopted as a reference for evaluating the transient temperature characteristics of CFRP composites after lightning strikes, the thermal responses of the samples under different combinations of multiple continuous uninterrupted lightning current components can be compared, as shown in Fig. 5.7.

From the experimental results shown in Fig. 5.7, it can be observed that as the number of applied continuous time-sequenced lightning current components increases, the surface temperature of the CFRP composites increases accordingly, and the higher the localized temperature at the lightning injection point, the more severe the lightning-induced damage to the CFRP composites. It can also be concluded that the sample temperature increases significantly under the action of multiple continuous uninterrupted lightning current components when the contribution of current component C is enhanced, i.e., the thermal effect of the lightning current C component is more prominent during the energy dissipation process within CFRP composites compared to components A and B.

5.2.3 Analysis of Lightning Damage Characteristics Under Multiple Lightning Current Components

Ultrasonic scanning imaging was employed to assess the extent and characteristics of lightning-induced damage in CFRP composites, as presented in Figs. 5.8 and 5.9.

(a) Type S4-1 (Mode A)

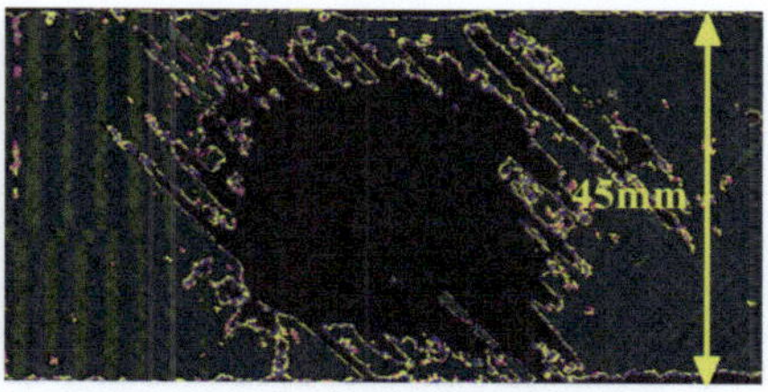

(b) Type S4-2 (Mode AB)

(c) Type S4-3 (Mode DBC)

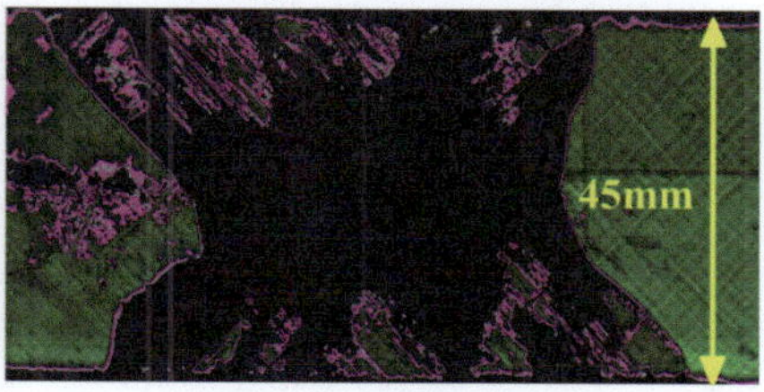

(d) Type S4-4 (Mode ABC)

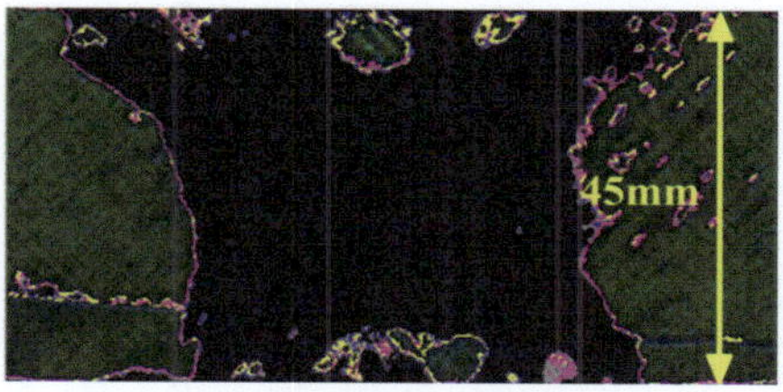

(e) Type S4-5 (Mode ABCD)

Fig. 5.8 Ultrasonic scan of damage area in CFRP composites after multiple lightning current components

Based on the ultrasonic scanning results in Figs. 5.8 and 5.9, it is evident that the damage region is primarily distributed along the carbon fiber lamination direction. Under the four-sequence ABCD lightning current component mode, the damage area extends to the edge of the composite material. In addition, due to the anisotropic structure and lamination process of CFRP composites, when subjected to lightning strikes, the lightning current preferentially propagates along the fiber lamination direction. The Joule heating effect generated during conduction induces pyrolysis of the resin and ablation or even sublimation of carbon fibers, resulting in damage that propagates outward along the axial direction of the carbon fibers within each ply. This eventually leads to the formation of a prominent fan-shaped, symmetric delamination pattern within the CFRP composites.

The lightning-induced damage in CFRP composites includes resin pyrolysis, fiber fracture, material ablation, and interlaminar delamination. To facilitate the quantitative analysis of damage severity in CFRP composites, the damage level of the samples was evaluated in terms of both damage depth and damage area. From the perspective of the macroscopic effects of lightning, damage depth and area reflect distinct

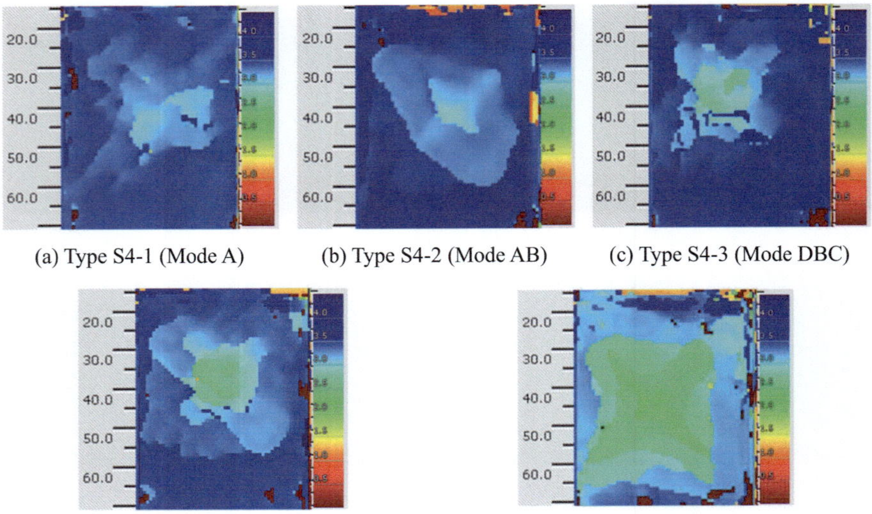

(a) Type S4-1 (Mode A) (b) Type S4-2 (Mode AB) (c) Type S4-3 (Mode DBC)

(d) Type S4-4 (Mode ABC) (e) Type S4-5 (Mode ABCD)

Fig. 5.9 Ultrasonic scan of damage depth in CFRP composites after multiple lightning current components

physical processes associated with lightning current application: damage depth is primarily influenced by shock wave and overpressure effects, electromagnetic forces, and localized thermal effects resulting from high-temperature arc contact, whereas damage area expansion is closely associated with the Joule heating effect caused by current dispersion within the CFRP composites. The measured damage area and depth for different experimental modes were obtained through pixel-based statistical analysis, as presented in Table 5.3.

Given that the lightning current components were applied in a sequential and continuous manner, the incremental increases in damage area and depth can be attributed to the additional contribution of newly applied current components beyond those used in previous test modes. For example, by comparing the lightning damage results under the "A" and "AB" test modes, the contribution of the lightning current

Table 5.3 Measured damage area and depth of samples under different combinations of multiple continuous lightning current components

Sample number	Experimental modes	Damage area/cm^2	Damage depth/mm
Type S4-1	Mode A	11.04	0.513
Type S4-2	Mode AB	13.23	0.810
Type S4-3	Mode DBC	14.53	0.900
Type S4-4	Mode ABC	24.00	0.930
Type S4-5	Mode ABCD	27.90	1.280

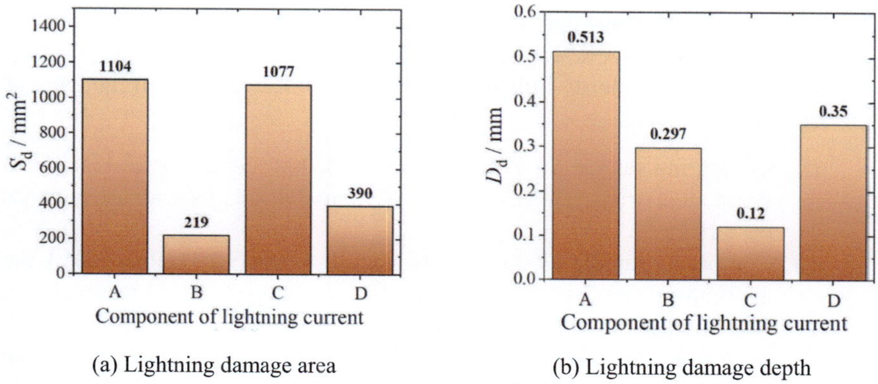

Fig. 5.10 Contribution of individual lightning current components to damage depth and area during multiple continuous lightning strikes

B component following the action of component A can be assessed. Similarly, by comparing the "AB" and "ABC" test modes, the incremental effect of the C component after sequential application of A and B components can be determined. Following this logic, the relative contribution of each individual lightning current component to the total damage of CFRP composites under multiple continuous lightning strike conditions can be quantitatively analyzed, as illustrated in Fig. 5.10 [1–4].

As can be seen in Fig. 5.10:

(1) Both the damage area and damage depth of CFRP composites increased with the number of continuously applied lightning current components. When the A, B, C, and D components were applied in sequence, the damage area reached 27.90 cm^2, and the damage depth reached 1.28 mm, with more than eight layers of the fiber laminate structure affected.
(2) The lightning current A, B, and D components, particularly the A component, had a more pronounced effect on damage depth, whereas the continuous current C component exhibited a comparatively weaker influence on depth.

   (i) Under test mode A (sample Type S4-1), the lightning current A component (peak value 40 kA) alone resulted in a damage depth of 0.5 mm.
   (ii) When the B component was added continuously and uninterruptedly after A (test mode AB, sample Type S4-2), the damage depth increased by 0.3 mm.
   (iii) For sample Type S4-4 under test mode ABC, which included the subsequent continuous application of the C component, the damage depth increased by only 0.12 mm.
   (iv) Finally, the addition of the pulse current D component (peak value: 20 kA) led to a further increase of 0.35 mm in damage depth for sample Type S4-4.

(3) Analysis of the damage depth contribution from each lightning current component indicates that the C component, despite having a long duration and high charge transfer, primarily influences the damage area rather than the damage depth.

  (i) Comparing sample S4-1 (mode A) with sample S4-2 (mode AB), the B component (transferred charge: 1.88 C) caused the damage area to increase by 2.19 cm^2.
  (ii) Comparing sample S4-2 (mode AB) with sample S4-4 (mode ABC), the addition of the C component (transferred charge: ~30 C) resulted in a 10.77 cm^2 increase in damage area.
  (iii) Comparing sample S4-4 (mode ABC) with sample S4-5 (mode ABCD), the D component (transferred charge 2.07 C) led to a 3.90 cm^2 increase in the damage area.

In the multiple continuous lightning current component impact experiments, each lightning current component contributed differently to the damage depth and damage area of the CFRP composites. This can be explained by the following analysis:

(1) Under the combined action of multiple continuous lightning current components, the characteristic parameters of each component differ significantly. For instance, the current amplitudes of components B and C differ from those of components A and D by more than two orders of magnitude, and the current rise rates differ by nearly four orders of magnitude. In terms of charge transfer, component C delivers approximately 20 times more charge than components A and D. These disparities suggest that, although all components contribute to CFRP damage, the mechanisms and effects by which they act differ substantially.
(2) When high-amplitude, fast-rising components A or D are applied to the CFRP composites, the stored energy in the capacitor is rapidly released through a narrow discharge channel. This causes rapid air compression and intense shock waves, exerting substantial pressure on the CFRP surface. The higher the current amplitude and faster the energy release, the stronger the resulting shock wave. These shock forces, along with dielectric breakdown under high electric field intensities, lead to the instantaneous rupture and exposure at the arc attachment point. Additionally, the high-temperature lightning pulse arcs contribute to increased damage depth. The instantaneous melting and fracture of carbon fibers at the lightning injection point expose the underlying laminate, allowing subsequent current components to propagate through both surface and interior fiber structures, thereby further increasing the damage depth.
(3) As discussed in Chap. 3 regarding the formation mechanisms of fiber fracture and resin pyrolysis-induced delamination, the expansion of the damage area is strongly correlated with the thermal effects of the lightning current. This inference is further supported by a comparison of the temperature profiles shown in

Fig. 5.7 and the damage area data in Table 5.3. The lightning current C component, due to its long duration and large transferred charge, plays a dominant role in enlarging the damage area.

5.2.4 Comparison of Damage Effects of Single and Multiple Continuous Lightning Current Components

The successive application of lightning current components modifies the thermal, electrical, and mechanical properties of CFRP composites, so that the damage caused by an individual component differs markedly from that produced by the same component applied after preceding pulses. For instance, the D component (peak 40 kA) applied in isolation produces the damage shown in Fig. 5.11.

From the surface observations and ultrasonic transmission scans, it is evident that when the D component acts alone, the damage is confined to the top few plies, yielding an overall damage area of approximately 9.0 cm^2 and a maximum depth of about 0.5 mm. By contrast, when the D component is applied after the A $\rightarrow$ B $\rightarrow$ C sequence (comparing Type S4-4 under ABC mode in Fig. 5.9d with Type S4-5 under ABCD mode in Fig. 5.9e), the deep-damage effects are substantially amplified. Under ABC alone, Type S4-4 exhibits a maximum depth of 0.93 mm, but

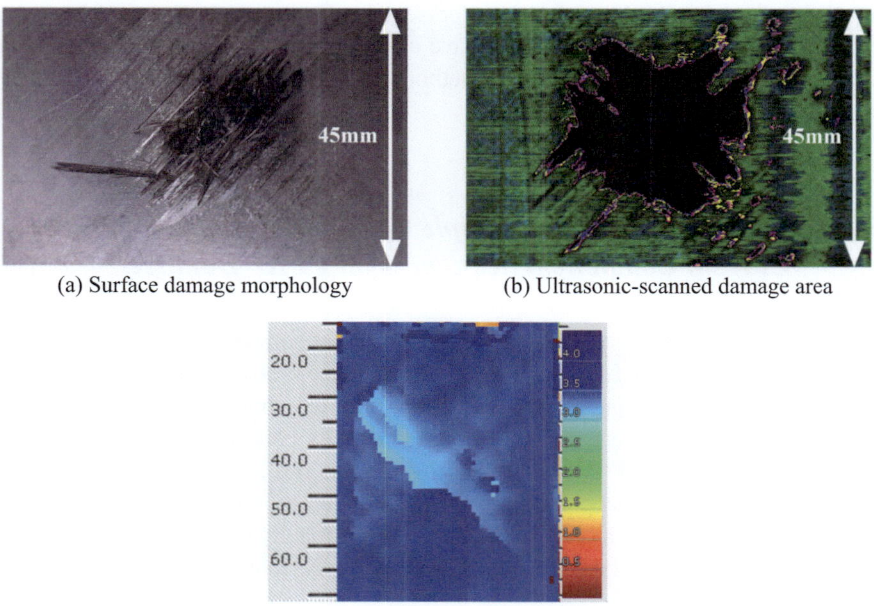

(a) Surface damage morphology (b) Ultrasonic-scanned damage area

(c) Ultrasonic-scanned damage depth

Fig. 5.11 Lightning damage of CFRP composites under the action of D-component of lightning current alone this depth is limited to an area under 1.00 cm^2, indicating that most damage remains near the surface. After addition of the D component, Type S4-5 shows only a 3.90 cm^2 increase in total damage area, yet the area exceeding 1.0 mm depth expands to over 10.00 cm^2 (green regions in Fig. 5.9e). Thus, the D component's contribution to deep damage is highly significant when acting on a material already thermally conditioned by components A, B, and C.

This behavior arises because, after continuous application of components A, B, and C, the subsequently applied D component encounters a much higher effective voltage/current amplitude, establishing a steep potential gradient through the CFRP thickness. On one hand, under high amplitude, the through-thickness resistance of the laminate decreases, facilitating current divergence into deeper plies and thus increasing damage depth. On the other hand, this high potential gradient can trigger interlaminar resin breakdown, enabling component D to conduct along deeper layup planes and thereby expand the deep-level damage area. Simultaneously, following A-B-C loading, the specimen's damage zone remains at elevated temperature and endures overpressure from resin-pyrolysis gases, while partial fiber sublimation and thinning weaken interlaminar bonding and reduce mechanical strength in the damaged region. When the high-rise-rate D component is then applied, its intense impact and electromagnetic forces further propagate deep-layer damage within the CFRP.

Consequently, the damage produced by an isolated current component does not fully represent its behavior within a multi-pulse sequence. Experimental and mechanistic studies of multiple continuous time-sequence lightning currents are therefore of critical theoretical importance and have direct application value for establishing CFRP lightning-damage assessment methods and designing optimized protective structures.

5.2.5 Effect of Timing in Multiple Continuous Lightning Current Component Sequences on CFRP Damage

In sequential application of multiple lightning current components, prior pulses not only modify the material state of CFRP composites but also influence the damage inflicted by subsequent pulses. Consequently, both the combination and timing of components critically determine the overall damage pattern. Comparison of the "DBC" (Type S4-3) and "ABC" (Type S4-4) test modes in Fig. 5.9c and d shows that, in both cases, the deep damage zone is centered around the current injection point. However, in DBC mode, damage propagates mainly along the −45° and 45° ply directions, whereas in ABC mode, damage extends along −45°, 45°, 0°, and 90° directions.

Considering the $[-45/45/0/90]_{3S}$ layup, this discrepancy can be attributed to the different damage depths induced by components A and D. Under 40 kA A-component-only loading (Type S4-1), the damage penetrates to approximately the fourth ply ($D_d \approx 0.513$ mm). When the C component follows the A component (ABC mode), the previously formed conductive channels allow current to propagate in all four principal ply directions, resulting in a nearly circular damage footprint. In contrast, in DBC mode, the D component's lower amplitude and energy produce a shallower damage depth compared to component A. As a result, the subsequent C component can only conduct along the first two plies (−45° and 45°), and the damage area remains concentrated along those directions.

5.3 Lightning Damage Assessment and Analysis of CFRP Composites

Lightning strikes act as a multi-component current process, following a specific time sequence, where each component exhibits distinct waveform characteristics. Among them, components A and D are marked by high peak current, steep front, and short duration, whereas component C stands out for its long-lasting discharge and large charge transfer. These differing characteristics lead to significant variation in how each component contributes to the damage depth and surface damage area in CFRP composites. The high energy density of A and D components causes rapid compression of the surrounding air and generates intense mechanical shock waves, exerting severe impact pressure on the CFRP surface. The greater the current amplitude, the more explosive the energy release, resulting in a stronger shock wave and deeper material damage. In contrast, although component C features a lower current amplitude and gentle front, its prolonged duration leads to substantial resistive heating, promoting lateral expansion of the damage region, even though depth damage remains limited due to the lack of electrical breakdown.

Based on experimental studies and simulation analyses, this section explores the influence patterns between rise rate, charge transfer, action integral, and the resultant lightning damage to CFRP composites. Through systematic evaluation of test data, the study identifies the dominant contributing factors and proposes a multifactorial assessment method for damage evaluation. These insights provide a scientific basis for future formulation design, processing optimization, structural layout, and performance enhancement of CFRP materials in lightning-prone applications.

5.3.1 Lightning Damage Assessment of CFRP Composites Under the Action of a Single Lightning Current Component

To investigate the influence of lightning current component parameters on the lightning damage behavior of CFRP (Carbon Fiber Reinforced Polymer) composites, this study evaluates the damage from three key perspectives: damage area, damage depth, and damage volume. Based on the CFRP lightning damage model established in Chap. 4, the total damage volume can theoretically be obtained by integrating the damage area of each ply with its corresponding depth. However, in practical experiments, the damage area is typically obtained via ultrasonic through-transmission scanning, which only provides a 2D projection of the damaged region and lacks layer-specific information. Therefore, the exact damage volume of the laminate cannot be precisely quantified. For the sake of consistency and comparative analysis in this study, the product of damage area and damage depth is adopted as an approximate representation of damage volume.

Using the experimental data (Table 3.4) from Chap. 3, which investigated the lightning strike response of CFRP laminates with unidirectional layup structures, the relationships between damage metrics, including resin pyrolysis delamination area (S_2), damage depth (D_d), and damage volume (V_d), and key parameters of the lightning current component A, including peak current (i_p), charge transfer (Q), and action integral (W_A), were analyzed. The results are illustrated in Figs. 5.12, 5.13, and 5.14 [2].

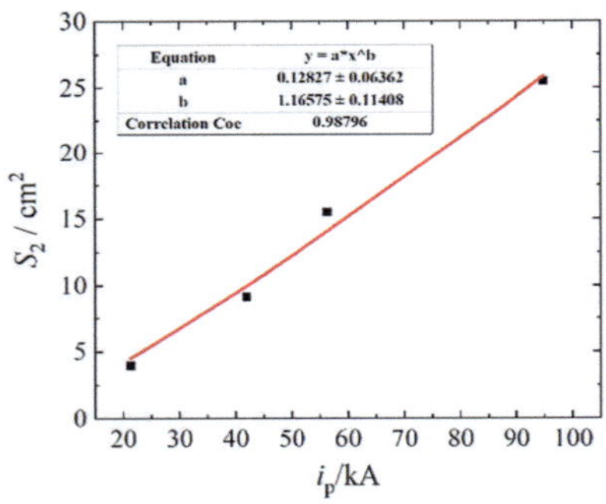

(a) Peak current vs. Damage area

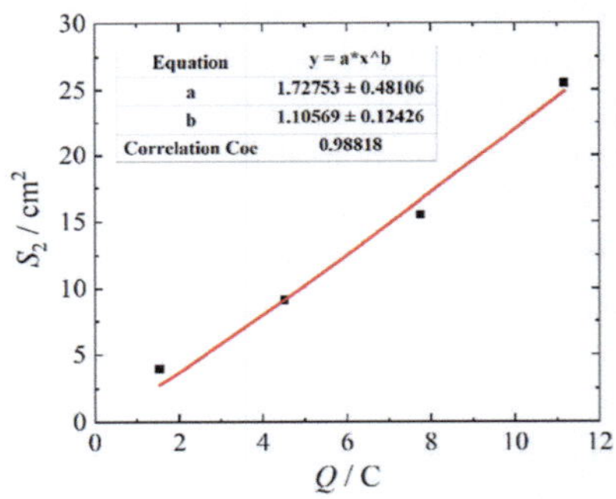

(b) Charge vs. Damage area

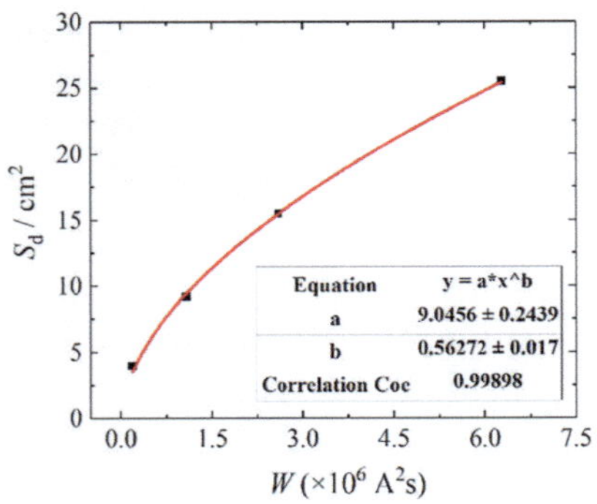

(c) Action integral vs. Damage area

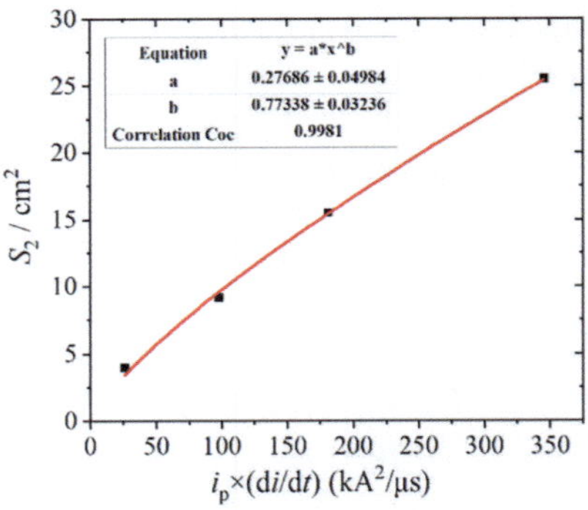

(d) Peak × Rise rate vs. Damage area

Fig. 5.12 Relationship between lightning damage area and lightning current parameters for CFRP composites

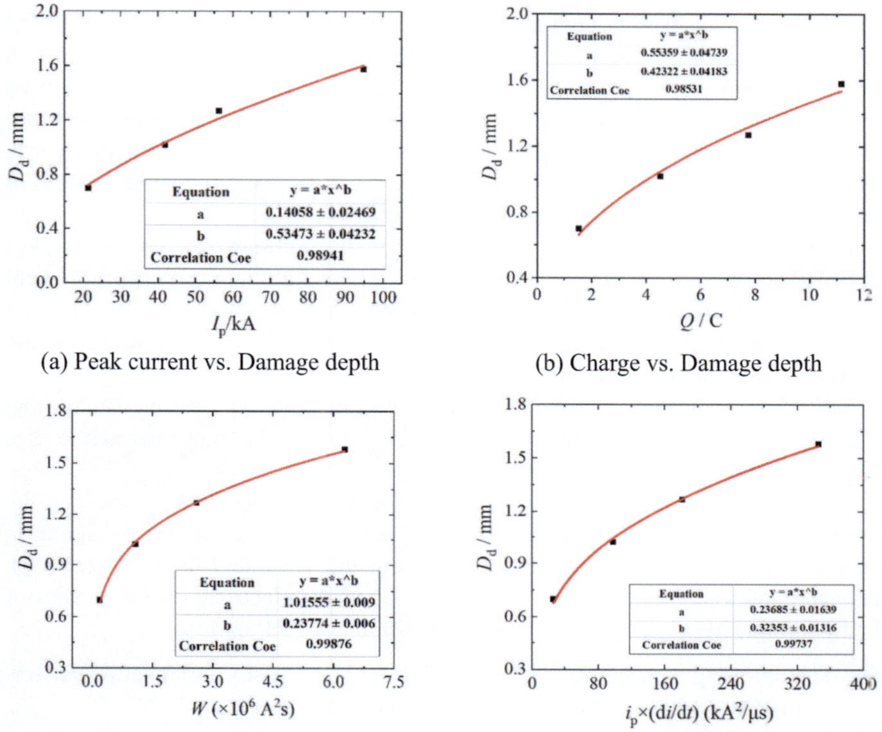

(a) Peak current vs. Damage depth

(b) Charge vs. Damage depth

(c) Action integral vs. Damage depth

(d) Peak × Rise rate vs. Damage depth

Fig. 5.13 Relationship between lightning damage depth (D_d) and lightning current parameters for CFRP composites

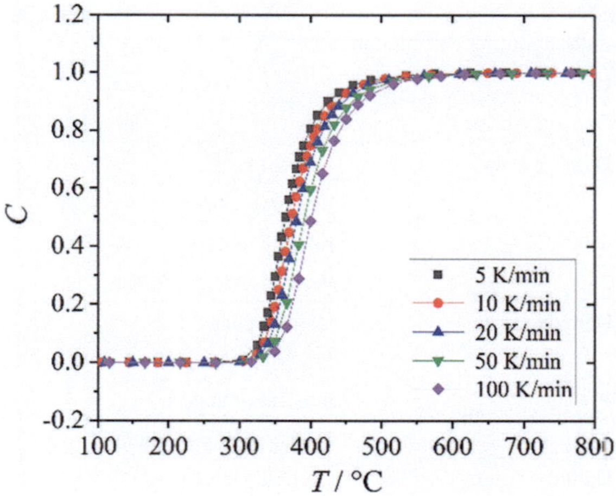

Fig. 5.14 Relationship between lightning damage volume (V_d) and lightning current parameters for CFRP composites

From Fig. 5.12, it is evident that the pyrolytic delamination area (S_2) of CFRP composites exhibits a nonlinear power-law relationship with the lightning current parameters. Similarly, as shown in Fig. 5.13, the damage depth (D_d) also follows a power-law dependence on the peak current, charge, and action integral of component A. These relationships indicate that lightning damage severity is not linearly proportional to input parameters, emphasizing the complex and coupled effects of electrical and thermal mechanisms during a strike.

Figure 5.14 illustrates the relationship between the lightning damage volume (V_d) of CFRP composites and the electrical parameters of lightning current component A, including peak current (i_p), charge transfer (Q), and action integral (W_A) [2]. As shown in the figure, the damage volume exhibits a nonlinear power-law correlation with each of these electrical parameters.

Table 5.4 presents the regression analysis between lightning-induced damage and lightning current electrical parameters of CFRP composites under the action of a single lightning current component A [2].

Based on the scatter plots in Figs. 5.12 through 5.14, the fitted curves, and the regression results in Table 5.4, the following conclusions can be drawn regarding the correlation between lightning damage parameters, namely damage area (S_2), damage depth (D_d), and damage volume (V_d), and the electrical characteristics of the lightning current component A for unidirectional CFRP laminates:

(i) The lightning damage parameters S_2, D_d, and V_d exhibit a nonlinear power-law relationship with the respective lightning current parameters.
(ii) Among all the evaluated parameters, the action integral (W_A) demonstrates the strongest correlation with lightning damage metrics, followed by the product of peak current and current rise rate, which also shows a significant influence.

Table 5.4 Regression analysis of lightning damage of CFRP composites under the action of a single lightning current component A

R^2		Power function (math.)
Damage metrics		$y = a \cdot x^b$
Damage area	Current peak	0.9880
	Electrical charge	0.9882
	Integral of action	**0.9990**
	Peak × rising gradient	0.9981
Damage depth	Current peak	0.9894
	Electrical charge	0.9853
	Integral of action	**0.9988**
	Peak × rising gradient	0.9974
Damage volume	Current peak	0.9903
	Electrical charge	0.9931
	Integral of action	**0.9997**
	Peak × rising gradient	0.9993

The empirical equations representing the relationships between lightning damage parameters and the action integral (W_A) of the lightning current component are provided in Eqs. (5.2) to (5.4).

$$S_2 = 9.0456 \times {W_A}^{0.5627} \tag{5.2}$$

$$D_d = 1.0155 \times {W_A}^{0.2377} \tag{5.3}$$

$$V_d = 0.8999 \times {W_A}^{0.8154} \tag{5.4}$$

5.3.2 *Lightning Damage Assessment of CFRP Composites Under Multiple Lightning Current Components*

Regression Analysis Between Lightning Damage Area and Electrical Parameters of Multiple Lightning Current Components

Based on the experimental results and analysis in Sect. 5.2.3, it is concluded that there is a strong correlation between the lightning damage area (S_d) of CFRP composites and the action integral (W). Therefore, the lightning damage under multiple lightning current components can be evaluated using the action integrals (W_A, W_B, W_C, W_D) of each component and the corresponding damage area (S_d) of the CFRP composites. The relevant experimental data are summarized in Table 5.5.

Figure 5.15 presents the scatter plot and fitted curve of the lightning damage area S_d versus the total action integral W (i.e., the sum of action integrals from all components) for CFRP composites subjected to multiple lightning current components. As shown in Fig. 5.15, the correlation between S_d and W under the influence of multiple consecutive lightning current components differs significantly from that under a single-component (A) lightning current. For the single-component case, the

Table 5.5 Data of action integrals for multiple lightning current components and corresponding composite damage areas

Sample number	W_A/A^2s	W_B/A^2s	W_C/A^2s	W_D/A^2s	S_d/cm^2
Type S4-1	79225	–	–	–	11.04
Type S4-2	81659	573.93	–	–	13.23
Type S4-3	–	638.73	4989.7	47968	14.53
Type S4-4	78359	643.67	4757.7	–	24.00
Type S4-5	76353	683.49	5548.1	18804	27.90

Note W_A, W_B, W_C and W_D represent the action integrals of the A, B, C, and D components of the lightning current, respectively relationship between S_d and W closely follows a nonlinear power function, with a correlation coefficient as high as 0.9990. However, for multiple lightning components in a continuous time sequence, the same fitting method yields a much lower correlation coefficient of only 0.42962, indicating that the power-law relationship no longer holds. Moreover, attempts to fit other models such as exponential functions also failed to establish a consistent relationship between S_d and W. This deviation is primarily attributed to the interdependent effects among the sequential lightning components. The damage caused by a later component is influenced by the cumulative effects of previous components, rather than acting independently. In addition, the distinct electrical characteristics of each lightning current component further complicate the relationship. For instance, the A component features a high current peak, steep front, and large action integral but transfers a relatively small amount of charge. In contrast, the C component involves larger charge transfer but a lower action integral.

These findings indicate that the lightning damage evaluation model based on single lightning current pulses is not suitable for multi-component lightning events. Therefore, it is crucial to develop a multifactor integrated assessment method that

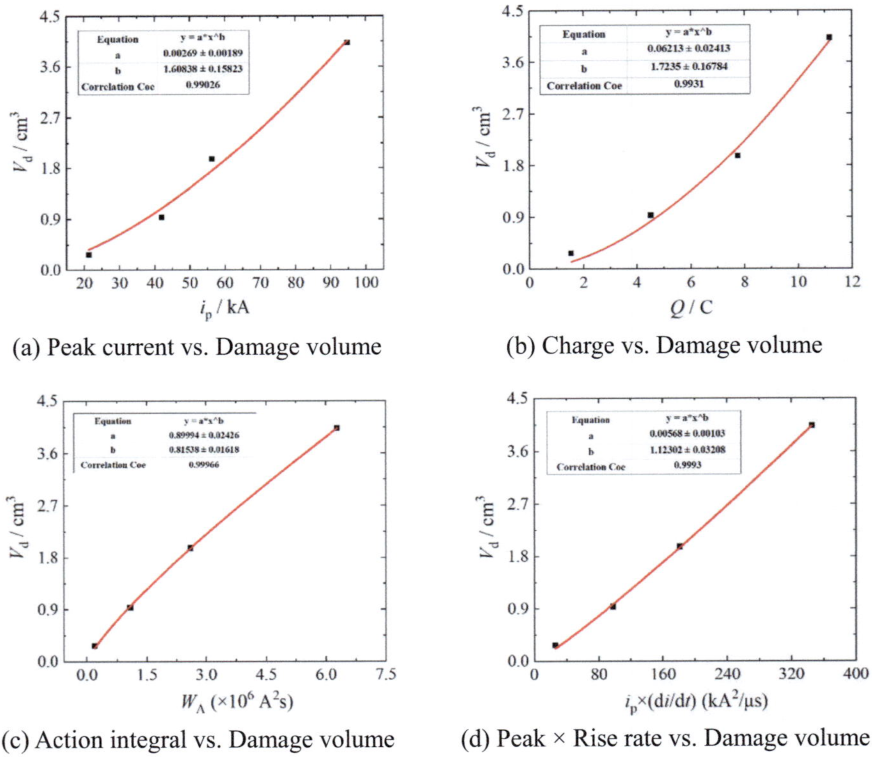

Fig. 5.15 Relationship between the damage area of CFRP composites and the total action integral of multiple lightning current components

accounts for the complex interactions among multiple sequential lightning current components.

Still taking the relationship between the lightning damage area S_d and the action integral W of CFRP composites as an example, it is assumed that the relationship between the lightning damage area S_d and the action integral W of each lightning current component follows the form given in Eq. (5.5).

$$S_d = k_{1SW} \times W_A + k_{2SW} \times W_B + k_{3SW} \times W_C + k_{4SW} \times W_D \tag{5.5}$$

In this equation, k_{1SW}, k_{2SW}, k_{3SW} and k_{4SW} represent the influence factors of the lightning damage area S_d contributed by the action integrals of lightning current components A, B, C, and D, respectively. Based on the experimental data from composite specimens Type S4-1, S4-2, S4-3, and S4-4 listed in Table 5.5, the values of these influence factors can be obtained through joint regression, as shown in Eq. (5.6).

$$\begin{cases} k_{1SW} = 1.3935 \times 10^{-2}(\text{mm}^2\text{A}^{-2}\text{s}^{-1}) \\ k_{2SW} = 3.2248 \times 10^{-1}(\text{mm}^2\text{A}^{-2}\text{s}^{-1}) \\ k_{3SW} = 2.3131 \times 10^{-1}(\text{mm}^2\text{A}^{-2}\text{s}^{-1}) \\ k_{4SW} = 1.9358 \times 10^{-3}(\text{mm}^2\text{A}^{-2}\text{s}^{-1}) \end{cases} \tag{5.6}$$

By substituting the calculated influence factors k_{1SW}, k_{2SW}, k_{3SW} and k_{4SW} into Eq. (5.5), the lightning damage area $S_{d\text{-ABCD}}$ of specimen Type S4-5 was estimated to be 26.04 cm^2, which shows a relative deviation of 6.67% compared to the experimental value of 27.9 cm^2. It is worth noting that the action integral W_D of the D component (with a peak current of 20 kA) is approximately 2×10^4 A^2s, while the action integral W_C of the C component (with an average current of 200 A over 200 ms) is about 5×10^3 A^2s. However, the resulting damage effects are not proportional to the magnitudes of these action integrals. According to the calculated influence factors, the C component contributes approximately 49.28% to the damage area of specimen S4-5, although its action integral accounts for only 5.47% of the total action integral W from all components. In contrast, the D component contributes just 1.40% to the damage area, despite its action integral making up 18.55% of the total. This inconsistency is attributed to the nonlinear dynamic impedance characteristics of CFRP composites [5–7]. When high-amplitude, high-rise-rate components such as A or D are applied, the equivalent impedance of the CFRP material is relatively low, resulting in reduced Joule heating. Conversely, under the action of the C component, the CFRP exhibits a higher equivalent resistance. Thus, although the current magnitude is lower, the Joule heating effect (I^2R) becomes more pronounced, causing a substantial increase in the surface temperature of the CFRP material and leading to a significantly larger lightning damage area.

Considering the nonlinear dynamic impedance characteristics of CFRP composites, and the fact that the transferred charge of the C component of the lightning current ($Q_C \approx 30$ C)—which is a low-voltage, small-current, long-duration waveform—is significantly greater than that of the A ($Q_A \approx 3$ C), B ($Q_B \approx 1.8$ C), and D ($Q_D \approx 2$ C) components, it is necessary to assess lightning damage not only from the perspective of action integral, but also in terms of the total transferred charge. The experimental data on the lightning-induced damage of CFRP composites, extracted from tests involving multiple successive lightning current components (Sect. 5.2), and corresponding to the transferred charge of each component, are listed in Table 5.6.

Figure 5.16 shows the scatter plots and fitted curves representing the relationship between the lightning damage area S_d and the transferred charge of each lightning current component for CFRP composites.

By comparing Figs. 5.12b and 5.16, a notable difference emerges. Under the action of a single A component, the lightning damage area S_d and the transferred charge exhibit a power-law relationship with a high correlation coefficient of 0.9882. In contrast, under the combined action of multiple lightning components, the correlation between the total transferred charge and the damage area drops significantly. When fitted with the same power function, the correlation coefficient is only 0.6056.

Table 5.6 Transferred charge of multiple lightning current components and corresponding damage areas in CFRP composites

Sample number	Q_A/C	Q_B/C	Q_C/C	Q_D/C	S_d/cm^2
Type S4-1	3.248	–	–	–	11.04
Type S4-2	3.226	1.883	–	–	13.23
Type S4-3	–	1.840	30.61	2.339	14.53
Type S4-4	3.118	1.843	30.00	–	24.00
Type S4-5	3.060	1.897	32.57	2.069	27.90

Note Q_A, Q_B, Q_C and Q_D denote the transferred charges of lightning current components A, B, C, and D, respectively

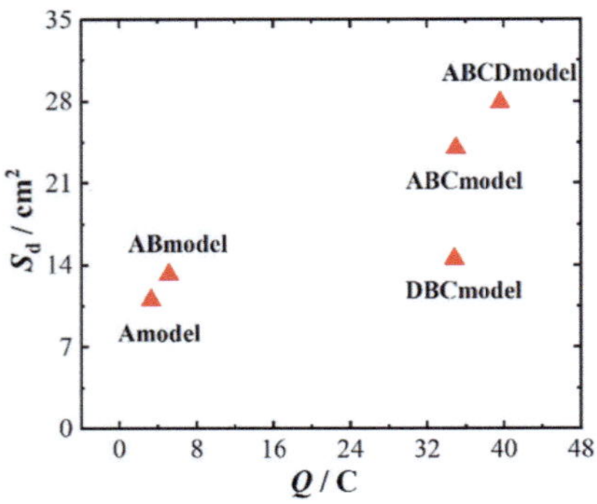

(a) Scatter plot

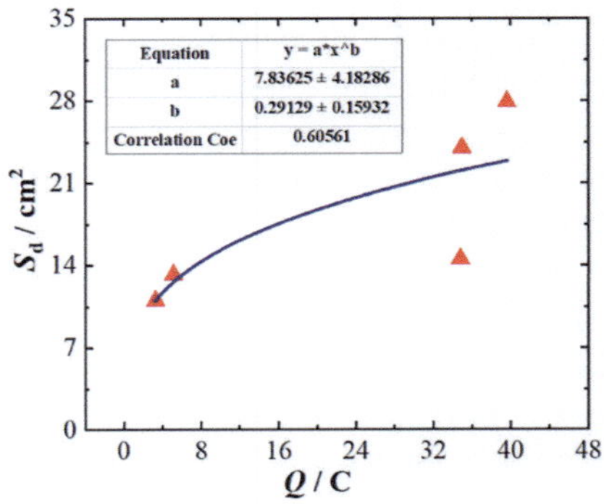

(b) Fitted curve

Fig. 5.16 Relationship between lightning damage area and transferred charge of multiple lightning current components for CFRP composites

Assuming that the relationship between the damage area and the transferred charge of each lightning component follows a similar form to Eq. (5.5), a new relational model is proposed Eq. (5.7), in which:

$$S_d = k_{1SQ} \times Q_A + k_{2SQ} \times Q_B + k_{3SQ} \times Q_C + k_{4SQ} \times Q_D \tag{5.7}$$

where, k_{1SW}, k_{2SW}, k_{3SW} and k_{4SW} represent the influence factors quantifying the contribution of the transferred charge of components A, B, C, and D to the total damage area of CFRP composites, respectively. By performing a joint fit using the lightning damage experimental data listed in Table 5.3, the values of these influence factors are obtained and presented in Eq. (5.8).

$$\begin{cases} k_{1SQ} = 339.9(\text{mm}^2\text{C}^{-1}) \\ k_{2SQ} = 120.3(\text{mm}^2\text{C}^{-1}) \\ k_{3SQ} = 37.28(\text{mm}^2\text{C}^{-1}) \\ k_{4SQ} = 38.69(\text{mm}^2\text{C}^{-1}) \end{cases} \tag{5.8}$$

Substituting the influence factors from Eq. (5.8) into the charge-based damage estimation model (Eq. (5.7)), the predicted damage area $S_{\text{d-ABCD}}$ of specimen Type S4-5 is 25.63 cm^2, with a relative deviation of 8.15% compared to the experimental value in Table 5.6. Furthermore, using the calculated influence factors and the damage estimation models based on both action integral (Eq. (5.5)) and transferred charge (Eq. (5.7)), the contribution percentage of each lightning current component to the total damage area of specimen Type S4-5 can be determined. These results are summarized in Table 5.7.

Based on the calculation results presented in Table 5.7, it can be concluded that the assessment of lightning damage area using both the action integral W and the transferred charge Q yields consistent and mutually supportive outcomes. The analysis further indicates that the C component of the lightning current plays a dominant role in determining the lightning damage area of CFRP composites.

The lightning damage area of CFRP composites is influenced by the combined effects of the action integral W and the transferred charge Q of the various lightning current components. For components A and D, characterized by short duration and high peak current, the action integral W serves as the primary parameter for evaluating their capacity to cause surface damage. In contrast, for the C component, which exhibits long duration and low average current, the transferred charge Q is the key factor governing the extent of lightning-induced damage. Therefore, in evaluating the

Table 5.7 Percentage contribution of lightning current components A, B, C, and D to the damage area of specimen Type S4-5

Component of lightning current	A (%)	B (%)	C (%)	D (%)
Equation (5.5): based on action integral	40.86	8.46	49.28	1.40
Equation (5.7): based on charge transferred	40.59	8.91	47.38	3.12

damage caused by multiple successive lightning current components, it is essential to comprehensively consider both the impulse energy (W) and the cumulative charge transfer (Q) to accurately assess the lightning damage area of CFRP composites.

It is also noteworthy that, although the calculated and experimental results show that the D component has a relatively limited contribution to the overall surface damage area, it exerts a significant effect on the depth and extent of internal damage. A comparison between specimens Type S4-4 and Type S4-5 reveals that the addition of the D component markedly enhances deep-layer damage. Specifically, specimen S4-4, exposed to components A, B, and C, exhibited only localized damage depths exceeding 1.5 mm within areas smaller than 1 cm^2, with most damage confined to the surface layers. In contrast, specimen S4-5, which was additionally subjected to the D component, showed a region with damage depth greater than 1.5 mm expanding to over 10 cm^2, despite the overall surface damage area increasing by only 3.9 cm^2. This phenomenon is attributed to the pre-heated and mechanically weakened state of the composite after the action of the C component. The high thermal loading causes resin pyrolysis, gas accumulation, fiber sublimation, and deterioration of inter-laminar bonding, rendering the composite more susceptible to subsequent damage. When the high-rise-rate D component is applied, its intense electromagnetic and mechanical shock effects further propagate the damage into deeper layers. These observations highlight that the measured damage area, as recorded in the experiments, primarily accounts for the surface defect regions, and may not fully capture the three-dimensional progression of damage. Therefore, in evaluating the lightning damage severity of CFRP composites, it is crucial to jointly consider both damage area and depth distribution, especially under complex multi-component lightning current sequences.

Regression Analysis of Electrical Parameters and Lightning Damage Depth in CFRP Composites

To investigate the correlation between the depth of lightning-induced damage in CFRP composites and the electrical parameters of multiple successive lightning current components, regression analyses were conducted based on three sets of electrical characteristics: the action integrals W, transferred charges Q, and current rise gradients of each lightning current component. The relevant data are summarized in Tables 5.8, 5.9, and 5.10, respectively.

Firstly, regarding the relationship between the lightning damage depth and the total action integral of multiple lightning current components in CFRP composites, Fig. 5.17 presents the scatter plot and fitted curve illustrating this relationship.

As shown in Fig. 5.17, the relationship between the lightning damage depth of CFRP composites and the action integral under multiple lightning current components differs significantly from that under a single lightning current component A. When subjected to only the A component, the damage depth exhibits a nonlinear power function relationship with the action integral, with a correlation coefficient as high as 0.9988. However, under the combined effect of multiple lightning current

Table 5.8 Action integrals of multiple lightning current components and the corresponding damage depth data of CFRP composites

Sample number	W_A/A^2s	W_B/A^2s	W_C/A^2s	W_D/A^2s	D_d/mm
Type S4-1	79225	–	–	–	0.513
Type S4-2	81659	573.93	–	–	0.81
Type S4-3	–	638.73	4989.7	47968	0.90
Type S4-4	78359	643.67	4757.7	–	0.93
Type S4-5	76353	683.49	5548.1	18804	1.28

Note W_A, W_B, W_C and W_D are the action integrals of the A, B, C, and D components of the lightning current, respectively

Table 5.9 Transferred charges of multiple lightning current components and the corresponding damage depth data of CFRP composites

Sample number	Q_A/C	Q_B/C	Q_C/C	Q_D/C	D_d/mm
Type S4-1	3.248	–	–	–	0.513
Type S4-2	3.226	1.883	–	–	0.81
Type S4-3	–	1.840	30.61	2.339	0.90
Type S4-4	3.118	1.843	30.00	–	0.93
Type S4-5	3.060	1.897	32.57	2.069	1.28

Note Q_A, Q_B, Q_C and Q_D are the transferred charges of the A, B, C and D components of the lightning current, respectively

Table 5.10 Current rise gradients of multiple lightning current components and the corresponding damage depth data of CFRP composites

Sample number	di_A/dt (kA/μs)	di_B/dt (A/ms)	di_C/dt (A/ms)	di_D/dt (kA/μs)	D_d/mm
Type S4-1	1.557	–	–	–	0.513
Type S4-2	1.603	1536	–	–	0.81
Type S4-3	–	1620	23.286	2.515	0.90
Type S4-4	1.612	1626	22.629		0.93
Type S4-5	1.600	1676	24.328	1.300	1.28

Note i_A, i_B, i_C and i_D are lightning current A, B, C and D component currents, respectively

components, the correlation coefficient drops to only 0.16783. This indicates that there is no significant nonlinear power function relationship between the total action integral and the damage depth under multi-component conditions. To establish a mathematical expression describing the relationship between the lightning damage depth of CFRP composites and the action integrals of each individual lightning current component, it is assumed that this relationship conforms to the form shown in Eq. (5.9):

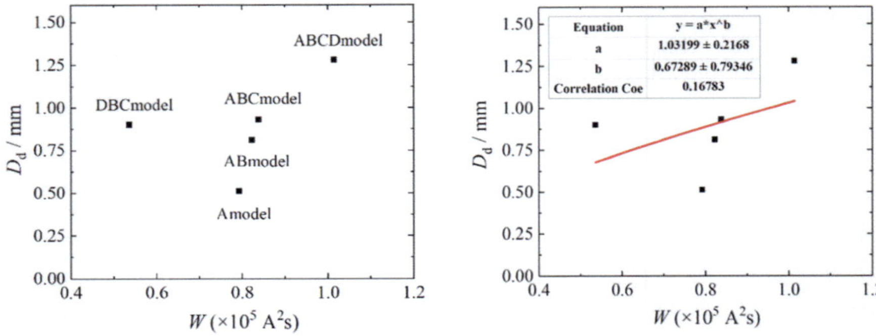

(a) Scatter plot (b) Fitted curve

Fig. 5.17 Relationship between the damage depth of the CFRP composites and the total action integral of the multiple lightning current components

$$D_d = k_{1\text{DW}} \times W_A + k_{2\text{DW}} \times W_B + k_{3\text{DW}} \times W_C + k_{4\text{DW}} \times W_D \tag{5.9}$$

where, $k_{1\text{DW}}$, $k_{2\text{DW}}$, $k_{3\text{DW}}$ and $k_{4\text{DW}}$ represent the influence coefficients corresponding to the contribution of each component's action integral to the overall lightning damage depth in CFRP composites. Based on the experimental data of composite specimens Type S4-1, S4-2, S4-3, and S4-4 in Table 5.8, the influence coefficients were obtained by solving the regression equation system, as expressed in Eq. (5.10).

$$\begin{cases} k_{1\text{DW}} = 6.475 \times 10^{-6}(\text{mm/A}^2\text{s}) \\ k_{2\text{DW}} = 4.900 \times 10^{-4}(\text{mm/A}^2\text{s}) \\ k_{3\text{DW}} = 9.893 \times 10^{-6}(\text{mm/A}^2\text{s}) \\ k_{4\text{DW}} = 2.254 \times 10^{-5}(\text{mm/A}^2\text{s}) \end{cases} \tag{5.10}$$

Substituting the calculated influence coefficients $k_{1\text{DW}}$, $k_{2\text{DW}}$, $k_{3\text{DW}}$ and $k_{4\text{DW}}$ into Eq. (5.9), the predicted lightning damage depth $D_{\text{d-ABCD}}$ of specimen Type S4-5 is 1.308 mm. This result shows a relative deviation of only 2.19% from the experimentally measured damage depth of 1.28 mm.

Based on the data in Table 5.9, the scatter plot and fitted curve illustrating the relationship between the lightning damage depth D_d and the charge transfer Q of the lightning current components for CFRP materials are shown in Fig. 5.18.

As shown in Fig. 5.18, under the action of multiple continuous time-sequence lightning current components, the relationship between the damage depth D_d of CFRP composites and the total transferred charge Q does not follow the same pattern as observed under the action of a single lightning current component A. In the case of the single A component, the damage depth D_d and the transferred charge Q follow a nonlinear power function relationship, with a correlation coefficient of 0.9853. However, when multiple components act in succession, the correlation coefficient

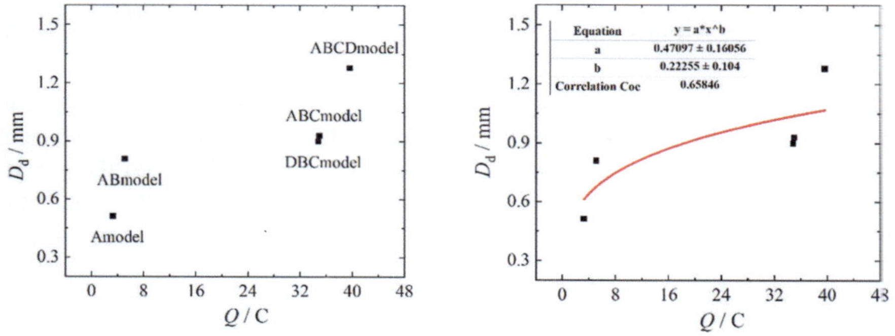

(a) Scatter plot (b) Fitted curve

Fig. 5.18 Relationship between damage depth and total transferred charge of multiple lightning current components for composite materials

between D_d and Q drops to 0.65846. This indicates that the power function relationship between the lightning damage depth and transferred charge observed under single-component conditions does not hold under multi-component excitation.

To establish a mathematical model for the relationship between the lightning damage depth D_d of CFRP composites and the transferred charge Q of each lightning current component, it is assumed that the relationship conforms to the form shown in Eq. (5.11):

$$D_d = k_{1\mathrm{DQ}} \times Q_A + k_{2\mathrm{DQ}} \times Q_B + k_{3\mathrm{DQ}} \times Q_C + k_{4\mathrm{DQ}} \times Q_D \tag{5.11}$$

where, $k_{1\mathrm{DQ}}$, $k_{2\mathrm{DQ}}$, $k_{3\mathrm{DQ}}$ and $k_{4\mathrm{DQ}}$ represent the influence coefficients that quantify the contribution of each component's transferred charge to the total lightning damage depth of the CFRP composites. Based on the experimental data from specimens Type S4-1, S4-2, S4-3, and S4-4 in Table 5.9, the influence coefficients were determined by solving the corresponding regression equation system.

$$\begin{cases} k_{1\mathrm{DQ}} = 0.15794(\mathrm{mm/C}) \\ k_{2\mathrm{DQ}} = 0.15957(\mathrm{mm/C}) \\ k_{3\mathrm{DQ}} = 4.78185 \times 10^{-3}(\mathrm{mm/C}) \\ k_{4\mathrm{DQ}} = 0.19667(\mathrm{mm/C}) \end{cases} \tag{5.12}$$

By substituting the obtained coefficients $k_{1\mathrm{DQ}}$, $k_{2\mathrm{DQ}}$, $k_{3\mathrm{DQ}}$ and $k_{4\mathrm{DQ}}$ into Eq. (5.11), the predicted lightning damage depth D_d for specimen Type S4-5 is 1.3486 mm. This result shows a relative deviation of 5.36% from the experimental damage depth of 1.28 mm provided in Table 5.9.

Finally, the relationship between the lightning damage depth of CFRP composites and the current rise gradient is investigated. It is assumed that the relationship between the damage depth and the rate of rise of each lightning current component conforms to the form shown in Eq. (5.13):

$$D_d = k_{1\mathrm{Ddi}} \times \frac{di(A)}{dt} + k_{2\mathrm{Ddi}} \times \frac{di(B)}{dt} + k_{3\mathrm{Ddi}} \times \frac{di(C)}{dt} + k_{4\mathrm{Ddi}} \times \frac{di(D)}{dt} \tag{5.13}$$

where, $k_{1\mathrm{Ddi}}$, $k_{2\mathrm{Ddi}}$, $k_{3\mathrm{Ddi}}$ and $k_{4\mathrm{Ddi}}$ are the influence coefficients representing the degree to which the damage depth D_d is affected by the current rise gradient of each corresponding lightning current component A, B, C, and D. Based on the experimental data of CFRP composite specimens provided in Table 5.10, the influence factors in Eq. (5.14) can be determined.

$$\begin{cases} k_{1\mathrm{Ddi}} = 3.295 \times 10^{-10} (\mathrm{mm\,s/A}) \\ k_{2\mathrm{Ddi}} = 1.835 \times 10^{-7} (\mathrm{mm\,s/A}) \\ k_{3\mathrm{Ddi}} = 4.442 \times 10^{-6} (\mathrm{mm\,s/A}) \\ k_{4\mathrm{Ddi}} = 1.988 \times 10^{-10} (\mathrm{mm\,s/A}) \end{cases} \tag{5.14}$$

By substituting the calculated coefficients $k_{1\mathrm{Ddi}}$, k_{2Ddi}, $\mathrm{k}_{3\mathrm{Ddi}}$ and $k_{4\mathrm{Ddi}}$ into Eq. (5.14), the predicted lightning damage depth $D_{\mathrm{d\text{-}ABCD}}$ for specimen Type S4-5 is 1.20 mm. Compared to the experimentally measured value of 1.28 mm in Table 5.10, this corresponds to a relative deviation of 6.25%. The lightning damage estimation models represented by Eqs. (5.9), (5.11), and (5.13) can also be used to evaluate the percentage contribution of each lightning current component to the total damage depth in the CFRP composite specimen Type S4-5. The results are summarized in Table 5.11.

From the data in Table 5.11, it can be observed that the predicted damage depths derived using the three different evaluation parameters, namely, action integral, charge transfer, and current rise gradient, are all relatively consistent with the experimental value. The deviations between the calculated and measured damage depths for specimen Type S4-5 are 2.19%, 5.36%, and 6.25%, respectively. These findings demonstrate the reliability of each model in approximating the damage depth from different physical perspectives. Furthermore, the individual contributions of the C component of the lightning current to the damage depth are relatively small across all three evaluation models, calculated as 0.0549 mm, 0.1557 mm, and 0.108 mm, respectively. This suggests that the C component, characterized by a long duration and large charge transfer, plays a comparatively minor role in the overall damage process. In contrast, the A, B, and D components, which are marked by higher peak values and faster current rise rates, exhibit a more pronounced impact on the damage depth of the CFRP composites.

Table 5.11 Percentage contribution of each lightning current component to the damage depth of specimen Type S4-5

Component of lightning current	A (%)	B (%)	C (%)	D (%)
Equation (5.9): based on action integral	37.80	25.60	4.196	32.40
Equation (5.11): based on transferred charge	35.84	22.44	11.55	30.17
Equation (5.13): based on current change rate	43.93	25.63	9.00	21.54

These results highlight the importance of the front-end components of the lightning current waveform in inducing severe thermal and mechanical effects on composite materials. Understanding the individual contributions of different lightning current components provides valuable guidance for optimizing lightning protection strategies for CFRP structures.

5.4 Summary

In this chapter, the direct effects of multiple consecutive and uninterrupted lightning current components on unidirectional layup structured CFRP composites were experimentally investigated. Through comprehensive characterization methods, including morphological observation of damage zones, infrared thermography, and ultrasonic scanning, the relationships between damage patterns, damage characteristics, and the parameters of multi-component, time-sequenced lightning currents were analyzed. Based on the experimental results, a multifactor evaluation of the lightning direct effects was conducted, considering both single-component lightning current (component A) and composite lightning currents (ABCD model). The key findings are summarized as follows:

(1) **Damage Mechanisms and Component Contribution**

The interaction between CFRP composites and multiple consecutive lightning current components involves complex physical processes. Both the damage area and depth increase with the cumulative energy of the applied lightning current components. However, different components contribute differently to the damage outcome due to variations in their electrical characteristics. Specifically, components A and D, with high peak amplitudes and steep current rise rates, are primarily responsible for damage in the depth direction. These components induce intense local thermal effects, mechanical impacts, and dielectric breakdown, causing abrupt rupture of carbon fibers and resin matrix at the strike point. Subsequent current dispersion within the layup structure leads to Joule heating, resin pyrolysis, and carbon fiber ablation or sublimation, contributing to elevated transient temperatures and enlarged damage areas. While the C component, characterized by long duration and high charge transfer, contributes relatively little to damage depth, it has a significant effect on damage area. In the ABCD current model, the C component alone accounts for approximately 40% of the total damage area.

(2) **Influence of Sequential Current Components and Temporal Effects**

The time sequence and combination of lightning current components significantly influence the damage morphology and extent. The pre-applied current components modify the electrical, thermal, and mechanical state of the CFRP, thereby altering its response to subsequent components. As a result, the damage caused by a given component in isolation differs notably from its effect when applied in combination with others. Moreover, the preceding components affect the dispersion path of subsequent currents within the composite, leading to variations in damage patterns and severity.

(3) **Nonlinear Correlations under Single Lightning Impact**

A detailed multifactorial analysis was conducted for the A component with varying peak amplitudes. The results revealed a strong nonlinear power-law relationship between the electrical parameters of the lightning current and the damage metrics (area, depth, and volume). Among the evaluated parameters, the action integral exhibited the strongest correlation with damage: the correlation coefficients were 0.9990, 0.9988, and 0.9997 for the damage area, depth, and volume, respectively.

(4) **Damage Evaluation Model for Multi-Component Lightning Impacts**

Under the action of multiple consecutive lightning current components, the original power-law relationship between damage metrics and current parameters (such as action integral and charge transfer) no longer holds. To address this, a multifactorial evaluation method was proposed to assess damage under complex lightning impact scenarios. Mathematical models were established to describe the relationship between lightning damage (area and depth) and key electrical parameters, including the integrated action and charge transfer of each component. The calculated results showed strong agreement with experimental data: the deviation of the predicted damage area was approximately 6.67%, and the deviation in damage depth was about 2.19%. These findings provide a reliable and practical foundation for future material design, process optimization, and lightning protection evaluation of CFRP composites.

References

1. Sun J.R., Yao X-L., Xu W.J., Chen J.L., & Wu Y. (2019). Effects of lightning current components on lightning-induced damage in carbon fiber-reinforced composites. High Voltage Engineering, 45(12), 3836–3843.
2. Sun J.R. (2019). Investigation of damage mechanisms in aircraft carbon fiber-reinforced polymer composites under lightning pulse currents (Doctoral dissertation). Xi'an Jiaotong University, Xi'an, China.
3. Sun J.R., Yao X.L., Li Y.F., Tian X.Y., & Chen J.L. (2019). Damage characteristics of carbon fiber–reinforced resin matrix composites under multiple sequential lightning current impacts. Journal of Composite Materials, 36(12), 2764–2770.
4. Sun J, Yao X, Tian X, Chen J, Wu Y. Damage characteristics of CFRP laminates subjected to multiple lightning current components[J]. Applied Composite Materials, 2019, 26: 745–762.
5. Sun J.R., Yao X.L., Xu W.J., et al. (2016). Dynamic characteristics of carbon fiber-reinforced polymer composites under nondestructive lightning current component A. Journal of Xi'an Jiaotong University, 50(6), 130–135.
6. Sun J, Yao X, Xu W, et al. Dynamic characteristics of carbon fiber reinforced polymer under nondestructive lightning current[J]. Polymer Composites, 2018. 39(5): 1514–1521.
7. Sun J, Li X, Tian X, et al. Dynamic electrical characteristics of carbon fiber-reinforced polymer composite under low intensity lightning current impulse[J]. Advanced Composites Letters, 2020, 29: 1–13.

Chapter 6
Lightning Protection Structure and Protection Performance of CFRP Composites

6.1 Overview of Lightning Strike Protection for CFRP Composites

Compared with metal and alloy materials, CFRP composites exhibit poor electrical and thermal conductivity, which prevents the efficient dissipation of lightning energy during a strike. As a result, lightning strike protection (LSP) has become a key technical challenge limiting the safe and large-scale application of CFRP composites in aircraft structures. Accordingly, the development of LSP materials, manufacturing processes, and structural integration techniques, as well as improvements in the intrinsic lightning resistance of CFRP materials, has become a major focus of both domestic and international research. In engineering practice, aircraft primarily constructed from composite materials must be equipped with effective LSP systems, and the protection strategies currently applied or under investigation can be broadly classified into the following categories.

6.1.1 Metal Lightning Protection

Metal-based LSP is a traditional and widely used approach in aircraft applications. It typically involves integrating high-conductivity metal materials, such as grids, foils, or conductive surface coatings, onto the outer layers of CFRP structures. These materials serve to divert lightning current across the aircraft surface, thereby reducing the thermal and mechanical damage to the underlying composites.

J. Sun et al., *Lightning Effects on Carbon Fiber-Reinforced Polymer Composites*,
https://doi.org/10.1007/978-981-95-3196-7_6

Metal Grid Lightning Protection

A metal grid refers to a conductive mesh layer embedded onto the composite surface during the curing process, which endows the outer layer with significantly improved electrical conductivity. This structure helps to minimize the localized ablation and delamination caused by lightning current. For example, Dexmet has proposed region-specific LSP configurations according to aircraft lightning zoning requirements [1]:

(i) In Lightning Zone I, a copper wire mesh can be applied as the surface protection layer. The mesh should have no fewer than 20 × 40 openings per square inch, with a wire diameter of at least 0.14 mm.
(ii) Zone I: In Lightning Zone II, the same mesh density standard applies, while the minimum wire diameter is reduced to 0.10 mm.

Dexmet conducted comparative lightning damage tests on CFRP samples with and without metal grid protection. The results indicated that under a simulated lightning current component with a 200 kA peak, composite panels protected by aluminum or copper grids experienced significantly less surface damage. In contrast, unprotected panels suffered severe burn-through and structural degradation, as illustrated in Fig. 6.1.

Kawakami [2] demonstrated that applying a copper grid effectively reduces the lightning-induced damage area in CFRP composites. Under lightning strike conditions, only minor surface damage is observed at the attachment point, while delamination damage is found along the outer layers of both the copper grid and the composite

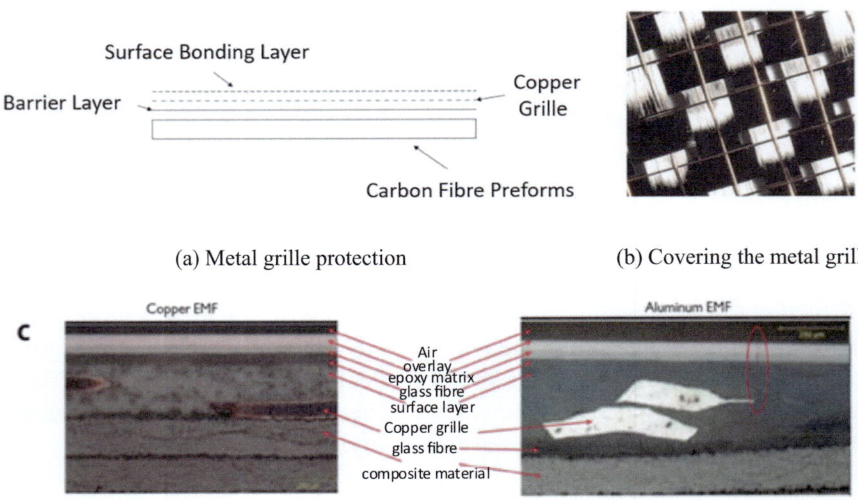

(a) Metal grille protection

(b) Covering the metal grille

(c) Metal Grating Thickness Interface

Fig. 6.1 Lightning protection structure for metal grilles

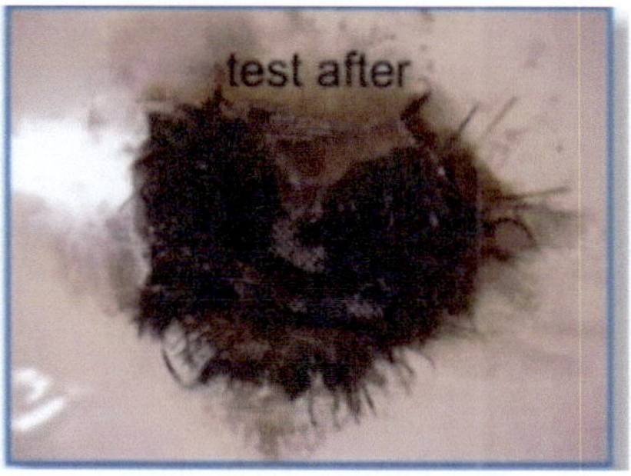

(a) Unprotected specimens

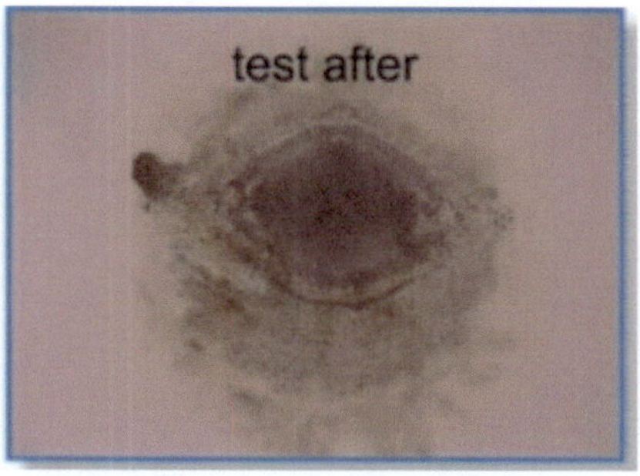

(b) Copper grid-protected specimens

Fig. 6.2 Comparison of lightning damage between protected and unprotected CFRP composites

material. The propagation of damage follows the orientation of the copper wires, and relatively severe damage occurs at the intersections of the grid wires, as shown in Fig. 6.2.

Metal Foil Lightning Protection

Compared with metal grids, metal foils offer more complete surface coverage and improved current conduction uniformity, thereby delivering a stronger lightning protection performance. The larger contact area enables better dispersion of the lightning current, resulting in enhanced suppression of thermal and mechanical damage. However, metal foils tend to be heavier than grids, exhibit poorer flexibility, and present challenges in achieving strong adhesion with CFRP surfaces. For these reasons, metal foils are typically applied to relatively flat structural regions, where higher demands exist for protection against both direct and indirect lightning effects.

In addition to physical bonding methods, thermal spray techniques can also be used to form conductive metallic coatings on CFRP surfaces. Through pulsed arc or flame spraying, molten metal particles are deposited onto the composite surface to form a lightning protection layer. For example, Liu et al. [3] employed a flame spraying process to form a continuous aluminum coating on the surface of CFRP composites. Lightning strike test results showed that the sprayed aluminum layer effectively enhanced surface conductivity and reduced the extent of lightning-induced damage. The comparison is illustrated in Fig. 6.3.

This aluminum spraying method is capable of covering complex and irregular surfaces where metal grids or foils are difficult to conform to, thus offering a certain degree of adaptability for composite structures with intricate geometries. However, this approach has not been widely applied in practice due to several inherent drawbacks: the high cost and added weight are significant limiting factors. Moreover, the coarse particle size of sprayed aluminum leads to rough surface finishes, which

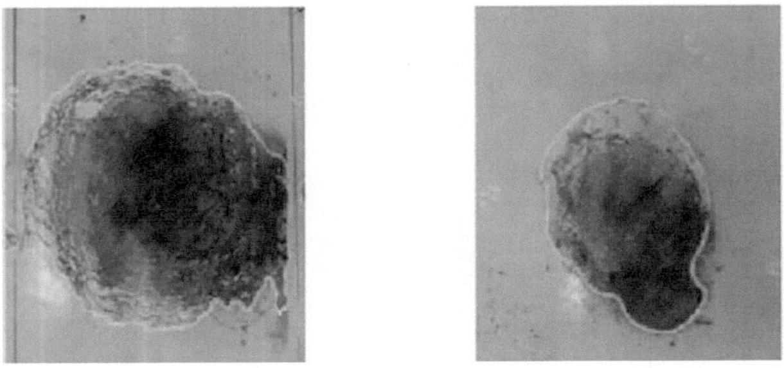

(a) 0.1 mm aluminium coating (b) 0.2 mm aluminium coating

Fig. 6.3 Lightning damage to CFRP composites with aluminum flame-sprayed protective coatings

may negatively affect aerodynamic and structural properties. Kawakami [4] investigated the repairability of CFRP composites protected with copper grids, conducting lightning strike tests based on the D component of lightning current. Through these experiments, the correlation between the repair process and the lightning strike resistance, including the damaged area and residual mechanical strength, was established, providing insights for post-strike restoration strategies.

When comparing different metal-based conductive protection techniques, the metal grid protection method is shown to deliver the best overall lightning strike protection (LSP) performance. As a result, it has been adopted as the preferred LSP solution by major aircraft manufacturers such as Boeing and Airbus. However, it suffers from a critical limitation: the significant additional weight introduced by high-density metal materials, which directly affects the aircraft's fuel efficiency and operational cost. According to the Dexmet lightning protection scheme mentioned previously, the mass of copper grids required for different lightning strike zones ranges from 73 to 195 g/m^2, while aluminum grids range from 66 to 136 g/m^2. These values correspond to approximately 9–30 wt% of additional lightning protection mass per square meter of CFRP composite, based on a 32-ply unidirectional CFRP laminate. For example, in the main nacelle of the Boeing 787, which is classified as Lightning Zone II, the protected surface area is approximately 3000 m^2. If a copper grid with a thickness of 0.1 mm and a surface density of 195 g/m^2 is used, the nacelle would gain an additional weight of 590 kg, significantly increasing the overall aircraft mass and fuel consumption.

In addition to the weight penalty, metallic materials are inherently susceptible to corrosion, especially aluminum, which exhibits higher electrochemical activity than carbon fibers. When bonded with resin systems, aluminum also shows relatively low interfacial strength, leading to the formation of voids or delamination at the resin-composite interface. During a lightning strike, electrical arcs or high-temperature sparks may penetrate through these defects, causing severe internal damage and, in extreme cases, igniting adjacent fuel tanks. This can result in explosions and catastrophic accidents. Consequently, airlines are required to conduct extensive maintenance and inspection routines to ensure the integrity of the protective layer and to prevent the presence of voids or delamination that might compromise the safety and performance of the aircraft.

6.1.2 Non-metallic Lightning Protection

In order to solve the problem of composite metal protective structures, many scholars at home and abroad have conducted research on non-metallic conductive composite protection mechanisms. The main objective of these studies is to enhance the electrical conductivity of composite materials while maintaining their lightweight and high-strength performance characteristics. Non-metallic conductive protection structures function either by forming a conductive layer on the surface of the composite material to shunt the lightning strike energy, or by enhancing the bulk conductivity of the composite to enable internal current dispersion.

Non-metallic conductive layer protection is typically achieved by applying a conductive layer composed of non-metallic conductive particles onto the surface of the composite material to form a lightning protection layer. Another method involves improving the electrical conductivity of the resin matrix itself, enabling the lightning current to be dispersed within the composite structure and thereby reducing lightning-induced damage. A commonly employed technique is the blending method, in which conductive nanoparticles are incorporated into the resin to increase its electrical conductivity, particularly in the direction perpendicular to the fiber orientation.

Hirano [5] developed a polyaniline-based conductive thermosetting resin, which significantly improved the electrical conductivity of the resin matrix. The in-plane conductivity and surface conductivity of the resulting polyaniline-based composite reached 148 and 0.73 S/cm, respectively, representing enhancements of 5.92 times and 27.4 times compared to those of conventional epoxy-based composites. Lightning strike tests under 40 and 100 kA A-component currents showed that the surface damage on CFRP composites was almost imperceptible, indicating that the polyaniline-based composite exhibited excellent lightning strike protection (LSP) performance. Kumar [6] utilized polyaniline-based conductive resin to fabricate four types of CFRP composites with varying levels of through-thickness conductivity. Test results indicated that an increase in through-thickness conductivity was directly correlated with a reduction in lightning-induced damage on the CFRP composites. In a separate study, Kumar [7] applied a staggered lay-up technique combining buckypaper and carbon fiber prepreg to enhance the electrical conductivity of CFRP composites, thereby improving their lightning strike resistance. Han [8] also utilized buckypaper for LSP of CFRP composites; experimental results showed that a 70 μm conductive layer combined with a 200 μm boron nitride-modified resin insulation layer could achieve a 30% weight reduction compared to traditional copper mesh while providing effective protection under 100 kA lightning current A-component conditions. Rajesh [9] investigated the use of heterogeneous material systems in conjunction with metallic conductive coatings to enhance the lightning strike resistance of CFRP composites. Results demonstrated that the metallic coatings provided the most effective protection performance. Guo [10] evaluated the LSP capability of nickel-plated carbon fiber nonwoven fabrics (Ni-CFNVs) under combined C- and D-component lightning currents, simulating Zone II lightning environments. Experimental findings indicated that the composite material incorporating 70 g/m^2 Ni-CFNVs exhibited superior lightning protection performance. Zhang [11] compared the LSP effectiveness of two strategies: interlayer addition of carbon nanotube (CNT) films and surface silver plating, under exposure to D-component lightning current. Results showed that incorporating CNT films between the composite layers significantly enhanced the dispersion capability of lightning current in both in-plane and through-thickness directions, thereby reducing in-plane lightning damage. Chakravarthi [12] incorporated Ni-decorated single-walled carbon nanotubes (Ni-SWNTs) into carbon fiber-reinforced bismaleimide composites, demonstrating that the addition of 4 wt% Ni-SWNTs led to a significantly higher electrical conductivity than that achieved with the same mass fraction of conventional single-walled carbon nanotubes (SWNTs). Simulated lightning tests revealed that the area of lightning-induced damage was reduced from 32.85 to 5.17 cm^2 following the addition of Ni-SWNTs, as illustrated in Fig. 6.4.

Although the non-metallic protection approach has become a prominent topic in current research and notable progress has been achieved, it still cannot fully replace traditional metallic protection structures. The primary limiting factors include the following.

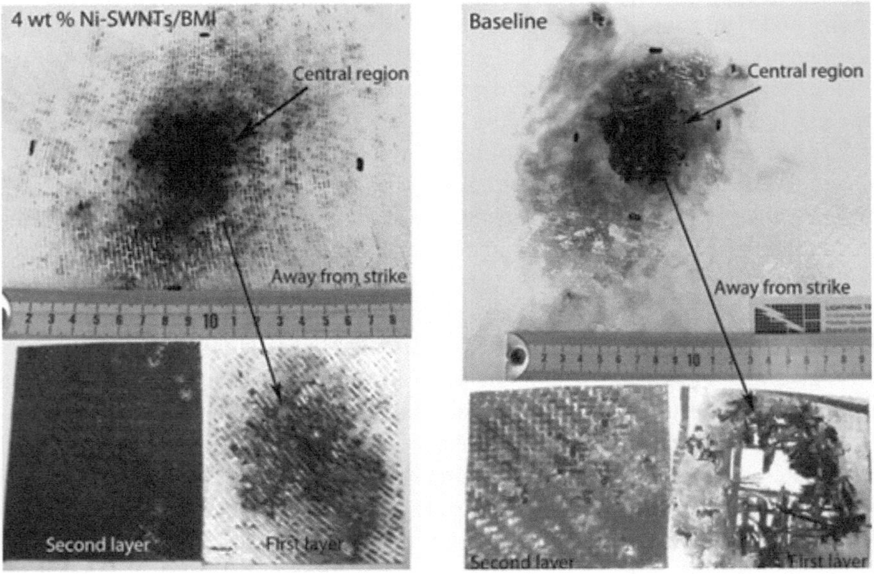

Fig. 6.4 Comparison of lightning damage areas in CFRP composites with and without protective layer

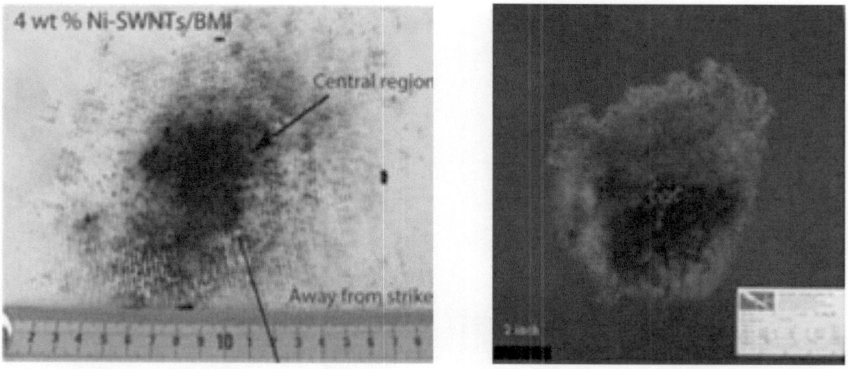

(a) 4% Ni-SWN added to composites

(b) Nanofibres and nickel nanowires used as protective layers

Fig. 6.5 Lightning damage morphology of CFRP composites under different lightning protection methods at a peak current of 40 kA

Insufficient Conductive Ability of Non-metallic Materials

Although the incorporation of conductive nanoparticles has significantly enhanced the conductivity of resin systems, the overall conductivity of nanoparticle-modified resins remains several orders of magnitude lower than that of metallic materials. For example, single- or multi-walled carbon nanotubes typically require a loading of 0.002–7 wt% to achieve conductivities in the range of 10^{-6}–10^{-2} S/cm [13], while graphene requires 0.1–0.5 vol.% to reach conductivities of 10^{-5}–10^{-2} S/cm [14]. In contrast, metallic copper-based protective structures exhibit conductivities as high as 3×10^7 S/cm. As a result, even under a peak current of 40 kA from a single lightning current D component, severe lightning-induced surface damage still occurs on non-metallic protective composite structures, as shown in Fig. 6.5.

Immature Moulding Process

The composite moulding process generally requires resins with high viscosity, which limits the feasible loading of non-metallic conductive particles into the resin matrix. Figure 6.6 presents the cross-sectional morphology of composites fabricated with varying carbon nanotube concentrations in the thickness direction [15].

In addition, the fibrous medium employed in the moulding process acts as a physical barrier to the mobility of non-metallic conductive particles. This hinders uniform dispersion, resulting in inhomogeneous particle distribution within the composite matrix. Such uneven distribution compromises the global lightning protection effectiveness of the composite structure and renders it insufficient for meeting lightning strike protection (LSP) performance requirements. Conductive polymer matrices such as polyaniline and polypyrrole also suffer from processability issues

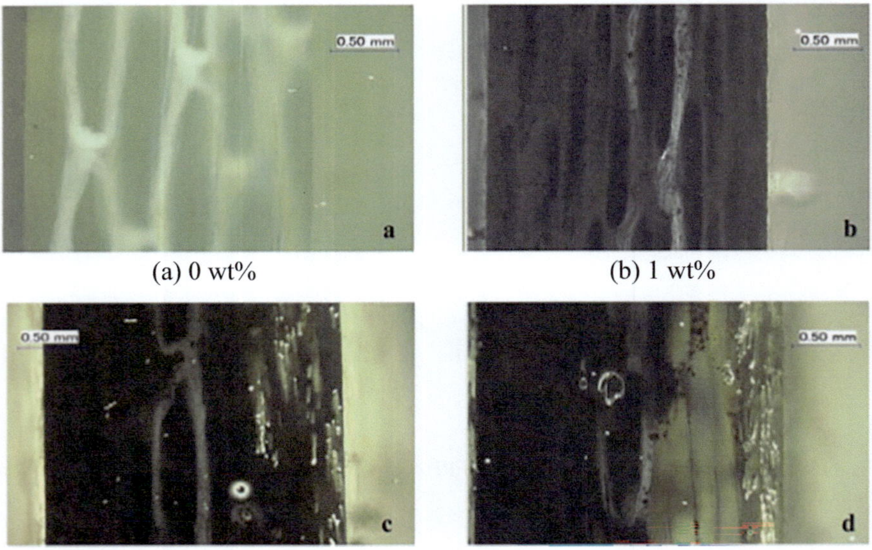

(a) 0 wt% (b) 1 wt%

(c) 1.5 wt% (d) 2 wt%

Fig. 6.6 Cross-sectional morphology of composites in the thickness direction fabricated by the RTM process using different carbon nanotube concentrations

due to their inherent high viscosities, making them incompatible with conventional composite moulding techniques. Moreover, the mechanical properties of conductive resins remain significantly inferior to those of epoxy resins, which further limits their applicability in the LSP of CFRP composites.

In summary, while various non-metallic protective structures have been shown to reduce lightning damage in CFRP composites to varying extents, they remain fundamentally constrained in performance due to their limited conductivity, particularly when compared to metal-based conductive layers. Notably, the majority of current research on LSP materials and structural configurations for CFRP composites has been conducted under isolated conditions, typically focusing on either the A or D component of the lightning current. Comprehensive investigations are still required to evaluate how these protective structures perform under multi-component, time-sequential lightning current profiles, which more accurately reflect real-world lightning strike environments.

6.2 Protective Performance of Metallic Copper Mesh Structures Against a Single Lightning Current

6.2.1 Experimental Methods for LSP of Copper Mesh Protective Structures

To investigate the influence of individual lightning current components on CFRP composites equipped with a metallic protective mesh, and to evaluate the lightning strike protection (LSP) efficiency of the mesh structure, a test configuration based on a one-dimensional unidirectional layup of CFRP composites was established. A 0.25 mm-thick metallic copper mesh with 1100 mesh count was designed and placed on the surface of the CFRP composite layup structure. The composite specimen has overall dimensions of 150 × 100 × 4.6 mm, as illustrated in Fig. 6.7. In the lightning damage test conducted on CFRP composites with the copper mesh protective layer, the injected current corresponds to the A component of the lightning waveform. The CFRP specimen is mechanically clamped at both ends, and the discharge gap is set to 2 mm. The detailed experimental parameters are summarized in Table 6.1.

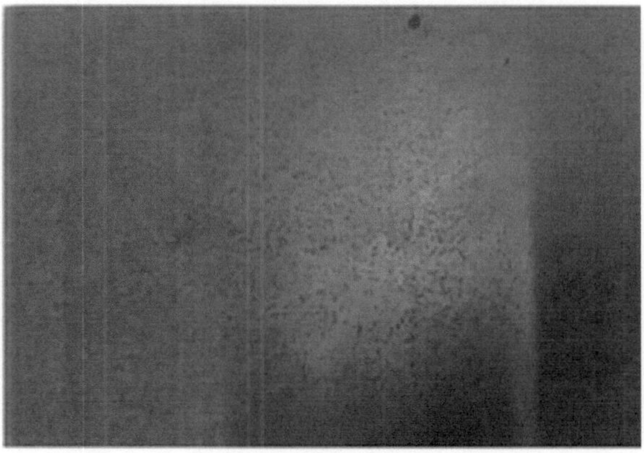

Fig. 6.7 CFRP composite specimen with copper mesh protective layer

Table 6.1 Experimental conditions for lightning-induced damage testing of CFRP composites protected by copper mesh

Specimens	I_p/kA	T_f/μs	T_t/μs	Q_C/C	W_A/ A^2s
Type Cu1-1	22.4 (A/10)	17.1	99.6	2.52	41,549
Type Cu1-2	42.4 (A/5)	17.5	102.5	5.02	117,747
Type Cu1-4	61.0 (3/10)	24.4	108.9	7.46	242,424
Type Cu1-4	98.4 (A/2)	25.6	117.4	11.87	681,940

6.2.2 Lightning Damage Morphology and Pattern of CFRP Composites with Copper Mesh Protective Structures

The lightning damage observed in unidirectional layup CFRP composites protected by a 0.25 mm-thick copper mesh after exposure to a simulated lightning current is shown in Fig. 6.8.

Analysis of the damage to CFRP composites with a copper mesh protective layer under the A component of the lightning current, as illustrated in Fig. 6.8, indicates the following:

(1) The copper mesh protective layer mitigates lightning-induced damage to the composite specimens to a certain extent when compared with unprotected CFRP composites. At a lightning current peak of 20 kA (corresponding to a charge transfer of approximately 2.52 C), only current dispersion marks and localized ablation traces appear on the copper mesh layer, with no significant damage observed in the CFRP substrate. When the lightning current peak increases to 40 kA, the copper mesh layer at the current injection point on the surface of specimen Type Cu1-2 is completely ablated, exposing the underlying CFRP and resulting in localized fiber breakage and slight warping. This behavior is attributed to the copper mesh layer's ability to effectively conduct the lightning

(a) Type Cu1-1, 22.4kA

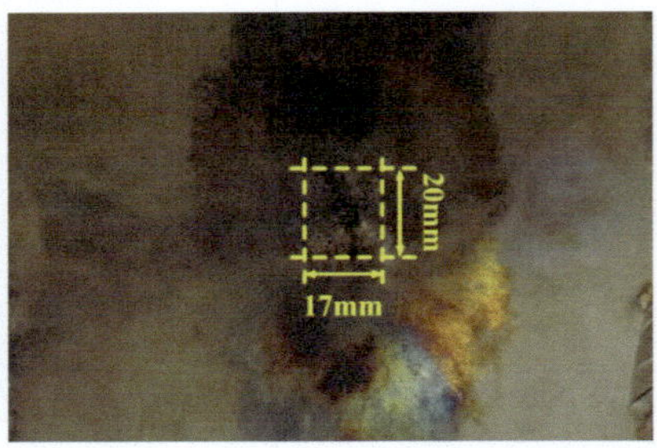

(b) Type Cu1-2, 42.4kA

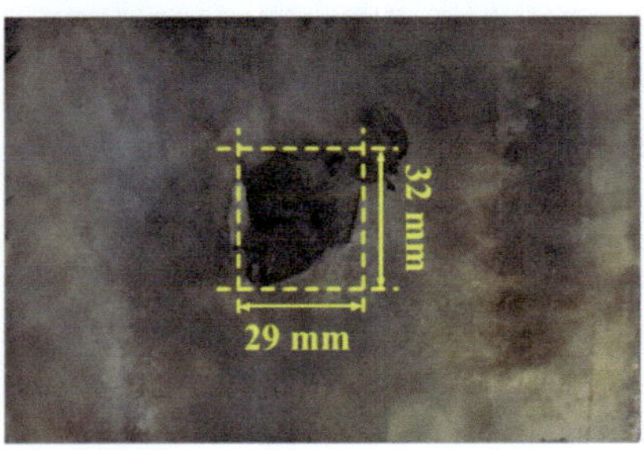

(c) Type Cu1-3, 61.0kA

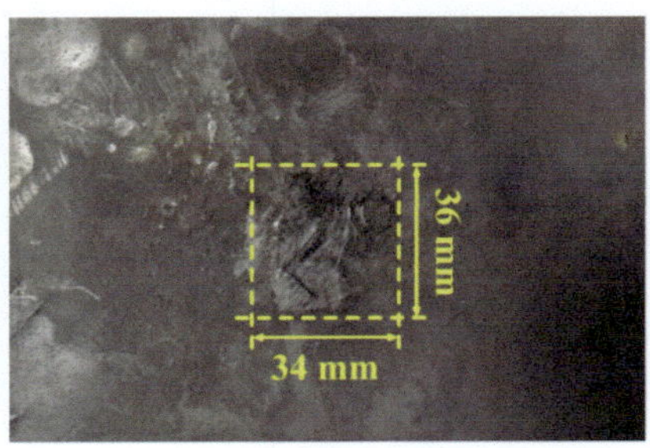

(d) Type Cu1-4, 98.4kA

Fig. 6.8 Surface damage pattern of unidirectional CFRP composites with copper mesh protective layer current when the peak current is moderate, thereby reducing current penetration in the thickness direction of the composite. As a result, the temperature rise within the fiber plies and the interlaminar resin is constrained, insufficient to reach the sublimation temperature of carbon fibers, and the thermal degradation of the resin matrix is limited.

(2) As the peak lightning current increases, the severity of damage to the CFRP composites with the copper mesh layer correspondingly intensifies. The ablation area of the copper mesh expands, exposing larger surface regions of the composite to direct impact. Pronounced fiber breakage becomes evident at the lightning attachment points. On specimens Type Cu1-3 and Type Cu1-4, the presence of flaky carbonaceous powder and small amounts of sintered residue on the composite surface can be clearly observed. These phenomena result from the melting and ablation of the copper mesh and the pyrolytic carbonization of the resin under the transient high temperatures generated by the lightning strike. Compared to the unprotected composite specimens (Fig. 3.11 in Chap. 3), the extent of resin pyrolysis, fiber ablation and fracture, and delamination damage is significantly reduced, demonstrating that the copper mesh offers a measurable degree of lightning strike protection.

6.2.3 *Analysis of Lightning Damage Results for CFRP Composites with Copper Mesh Protection Structures*

Ultrasonic Scanning Results of Lightning Damage on Copper Mesh-Protected CFRP Composites

The ultrasonic scan results for lightning-damaged CFRP composites with a unidirectional layup structure and a copper mesh protective layer, subjected to different peak values of the A component of the lightning current, are shown in Figs. 6.9, 6.10, 6.11 and 6.12.

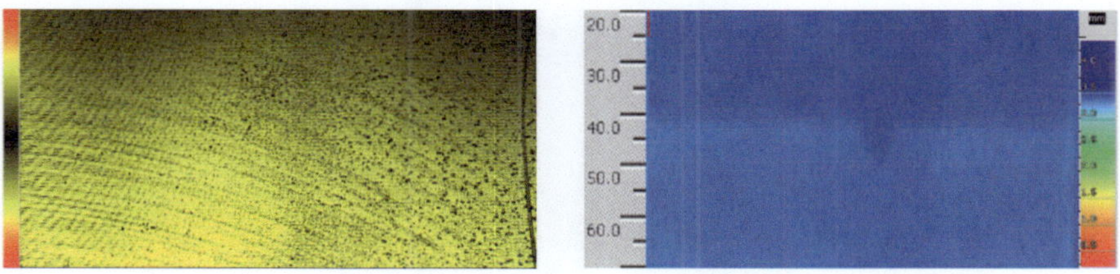

(a) Surface scan (80×40mm) (b) Depth scan map

Fig. 6.9 Lightning damage scan results for specimen Type Cu1-1 (22.4 kA)

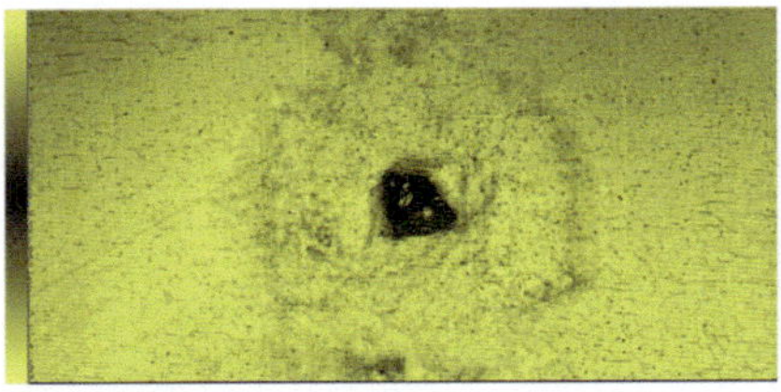

(a) Surface scan (80×40mm)

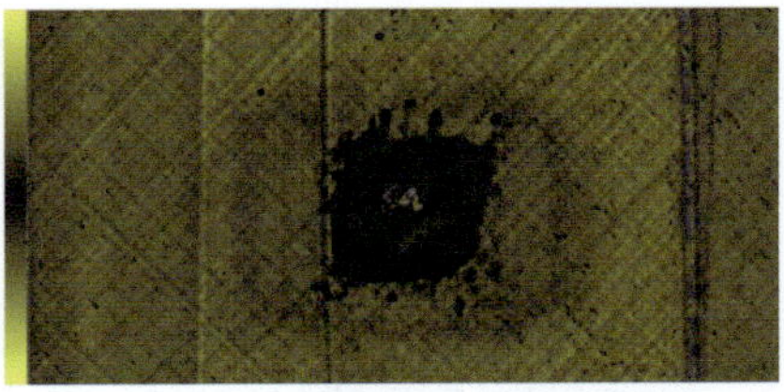

(b) Transmission scan (80×40mm)

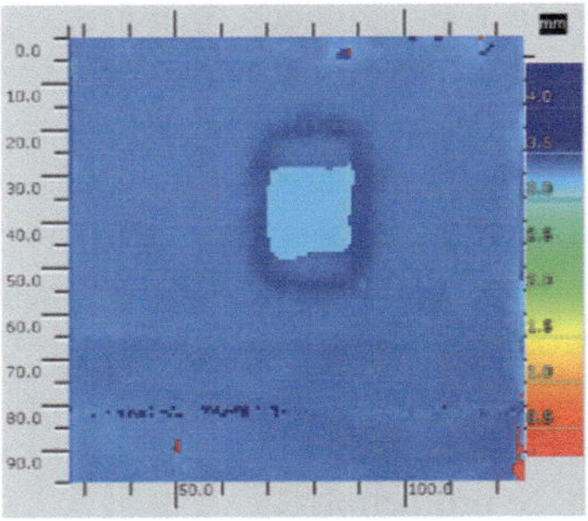

(c) Depth scan map

Fig. 6.10 Lightning damage scan results for specimen Type Cu1-2 (42.4 kA)

(a) Surface scan (80×40mm)

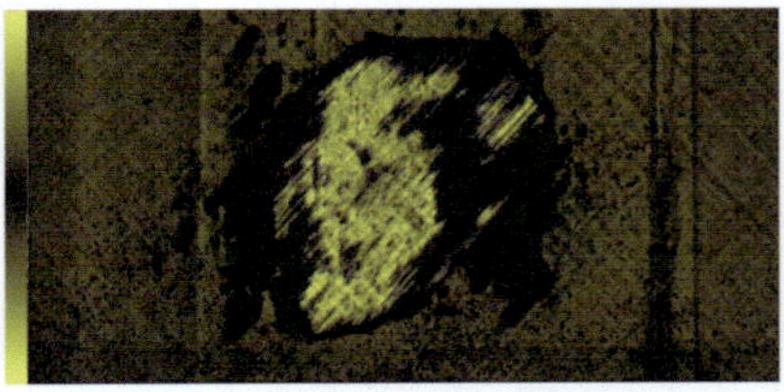

(b) Transmission scan (80×40mm)

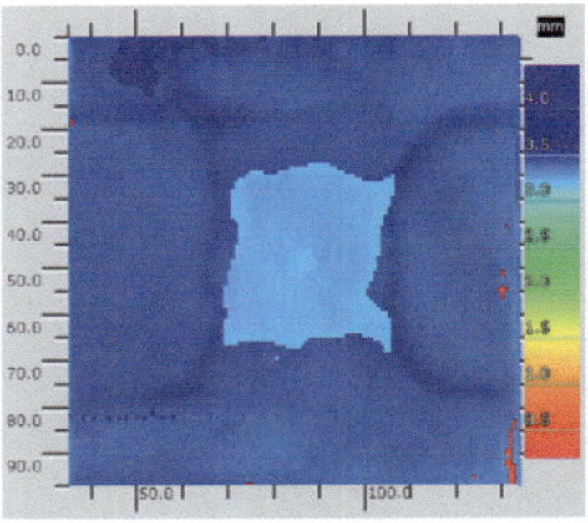

(c) Depth scan map

Fig. 6.11 Lightning damage scan results for specimen Type Cu1-3 (61.0 kA)

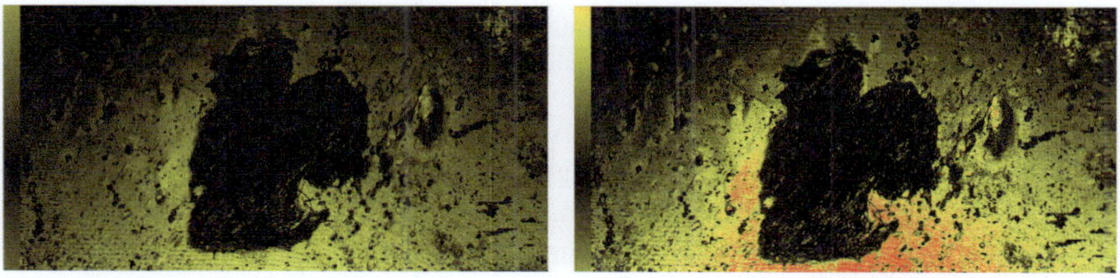

(a) Surface scan (80×40mm) (b) Transmission scan (80×40mm)

Fig. 6.12 Lightning damage scan results for specimen Type Cu1-4 (98.4 kA)

Damage Area and Depth Analysis

Based on the ultrasonic scanning results in Figs. 6.9, 6.10, 6.11 and 6.12, the lightning damage area and depth were quantified for the unidirectional CFRP composites with copper mesh protection. The calculated results are summarized in Table 6.2.

To further evaluate the effectiveness of the copper mesh protection, Fig. 6.13 presents a comparative analysis of lightning damage area and depth between copper mesh-protected CFRP composites and unprotected CFRP composites, as previously reported in Sect. 3.3.1 [16].

Table 6.2 Lightning damage area and depth for unidirectional CFRP composites with copper mesh protective layer

Specimens	I_p(kA)	Rising rate	Q_C (C)	W_A(A^2s)	S_Z (cm^2)	D_d (mm)
Type Cu1-1	22.4	1.404	2.52	41,549	0	0
Type Cu1-2	42.4	2.424	5.02	117,747	2.44	2.44
Type Cu1-3	61.0	2.618	7.46	242,424	7.25	7.25

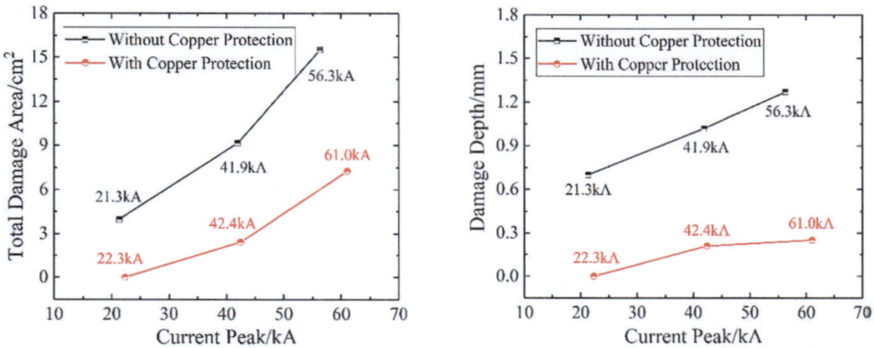

(a) Transmission damage area (b) lightning damage depth

Fig. 6.13 Comparison of lightning damage in CFRP composites with and without copper mesh protection

Based on the comparison curves presented in Fig. 6.13, the following conclusions can be drawn:

(i) Under identical peak values of the A component of the lightning current, the copper mesh protective layer can significantly reduce the extent of lightning-induced damage in CFRP composites. When the lightning current A component reaches a peak of 20 kA, the 0.25 mm-thick copper mesh is capable of effectively diverting the energy from the lightning current, thereby preventing its penetration into the depth of the CFRP composite. As a result, the underlying composite layers remain entirely unaffected by the lightning impact.
(ii) As the peak value of the lightning current increases, the CFRP composites protected by copper mesh still experience damage, but both the area and depth of lightning-induced damage remain substantially smaller compared to those of unprotected composites. For example, under a lightning current peak of 56.4 kA, the unprotected CFRP composite exhibits a damage area of 15.5 cm^2 and a damage depth of 1.27 mm. In contrast, under a slightly higher peak current of 61 kA, the copper mesh-protected composite sustains a much smaller damage area of 7.25 cm^2 and a shallower depth of 0.25 mm.

This difference arises because, with increasing lightning current amplitude, the thermal effect, which grows proportionally to the square of the current, intensifies. Consequently, the copper mesh at the lightning attachment point may melt due to the high temperature, exposing the underlying CFRP material to the subsequent energy of the strike. This results in localized sublimation and fracture of the carbon fiber layers, as well as pyrolysis of the resin matrix in the first and second plies, potentially extending even into the third layer. However, the presence of the copper mesh continues to serve a current-shunting and heat-dissipating role. Compared to unprotected specimens, this layer effectively reduces the transmission of both current and thermal energy into deeper regions of the composite, thereby mitigating the extent and depth of structural damage.

6.3 Protective Performance of Metallic Copper Mesh Structures Under Multiple Lightning Current Components

6.3.1 Experimental Method for Evaluating Copper Mesh Protection of CFRP Composites Against Multiple Lightning Strikes

Types of CFRP Composites

Unidirectional layup CFRP composite specimens with a metallic copper mesh protective layer were fabricated using the same material system. Each specimen measures 250 × 45 × 3.6 mm, with a total of 24 layers, each 0.15 mm thick. The layup sequence follows the $[-45/45/0/90]_{3S}$ configuration. A 0.25 mm-thick copper mesh with a density of 1100 mesh is laminated on the surface of each composite to serve as the lightning protection layer.

Experimental Conditions

The test setup is identical to that shown in Fig. 3.3, where specimens are clamped at both ends and the arc injection gap is set to 2 mm. The prepared specimens are divided into five groups, with each group subjected to a specific multi-component continuous lightning current test pattern as outlined in Table 6.3.

Under the ABCD continuous waveform mode, the waveforms and timing relationships of each lightning current component are illustrated in Fig. 5.1 (Chap. 5). The corresponding electrical parameters for each lightning current component applied to the copper mesh-protected CFRP composites in the multiple-strike tests are detailed in Table 6.4.

Table 6.3 Test models for multiple continuous uninterrupted lightning current components

Test models	Multiple sequential lightning current component combinations	Specimens
Model A	Application of component A only	Type Cu2-1
Model AB	Continuous application of components A and B	Type Cu2-2
Model DBC	Continuous application of components D, B and C	Type Cu2-3
Model ABC	Continuous application of components A, B and C	Type Cu2-4
Model ABCD	Continuous Application of components A, B, C and D	Type Cu2-5

Table 6.4 Electrical parameters of multiple lightning current components applied to unidirectional-layer CFRP composites with copper mesh protection

Specimens	Component A			Component B			Component C			Component D		
	$I_{p\text{-}A}$ /kA	T_f /μs	W_A/A^2s	$I_{ave\text{-}B}$/A	T_L/ms	Q_B/C	$I_{ave\text{-}C}$/A	T_L/ms	Q_C/C	$I_{p\text{-}D}$/kA	T_f/μs	W_D/A^2s
Type Cu2-1	41.24	26.43	76,100	–	–	–	–	–	–	–	–	–
Type Cu2-2	45.08	24.88	112,898	361.3	5.074	1.833	–	–	–	–	–	–
Type Cu2-3	–	–	–	361.7	5.061	1.830	166.1	191.2	31.76	34.83	13.50	46,666
Type Cu2-4	45.17	25.00	123,303	361.6	5.082	1.838	162.3	188.9	30.66	–	-	-
Type Cu2-5	44.95	24.50	119,161	373.7	5.050	1.887	167.9	188.5	31.65	19.87	14.00	15,745

6.3.2 *Morphology and Patterns of Multiple Lightning Damage in Copper Mesh-Protected CFRP Composites*

The lightning damage morphology of unidirectional layup CFRP composites protected by a 0.25 mm thick copper mesh after exposure to multiple continuous time-sequence lightning current components is shown in Fig. 6.14.

From the analysis of the lightning damage shown in Fig. 6.14, the following observations can be made:

(i) Under the action of the A component of the lightning current with a peak value of 40 kA and a charge of approximately 3 C, specimen Cu2-1 exhibited a distinct ablation crater at the lightning attachment point on the copper mesh. The ablation morphology is predominantly circular. Upon removal of the locally ablated copper mesh, the underlying CFRP composite shows clear surface damage, indicating localized fiber ablation. The damage characteristics

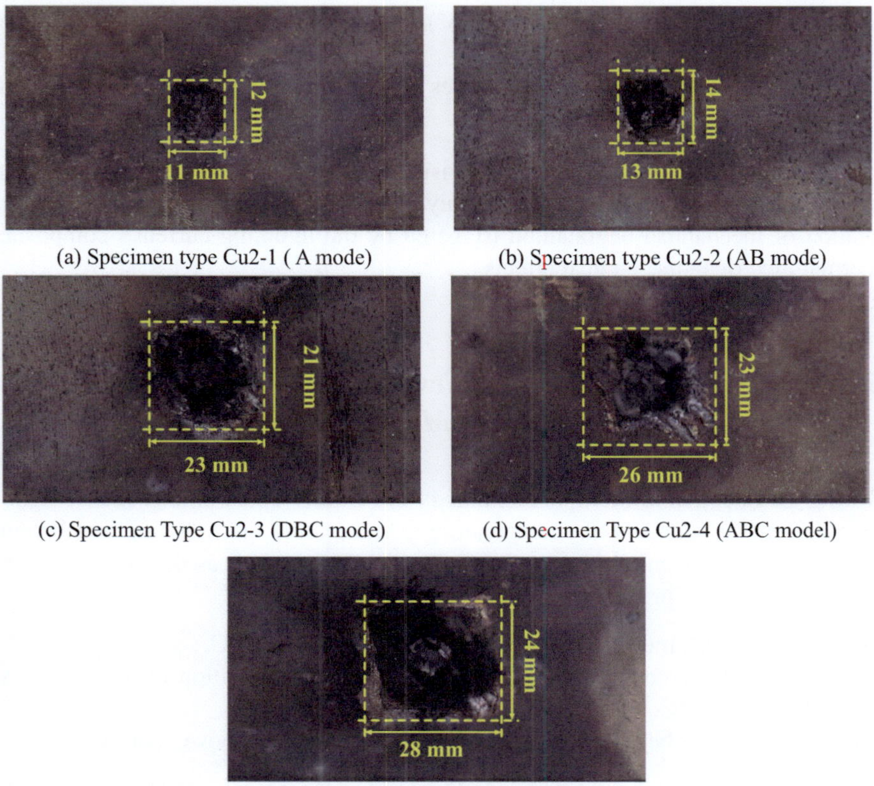

(a) Specimen type Cu2-1 (A mode)

(b) Specimen type Cu2-2 (AB mode)

(c) Specimen Type Cu2-3 (DBC mode)

(d) Specimen Type Cu2-4 (ABC model)

(e) Specimen Type Cu2-5 (ABCD mode)

Fig. 6.14 Lightning damage morphology of CFRP composite specimens with copper mesh protection layer of Cu2-2 (AB mode) are similar to Cu2-1; however, in addition to surface fiber ablation, resin matrix pyrolysis, fiber filament breakage, and local fiber uplift can be observed, indicating a more severe degree of degradation due to the additional B component.

(ii) For specimens Cu2-3, Cu2-4, and Cu2-5, subjected to more complex combinations of lightning current components including the high-charge C component (approximately 30 C), the overall damage is significantly aggravated. As shown in Fig. 6.14c–e, the degree of lightning damage increases progressively with the complexity of the current combination. The inclusion of the C component causes the initially circular ablation crater at the lightning attachment site to expand significantly, with a wider area of copper mesh melting and material removal around the arc attachment point. The intensified thermal effects result in:

   (a) Fiber ablation and matrix pyrolysis at the lightning attachment point;
   (b) Subsurface gas pyrolysis, accumulation, and explosive expansion inside the composite, producing internal pressure and mechanical stress;
   (c) Fiber fracture and upward delamination in the direction of the surface (−45°) and second-layer (+45°) fiber orientations due to the explosive outgassing;
   (d) Carbon black flake-like residues and sintered copper nodules distributed around the damaged zone.

Moreover, the lightning damage zone visibly extends toward the peripheral regions of the CFRP composite, indicating not only localized thermal damage but also propagation of mechanical degradation triggered by the lightning current's composite thermal-electrical–mechanical effects.

6.3.3 Temperature Characteristics of Copper Mesh-Protected CFRP Composites Under Multiple Lightning Strikes

During the injection of multiple continuous and uninterrupted time-sequence lightning current components, the temperature at and around the lightning attachment point of the copper mesh-protected CFRP composites increases rapidly. Based on the thermophysical properties of the constituent materials:

(i) Copper mesh melts at 1083 °C and sublimates at 2562 °C;
(ii) Resin matrix undergoes significant thermal decomposition at temperatures close to 200 °C;
(iii) Carbon fibers begin to sublimate only at temperatures above 3000 °C.

Therefore, under lightning strike conditions, the copper mesh is the first to undergo melting and sublimation. If the protective copper mesh fails to adequately conduct the lightning current, resulting in excessive temperature rise—the underlying CFRP

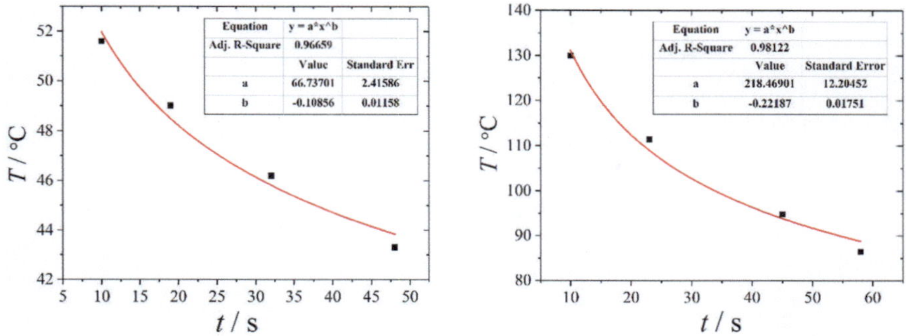

(a) Type Cu2-2 (b) Type Cu2-4

Fig. 6.15 Surface temperature trends of copper mesh-protected CFRP composite specimens after multiple uninterrupted lightning strikes

composite will inevitably suffer from severe lightning-induced thermal and mechanical damage. Infrared thermography was used to capture the surface temperature distribution of specimens Type Cu2-2 and Type Cu2-4 at various time intervals following lightning exposure under different current component combinations. As shown in Fig. 6.15:

(i) The high-temperature region is consistently located at the lightning attachment point, closely matching the nearly circular ablation crater observed in the physical damage profile.
(ii) For specimen Type Cu2-4, the maximum surface temperature reached 129.9 °C at $t = 10$ s after exposure to multiple sequential lightning current components.
(iii) The surface temperature of the specimens decreases rapidly over time due to convective and radiative heat exchange with the ambient environment.

Based on the post-strike surface temperature measurements and considering the thermal properties of the resin, carbon fibers, and the characteristic arc temperatures (on the order of 10^3–10^4 K), the time-dependent temperature variation of specimens Cu2-2 and Cu2-4 was obtained through curve-fitting techniques. The fitted temperature decay profiles are presented in Fig. 6.15, offering insight into the thermal response characteristics of the CFRP composites with copper mesh protection following severe lightning exposure.

$$T = 66.727 \times t^{-0.1085} \tag{6.1}$$

$$T = 218.469 \times t^{-0.2218} \tag{6.2}$$

According to Eq. (6.2), the instantaneous surface temperature of the Type Cu2-4 CFRP composite specimen with copper mesh protection can be estimated to reach

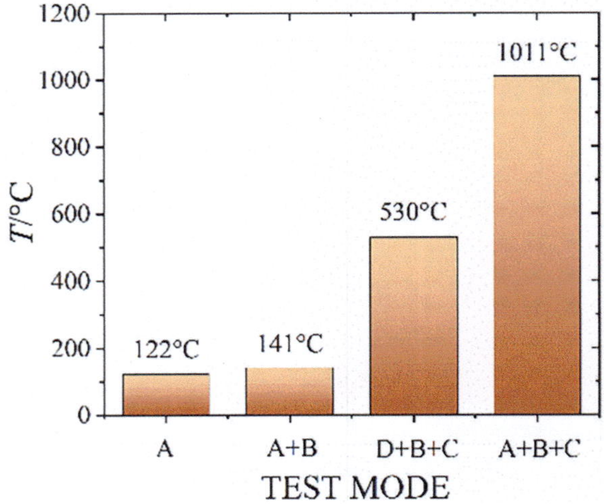

Fig. 6.16 Temperature characteristics of copper mesh-protected CFRP composites under different lightning current component combinations

1011 °C at $t = 1$ ms after exposure to multiple consecutive time-sequence lightning current components. This rapid and intense thermal effect induces carbon fiber sublimation and resin pyrolysis, resulting in significant internal and surface structural degradation. The temperature response characteristics of CFRP composites with copper mesh protection under different combinations of sequential lightning current components are illustrated in Fig. 6.16.

From the experimental results in Fig. 6.16, it is evident that the surface temperature of the CFRP composites increases with the cumulative energy (or total charge) of the applied lightning current sequence. A higher local temperature at the lightning attachment point leads to more severe material damage. Among the current components, the C component, with its larger charge transfer and thermal energy, has the most pronounced heating effect during current dissipation within the composite. This is particularly evident in specimens Type Cu2-2 and Cu2-4, which show steep temperature increases following exposure to sequences containing the C component, compared to sequences containing only A, B, or D components.

To further assess the effectiveness of the copper mesh as a thermal protection strategy, a comparison of temperature characteristics between unprotected and copper mesh-protected CFRP composites under multiple uninterrupted lightning strikes is shown in Fig. 6.17.

As seen from the comparative temperature characteristics shown in Fig. 6.17:

(i) The transient peak temperatures of CFRP composite specimens Type Cu2-1 and Type Cu2-2 with copper mesh protection do not show significant differences from those of unprotected CFRP composites Type S1-1 and Type S1-2 when subjected to a single lightning current component A (peak value 40 kA) or to the sequential action of components A and B of the lightning current. This indicates that under low-energy or short-duration lightning current components,

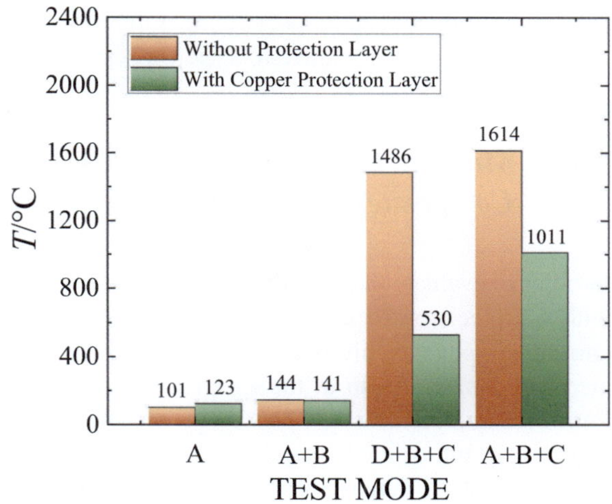

Fig. 6.17 Comparison of surface temperature response of CFRP composites with and without copper mesh protection after multiple sequential lightning current components

the presence of copper mesh does not markedly alter the surface temperature response.

(ii) With the addition of the lightning current component C, which is characterized by a large charge and long duration, the transient surface temperatures of copper mesh-protected specimens Type Cu2-3 and Type Cu2-4 are significantly lower than those of unprotected specimens Type S1-3 and Type S1-4. The primary reason lies in the inherent current conduction characteristics of component C, which features a relatively low average current magnitude and lower action voltage compared to components A, B, and D. As a result, component C tends to propagate along the plane of the composite rather than through its thickness. At the same time, due to the much higher electrical conductivity of copper mesh relative to the CFRP substrate, the copper mesh effectively diverts a substantial portion of the component C current, thereby shielding the underlying CFRP material. Moreover, although component C imparts a high overall thermal effect due to its long action time, the energy density is insufficient to induce melting or thermal damage to the copper mesh, owing to its favorable thermal conductivity. Therefore, from a temperature behavior perspective, the copper mesh layer significantly suppresses the transient temperature rise and mitigates thermal damage induced by the C component of the lightning current.

(iii) When comparing Type Cu2-3 and Type Cu2-4, the transient peak temperature of specimen Type Cu2-4 is notably higher than that of Type Cu2-3. This difference is primarily attributed to the higher peak amplitude of the A component of the lightning current in Type Cu2-4. Specifically, the A component at 40 kA in Cu2-4 causes deeper initial damage, compared to the D component at 20 kA applied to Cu2-3. As a result, during subsequent exposure to the C component, the current in Cu2-4 is able to penetrate more deeply into the layup, producing more extensive thermal damage. Additionally, the total energy (integral) and duration of component A are greater than those of component D, leading to a more pronounced transient temperature increase in Type Cu2-4 following the lightning strike, relative to Type Cu2-3.

6.3.4 Multiple Lightning Damage Characterization of CFRP Composites with Copper Mesh Protection

In order to evaluate the cumulative lightning-induced damage on CFRP composites with a protective copper mesh under the action of multiple, sequential lightning current components, ultrasonic C-scan imaging was employed to detect the damage area and damage depth of the specimens after exposure. The results are shown in Figs. 6.18 and 6.19.

From the ultrasonic C-scan imaging results shown in Figs. 6.18 and 6.19, it can be observed that the damage area of CFRP composites with copper mesh protection is

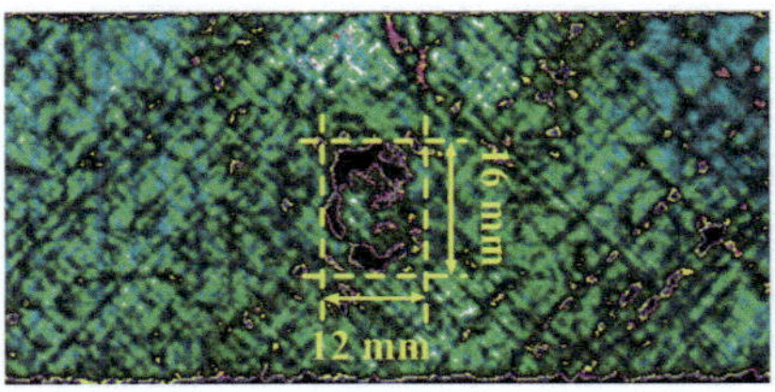

(a) Specimen Type Cu2-1 (A mode)

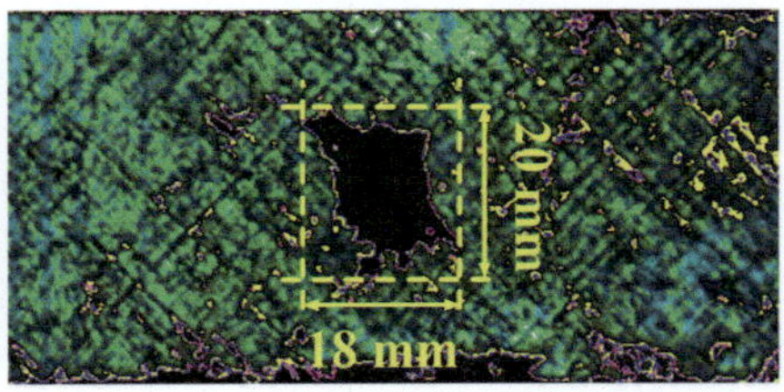

(b) Specimen Type Cu2-2 (AB mode)

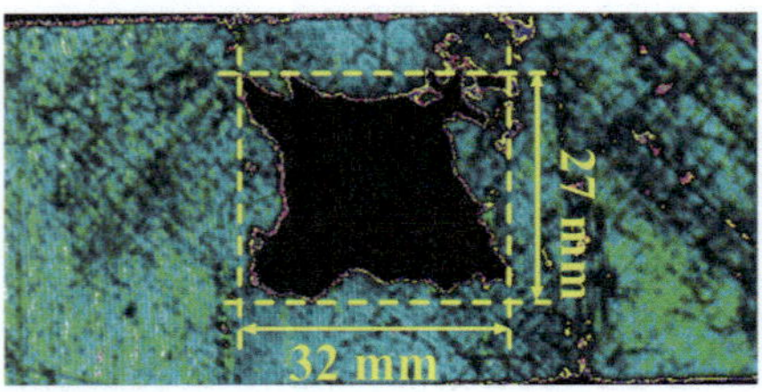

(c) Specimen Type Cu2-3 (DBC mode)

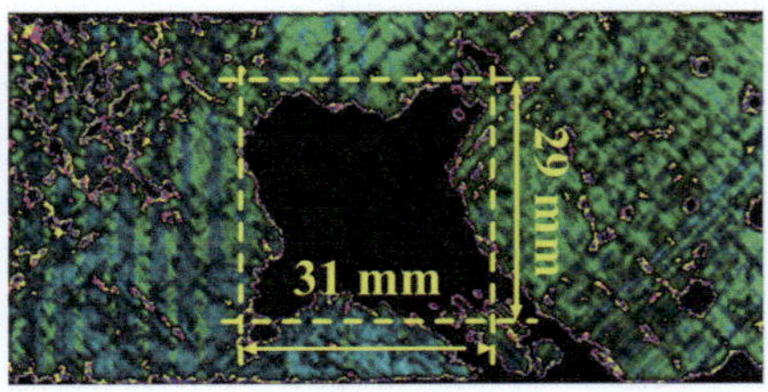

(d) Specimen Type Cu2-4 (ABC mode)

(e) Specimen Type Cu2-5 (ABCD mode)

Fig. 6.18 Ultrasonic C-scan images of damage areas in CFRP composites with copper mesh protection after multiple lightning current components

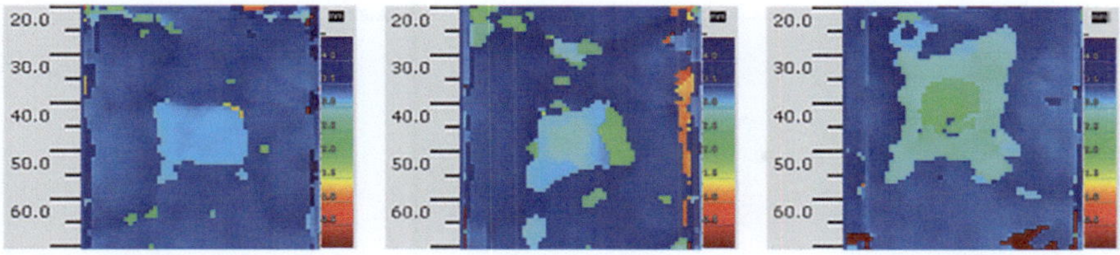

(a) Specimen Type Cu2-1 (A mode) (b) Specimen Type Cu2-2 (AB mode) (c) Specimen Type Cu2-3 (DBC mode)

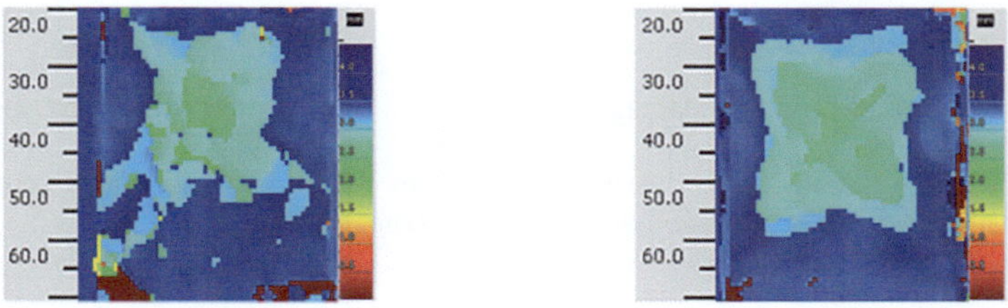

(d) Specimen Type 5-4 (ABC mode) (e) Specimen Type 5-5 (ABCD mode)

Fig. 6.19 Ultrasonic C-scan images of damage depth in CFRP composites with copper mesh protection after multiple lightning current components

primarily concentrated around the lightning attachment point, and exhibits a roughly circular distribution. Compared with the lightning-induced damage in unprotected CFRP composites, the fiber fracture warpage along the ply direction is significantly less pronounced, indicating that the presence of the copper mesh layer effectively suppresses the formation and accumulation of pyrolysis gas within the composite plies. This suppression reduces the fiber fracture damage induced by gas expansion and explosion effects. By applying quantitative methods such as pixel-based statistical analysis, the damage area and depth for CFRP composites with copper mesh protection under various experimental lightning current modes were obtained, as summarized in Table 6.5 and illustrated in Fig. 6.20.

Analyzing the experimental results presented in Table 6.5 and Fig. 6.20, it is evident that the influence of each lightning current component on the damage of CFRP composites protected by a copper mesh is cumulative and synergistic.

Table 6.5 Damage area and depth of CFRP specimens with copper mesh protection under different combinations of continuous lightning current components

Specimens	Test modes	Damage area/cm^2	Damage depth/mm
Type Cu2-1	Mode a	1.76	0.37
Type Cu2-2	Mode ab	2.53	0.59
Type Cu2-3	Mode dbc	8.00	0.91
Type Cu2-4	Mode ABC	8.63	0.97
Type Cu2-5	Mode ABCD	9.99	1.20

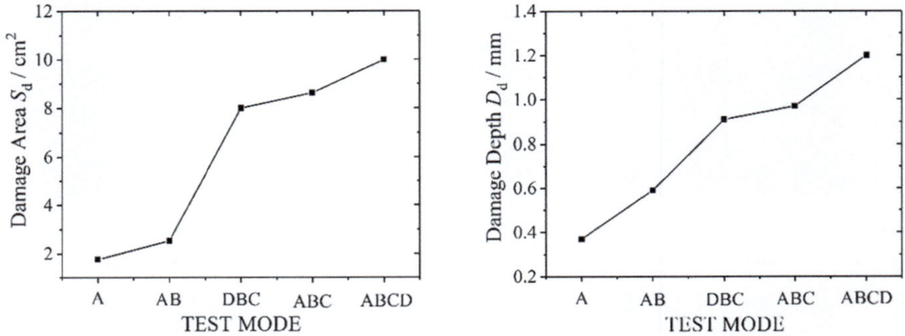

(a) Area of lightning damage (b) Depth of lightning damage

Fig. 6.20 Damage area (S_d) and damage depth (S_d) of CFRP composites with copper mesh protection after exposure to multiple lightning current components

When the energy of the applied lightning current component is relatively low, the copper mesh effectively conducts the majority of the current, thereby protecting the composite structure. As such, for specimens Cu2-1 and Cu2-2, the copper mesh protection layer effectively suppresses both the damage area and depth. However, as the number of lightning current components increases, especially with the introduction of the long-duration component C, a notable escalation in damage is observed in specimens Cu2-3, Cu2-4, and Cu2-5. These specimens exhibit significantly increased damage areas and depths, highlighting the reduced shielding effectiveness of the copper mesh under high cumulative current and energy loads.

To further assess the protective effect of the copper mesh layer, Fig. 6.21 compares the damage area and depth of CFRP composites with and without copper mesh protection under different combinations of multiple continuous lightning current components.

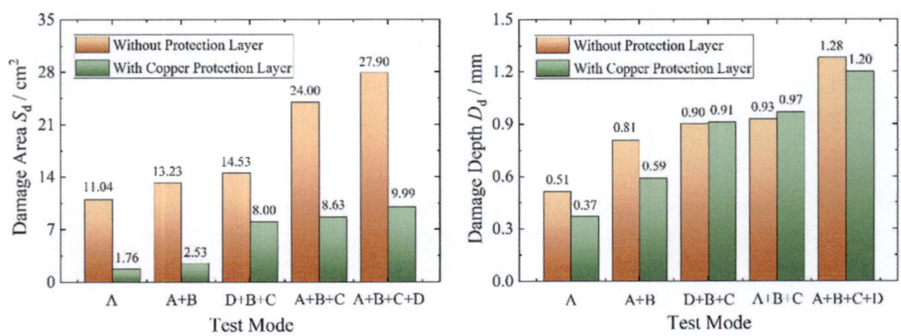

(a) Area of lightning damage (b) Depth of lightning damage

Fig. 6.21 Damage area (S_d) and damage depth (D_d) of CFRP composites with and without copper mesh protection under multiple lightning strikes

(1) **Effectiveness of the copper mesh protection layer on lightning-induced damage area**

From the comparison results shown in Fig. 6.21, the following observations can be made. As the number of lightning current components increases, the damage area in both protected and unprotected composites tends to grow. However, the copper mesh protection layer significantly mitigates surface damage. Under the action of a single lightning component (A) or two components (A and B), the copper mesh reduces the damage area by more than 80% compared to unprotected CFRP composites. This protective effect diminishes as the number and duration of current components increase, particularly when the C component (long-duration current) is involved. In such cases, the current conduction capacity of the copper mesh becomes insufficient to divert all the energy, leading to increased internal current shunting and greater damage propagation into the composite. As a result, specimens Cu2-3, Cu2-4, and Cu2-5 exhibit significantly enlarged damage regions, both in area and depth, indicating the limitations of copper mesh protection under complex, multi-component lightning strikes.

(2) **Effectiveness of the copper mesh protection layer on lightning-induced damage depth**

The presence of a copper mesh protection layer also contributes to a reduction in the damage depth of CFRP composites, although this effect is generally less pronounced than its suppression of surface damage area. When the energy of the applied lightning current component is relatively low, the copper mesh demonstrates a noticeable ability to limit the penetration depth of thermal and mechanical damage. For example, under the action of the single lightning current component A, the damage depth is reduced from 0.513 (unprotected) to 0.37 mm (protected). Similarly, under the combined action of two components (A and B), the damage depth decreases from 0.81 to 0.59 mm, representing a reduction of approximately 30%. These results indicate that at low to moderate energy levels, the copper mesh retains its integrity and serves as an effective current shunt, thereby protecting the underlying composite layers from deep thermal and arc-induced damage. However, as the number and duration of lightning current components increase, particularly with the introduction of component C (long-duration current), the protective effectiveness of the copper mesh with respect to damage depth is drastically diminished. In specimens Cu2-3, Cu2-4, and Cu2-5, subjected to three or more lightning current components, the damage depth is nearly comparable to that of the unprotected specimens (Types 4.3, 4.4, and 4.5).

This degradation in protective performance can be attributed to thermal and mechanical failure of the copper mesh itself. When the cumulative lightning energy exceeds a critical threshold, the energy density at the lightning attachment point becomes sufficient to melt or vaporize portions of the surface copper mesh. This loss of conductive material not only disables the mesh's current-shunting function, but also exposes the underlying CFRP layers to direct arc ablation and joule heating from subsequent current components. In this scenario, the later lightning components effectively act on the composite material without any meaningful protection, leading to deep internal damage. At present, most modern aircraft employ metal foils or metal meshes as primary elements of lightning protection systems. Under single-component impulse currents, commonly used copper meshes offer moderate protection to CFRP structures. Their effectiveness is influenced by factors such as mesh density, thickness, and the specific waveform characteristics of the lightning current. However, the real-world lightning environment typically involves multiple, continuous, time-sequenced current components. Therefore, the suitability of single-layer metallic protection schemes must be re-evaluated under these more realistic multi-component lightning exposure scenarios.

6.4 Non-metallic Protective Structures

6.4.1 3D Printing of Highly Conductive Grids

To address the limitations of conventional copper mesh protective structures, namely their high density and poor ablation resistance, there is significant engineering value in developing lightweight, thermally stable lightning protection layers for CFRP composites. In contrast to metal meshes, which tend to melt rapidly under high-temperature conditions, carbon-based materials exhibit superior thermal stability. These materials do not melt or ignite under high temperatures; instead, they undergo oxidative reactions with ambient oxygen, which occur more slowly and typically only affect the surface. Due to the short duration of a typical lightning strike, carbon-based protective structures can retain their structural integrity during the current pulse, thereby maintaining electrical continuity and effectively diverting lightning currents throughout the event.

As illustrated in Fig. 6.22, nickel-plated carbon fiber filaments are composite fibres created by depositing a highly conductive nickel coating, via electrophoretic deposition or electroplating, onto traditional carbon fiber substrates. The average thickness of the metallic coating is approximately 226 nm. The untreated 3K carbon fiber tows exhibit a linear density of 0.2 g m^{-1} and a resistance of 157.26 $\Omega\,m^{-1}$. After nickel plating, the filament density increases to 0.35 g m^{-1}, while the linear resistance decreases dramatically to 4.51 $\Omega\,m^{-1}$. This demonstrates that the nickel-plated filaments maintain the lightweight characteristics of carbon fibres while achieving electrical conductivity levels approaching those of metals.

To integrate these filaments into composite structures, continuous fiber 3D printing is employed, which is a form of automated additive manufacturing that enables the deposition of continuous fibres in predefined patterns. This method leverages fiber filaments as the reinforcement medium and uses a resin system identical to that of the composite prepreg, ensuring excellent interfacial compatibility and enabling

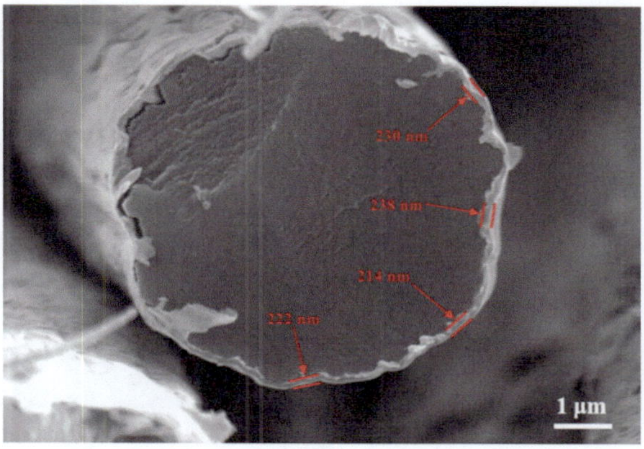

Fig. 6.22 Microscopic cross-sectional morphology of nickel-plated carbon fiber filaments

co-curing during the composite lay-up process. This integration effectively eliminates delamination or interfacial debonding between the conductive grid and the host composite. As shown in Fig. 6.23, nickel-plated carbon fiber filaments were printed in-situ onto the surface of the composite prepreg using this technique. The printed grid had a spacing of 3.59 mm, carefully selected to achieve a surface mass density (~195 g m^{-2}) comparable to that of conventional copper mesh. This design facilitates a direct and meaningful comparison of lightning protection effectiveness between the novel and traditional protective structures.

Figure 6.24 presents the arc evolution process observed on three types of specimens, which are unprotected CFRP panels, CFRP panels protected with nickel-plated carbon fiber grids, and those protected with copper mesh, when subjected to a 100 kA peak amplitude A-component lightning current. At 5 μs, initial charge attachment occurs, forming a localized hot spot on the specimen surface. Between 25 and 50 μs, as the current rises, the arc begins to expand significantly. Notably, a clear difference in arc propagation behaviour is observed among the three configurations. For specimens equipped with either copper mesh or nickel-plated carbon fiber grids,

Fig. 6.23 In-situ 3D printing of nickel-plated carbon fiber filament grid

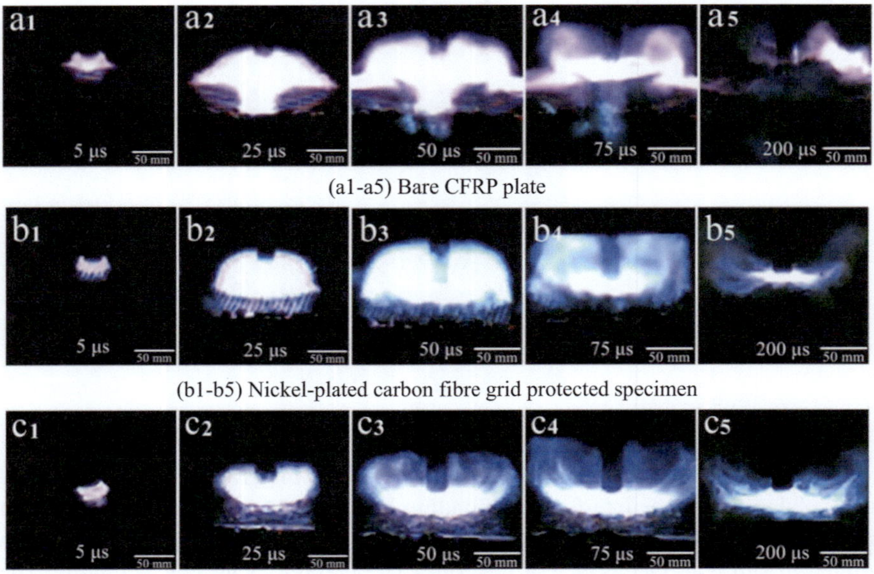

(a1-a5) Bare CFRP plate

(b1-b5) Nickel-plated carbon fibre grid protected specimen

(c1-c5) Copper mesh protected specimen

Fig. 6.24 Expansion process of lightning arcs on various specimen surfaces

the arcs exhibit hemispherical expansion patterns, extending laterally toward the surrounding ground plane. This is attributed to the high surface conductivity of these protective layers, which promotes efficient current diffusion and arc root mobility. In contrast, the unprotected composite specimens show constrained arc development in the central region, indicating limited current dispersion and greater energy concentration. As the lightning pulse enters the decay phase (75–200 μs), the arcs progressively weaken and dissipate, following the current waveform's decline.

Figure 6.25 shows the surface and internal damage morphologies of the CFRP composites after the lightning test. The bare specimens exhibit the most severe damage, characterized by complete penetration, severe fiber fracture, and visible interlaminar delamination. In contrast, both nickel-plated carbon fiber grid and copper mesh protections significantly reduce the extent of damage. The affected area in these two cases can be categorized into two zones: Zone I (arc attachment and injection zone) and Zone II (thermal radiation zone). For the nickel-plated grid, fiber bulging and breakage are observed in Zone I, while Zone II primarily exhibits slight resin ablation, and the overall grid structure remains largely intact. In the case of copper mesh, although Zone I damage is controlled, the entire grid structure in Zone II is destroyed, attributed to the low melting point of copper. CT scan results confirm that both protection schemes effectively suppress delamination and porosity, with no obvious internal interlayer separation.

The primary damage mechanism in CFRP composites under lightning strikes is the instantaneous vaporization of the matrix resin due to Joule heating, which generates

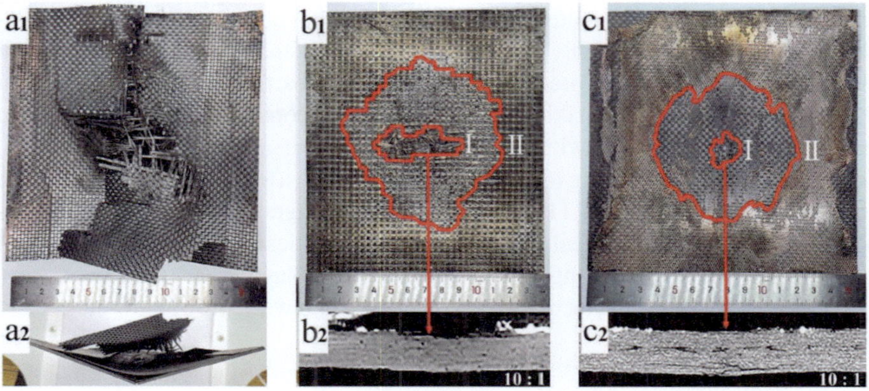

(a1-a2) Bare CFRP specimens

(b1-b2) Specimens with nickel-plated carbon fibre grid protection

(c1-c2) Specimens with copper mesh protection

Fig. 6.25 Damage morphology post-lightning strike

a high-pressure recoil force leading to fiber fracture and interlaminar delamination. Effective protective layers with high electrical conductivity, such as nickel-plated carbon fiber grids and copper meshes, can rapidly conduct the lightning current away from the composite surface, thereby preventing deep current penetration and mitigating thermal/mechanical damage. However, in terms of thermal endurance and structural retention, the nickel-plated carbon fiber grid exhibits superior performance. The carbon fibres in the core of the filaments possess significantly higher melting point (3316 °C vs. 1083 °C), specific heat capacity (712 J $(kg\ K)^{-1}$ vs. 385 J $(kg\ K)^{-1}$), and latent heat of vaporization (355.8 kJ mol^{-1} vs. 300.3 kJ mol^{-1}) compared to copper. These properties enable the nickel-plated carbon fiber grids to withstand stronger arcs and higher temperatures, maintaining structural integrity even under a single 100 kA lightning pulse. Additionally, the 3K nickel-plated carbon fiber tows, composed of approximately 3,000 filaments per strand, are 3D printed into an interconnected parallel network, creating multiple conductive pathways. This redundant conduction architecture further enhances current dissipation efficiency and contributes to the system's robust lightning protection capability.

6.4.2 Lightning Damage to Graphene Film-Protected Structures

Graphene differs from carbon fiber in that it is a two-dimensional planar film composed of carbon atoms in sp^2-hybridized orbitals, arranged in a hexagonal honeycomb lattice with a single-atom thickness of 0.335 nm. Each carbon atom participates in three π-bonds and contributes its remaining p-orbital electron to a delocalized π-bond network, enabling highly mobile charge carriers and resulting in exceptional electrical conductivity. Theoretical conductivity values for pristine graphene can reach up to 1×10^8 S/m, making it among the most conductive materials at room temperature. Additionally, with a C–C bond length of approximately 0.142 nm, graphene exhibits outstanding mechanical strength, with a tensile strength of 125 GPa and an elastic modulus of 1.1 TPa, ranking it as the strongest known crystalline material. These combined electrical and mechanical properties render graphene an ideal candidate for use as a lightning strike protection (LSP) layer on CFRP composites.

Graphene can be produced by several methods, each with its own advantages and limitations:

(i) Mechanical Exfoliation

A physical delamination technique, originally employing adhesive tape to peel off graphene layers from bulk graphite. This approach yields high-quality monolayers but suffers from extremely low throughput, making it unsuitable for industrial scale. Automated "knife-edge" exfoliation has improved yield but remains cost-prohibitive.

(ii) Chemical Vapour Deposition (CVD)

A chemical synthesis method in which hydrocarbon gases (e.g., methane) are thermally decomposed on a metal substrate (commonly copper or nickel) to form a continuous graphene film. Subsequent substrate etching releases the graphene layer. CVD offers large-area coverage, relatively low cost, and compatibility with aerospace-grade production requirements, and thus represents the most prevalent industrial technique.

(iii) Chemical Reduction

Involves oxidation of graphite to graphene oxide, followed by chemical or thermal reduction to restore the sp^2 carbon network. Although this method enables bulk graphene powder production, the resulting films often contain residual defects and lower electrical conductivity compared to CVD-grown graphene.

Preparation of Highly Conductive Graphene Layers via Resin Infiltration

To enhance the lightning damage resistance of CFRP composites, a highly conductive surface layer must be established to rapidly channelize lightning current, thereby preventing current and heat penetration through the composite thickness. Although graphene exhibits exceptional electrical and thermal conductivity, its precursor graphene oxide (GO) flakes typically measure less than 50 μm in lateral dimension, as shown in Fig. 6.26. Consequently, large-area conductive graphene layers must be fabricated through material self-assembly techniques, such as resin infiltration or standalone graphene film deposition.

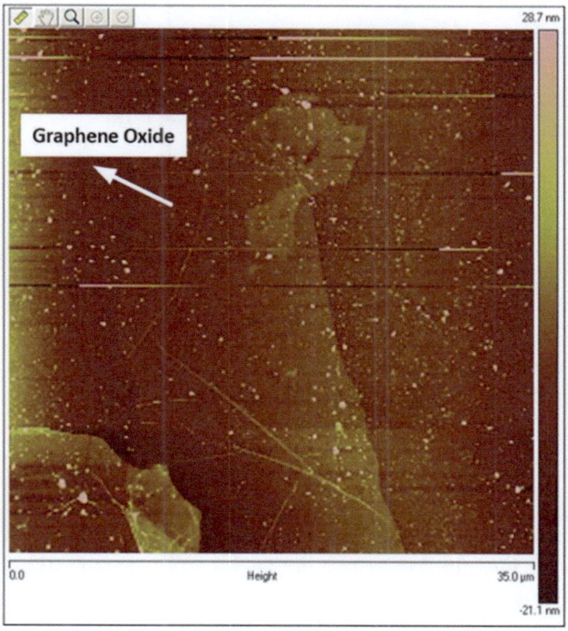

Fig. 6.26 Atomic force microscopy image of graphene oxide sheets (flake size < 50 μm)

As depicted in Fig. 6.27, the resin infiltration approach entails premixing graphene nanosheets into an epoxy resin, followed by ball-milling to ensure uniform dispersion and elimination of large-area agglomerates. The resulting graphene-filled resin is then calendared into a thin film. Microscopic and macroscopic observations confirm a homogeneous distribution of graphene within the resin matrix. Figure 6.28 illustrates the resin dissolution and infiltration mechanism. A graphene-laden resin film is placed at the base of the fiber preform, with filter paper beneath to prevent graphene loss during curing. Upon heating, the resin's viscosity decreases and, under vacuum-induced negative pressure, infiltrates the fiber bundles. Due to the dense fiber architecture, graphene nanosheets experience flow resistance, impeding their penetration into the fiber interior. Consequently, graphene gradually deposits onto the composite surface during resin curing, forming the desired surface conductive layer.

Figure 6.29 presents the microscopic cross-sectional morphology of a cured CFRP specimen, confirming the formation of a continuous graphene layer on its surface. The images clearly demonstrate a uniform, dense distribution of graphene, which enhances the electrical conductivity and overall performance of the composite. For a 150 × 150 mm specimen with a $[0/90]_{4S}$ layup, the average graphene layer thickness is measured as 84 μm along the fiber direction and 37 μm perpendicular to the fibers. This anisotropy in thickness reflects the fiber architecture and resin flow behavior during cure, and indicates the establishment of a crosslinked conductive network on the composite surface.

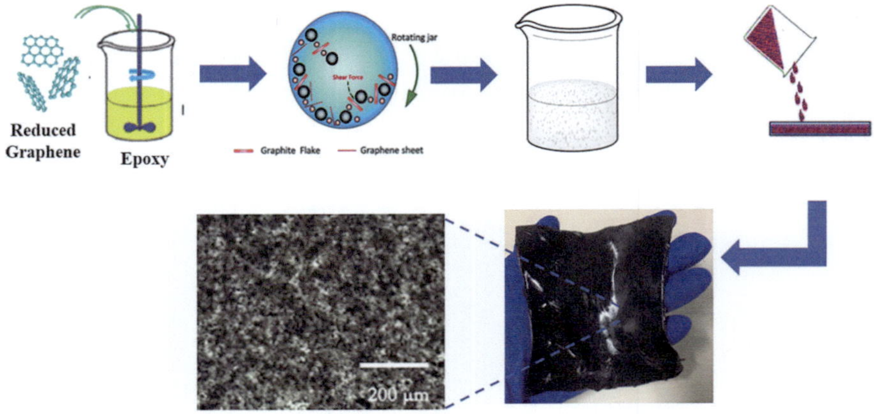

Fig. 6.27 Process flow for graphene dispersion in epoxy resin

As shown in Fig. 6.30, surface resistivity measurements reveal a dramatic drop from 104.2 to 0.55 Ω/sq in the perpendicular direction when 0.05 g of graphene is added. This sharp decrease underscores the significant improvement in conductive properties imparted by the graphene layer. Moreover, the conductivity along and across the fiber directions becomes nearly identical, indicating excellent isotropic conductivity. Such isotropy is critical for the rapid evacuation of lightning-strike currents within the graphene layer, thereby minimizing Joule heating and mitigating thermal damage. However, beyond a certain graphene loading, surface conductivity plateaus, implying the existence of a percolation threshold. As graphene content increases, nanosheets coalesce into a continuous film, shifting the primary electron conduction path to the graphene layer. Once fully percolated, additional graphene yields diminishing returns in conductivity while increasing the protective-layer mass.

Figure 6.31 illustrates the lightning damage morphology and CT inspection images of CFRP composites with and without a graphene surface layer under a 40 kA A-component lightning current. In the unprotected specimen (Fig. 6.31a, c), the damage concentrates in the central region, with an area of approximately 751 mm^2, encompassing fiber fracture and resin ablation. This severe damage results from the inability to evacuate the lightning current rapidly, leading to excessive Joule heating at the strike attachment site, instantaneous resin vaporization, and vapor recoil–induced fiber breakage. By contrast, the graphene-protected specimen (Fig. 6.31b, d) exhibits a markedly reduced damage area of 211 mm^2, primarily due to the graphene layer's effectiveness in lowering Joule heat generation.

To further evaluate internal damage, both unprotected and graphene-protected specimens were subjected to CT scanning (Fig. 6.32). In the CT images, dark regions denote severe lightning damage. The unprotected composite shows extensive delamination spanning ~43.8 mm in length and ~0.7 mm in depth (3–4 plies). In contrast, the graphene-protected composite exhibits delamination reduced to 19.4 mm in length and 0.3 mm in depth (1–2 plies). Porosity analysis via VG Studio Max 2.1 and

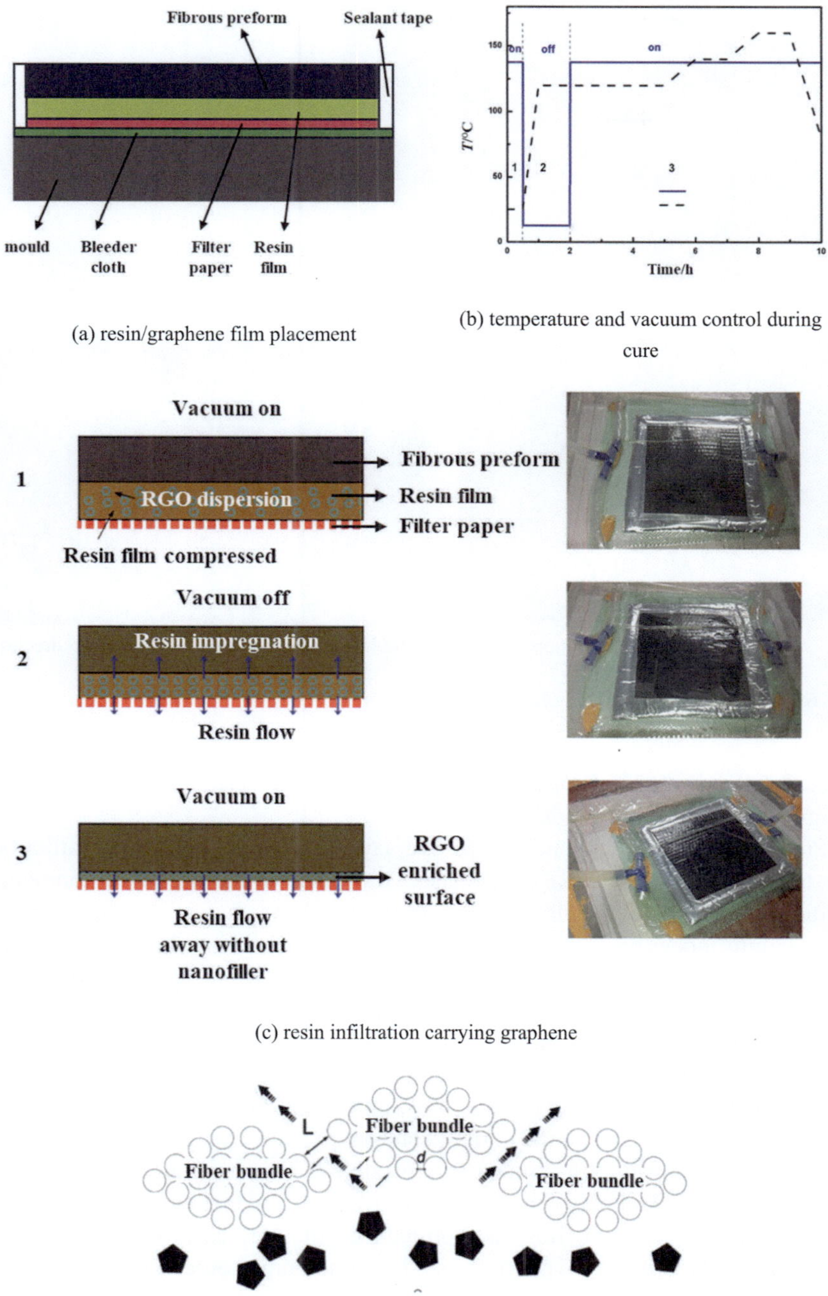

(a) resin/graphene film placement

(b) temperature and vacuum control during cure

(c) resin infiltration carrying graphene

(d) graphene impeded by fiber bundles

Fig. 6.28 Resin dissolution and infiltration schematic

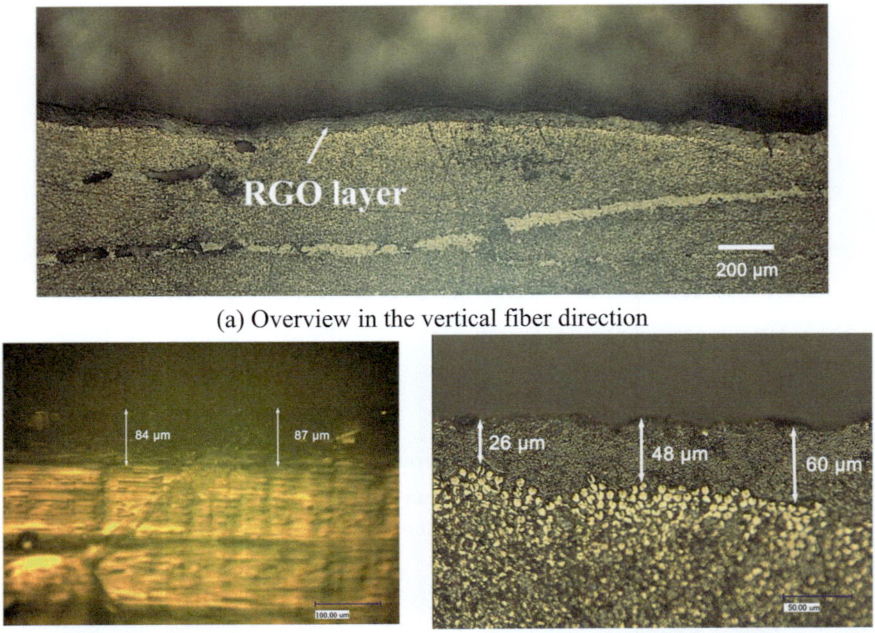

(a) Overview in the vertical fiber direction

(b) along the fiber direction

(c) Magnified view in the vertical fiber direction

Fig. 6.29 Cross-sectional morphology of cured specimens

Avizo® 6.0 indicates 16.93% porosity in the unprotected specimen versus 4.08% in the graphene-coated specimen.

Preparation of a graphene conductive layer via resin dissolution-infiltration substantially enhances CFRP composites' resistance to the 40 kA A-component lightning current. Nevertheless, as the peak current increases, delamination and other failure modes still occur. This behavior primarily arises because insulating epoxy remains interspersed between graphene sheets within the protective layer, impeding efficient current conduction and causing localized thermal hotspots and current crowding, which in turn damage the material. Figures 6.33 and 6.34 present SEM images of the protective layer's surface and cross section. The surface micrograph reveals overlapping graphene sheets, indicating the formation of continuous conductive pathways and partial current conduction. However, cross-sectional SEM still shows large resin-rich regions (bright areas), demonstrating that significant epoxy remains trapped between sheets, hindering full conductivity.

Therefore, for graphene film-based protective structures, ensuring effective inter-sheet contact is essential. This not only facilitates rapid evacuation of lightning currents at high temperatures but also maximizes bulk conductivity. Graphene films, which are macroscopic assemblies of nanosheets, are typically produced by ultra-high-temperature reduction of graphene oxide films. During this process, thermal energy not only restores sp^2 bonding but also promotes covalent cross-linking between adjacent sheets. Such cross-linking enhances sheet adhesion and stabilizes

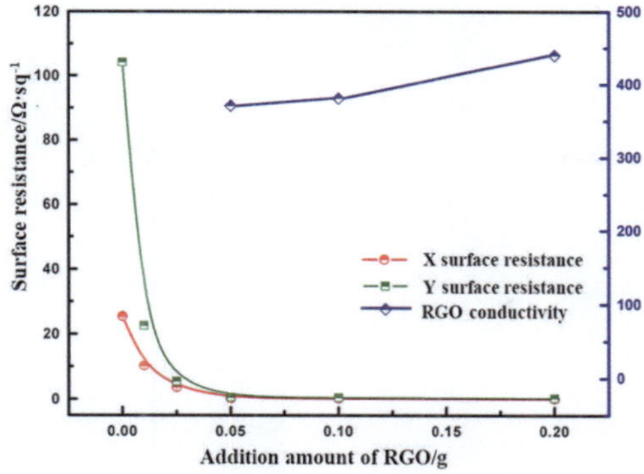

(a) surface conductivity vs. graphene loading

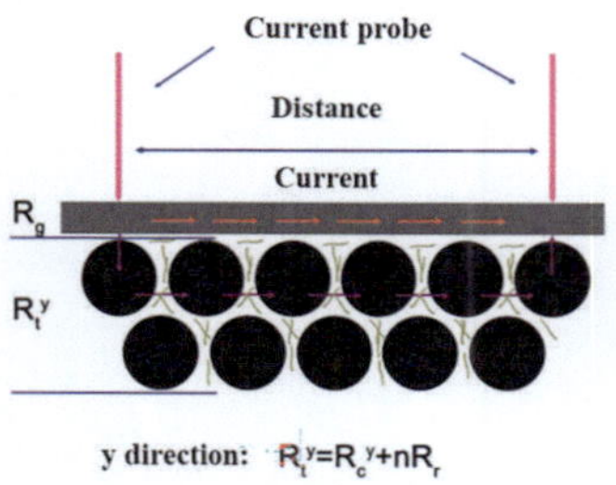

(b) conduction mechanism perpendicular to fibers

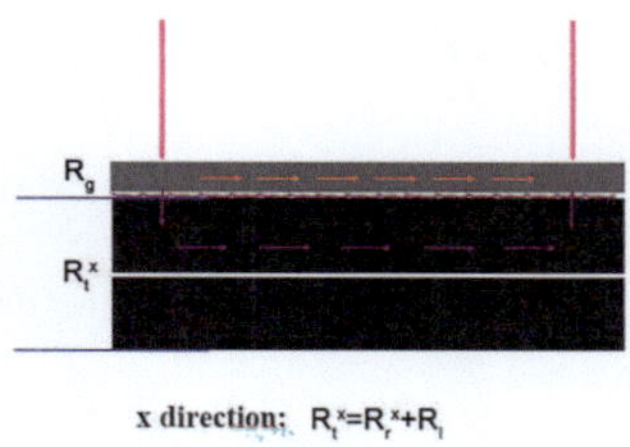

(c) conduction mechanism along fibers

Fig. 6.30 Influence of graphene content on surface conductivity and conduction mechanisms

the conductive network, thereby minimizing localized heat accumulation and current concentration under lightning current impact.

Preparation and High-Temperature Reduction of Graphene Films

Common methods for preparing graphene oxide films include the vacuum filtration and coating techniques, both favored for their simplicity and low cost. Figure 6.35 displays graphene oxide films produced by these methods. However, the films contain numerous oxygen-containing functional groups and structural defects, which severely impede electron transport between sheets. Consequently, as-prepared graphene oxide films are electrically insulating, limiting their utility, particularly as a lightning strike protection layer for CFRP composites. To address this, further reduction, either chemical or thermal, is required.

Chemical reduction commonly employs strong reducing agents (e.g., hydriodic acid, hydrazine hydrate) that remove oxygen functionalities and convert them to water. However, these reagents are highly corrosive, pose environmental hazards, and are incompatible with large-scale production. Traditional high-temperature furnaces

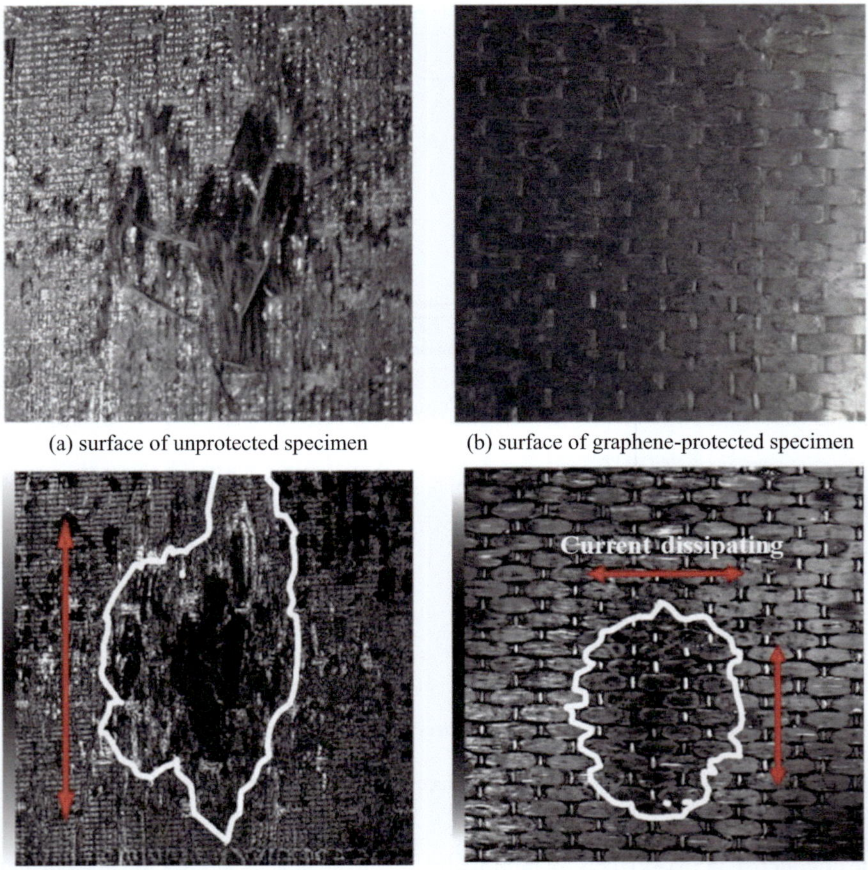

(a) surface of unprotected specimen (b) surface of graphene-protected specimen

(c) CT images of unprotected specimen (d) CT images of graphene-protected specimen

Fig. 6.31 Lightning damage and CT scan morphology under 40 kA A-component current

offer an alternative by heating films under inert gas flow to prevent re-oxidation. Yet such furnaces suffer from slow temperature ramp rates, high energy consumption, and limited chamber size, constraining industrial throughput. An emerging approach uses ascorbic acid as a green reducing agent, combining environmental friendliness with effective deoxygenation to impart initial conductivity and a modest Joule heating effect. Subsequent ultra–high-temperature reduction is achieved via electric heating, greatly enhancing reduction efficiency and uniformity, as shown in Fig. 6.36.

The reduction temperature profoundly influences graphene's electrical conductivity, mechanical strength, thermal conductivity, and surface chemistry. Graphitization near 2800 °C effectively eliminates oxygen groups, restoring graphene's exceptional properties. Raman spectroscopy of films after Joule heating (Fig. 6.37) shows the absence of the D-peak, indicating minimal defects and recovered graphene lattice integrity.

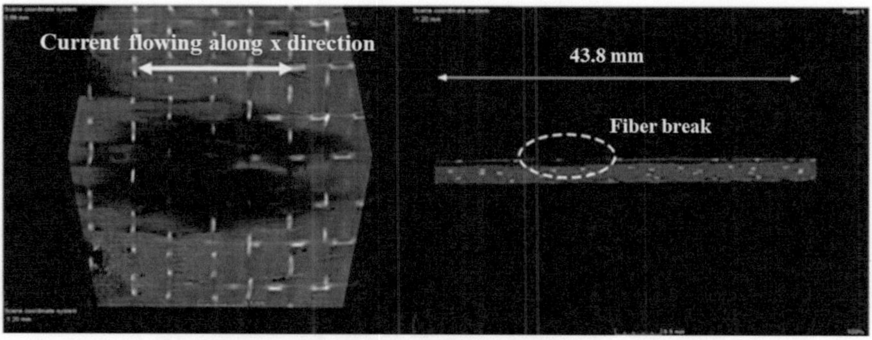

(a) unprotected specimen

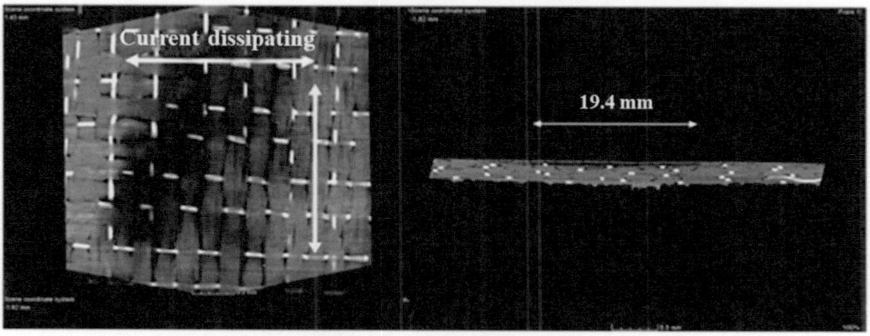

(b) graphene-protected specimen

Fig. 6.32 CT images showing internal delamination damage after lightning strike

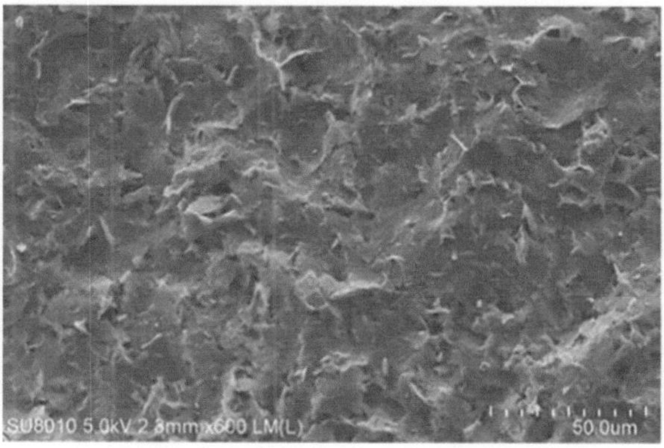

Fig. 6.33 SEM micrograph of graphene layer surface morphology

Lightning Protection Performance Testing of Graphene Films

Figure 6.38 shows both the schematic and photographic views of a graphene film co-cured with CFRP prepreg, laid up in the $[-45/90/45/0]_{2S}$ sequence. The cured

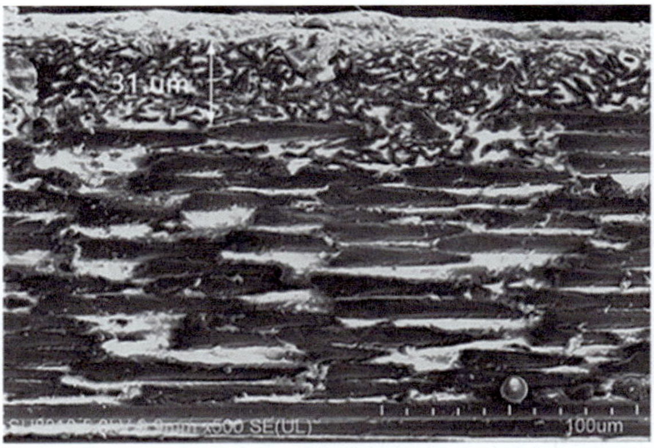

Fig. 6.34 SEM micrograph of graphene layer cross-sectional morphology

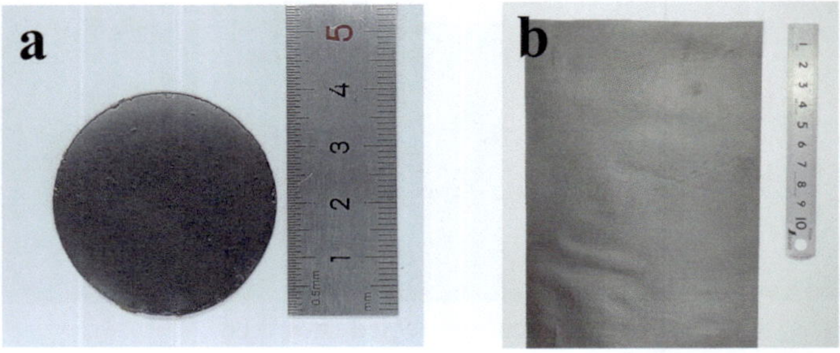

(a) Vacuum filtration (b) Coating method

Fig. 6.35 Preparation of graphene oxide films

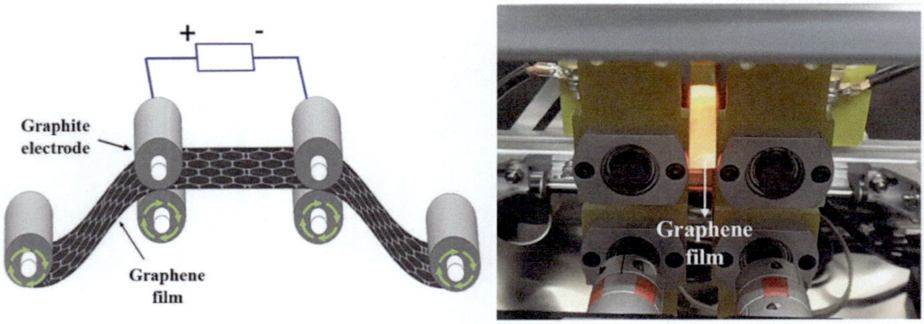

(a) Principle of electric heating device (b) Photograph of electric heating apparatus

Fig. 6.36 Ultra-high-temperature reduction of graphene films via electric heating graphene film is intimately bonded to the composite substrate via the resin, exhibiting a flat, wrinkle-free surface, and is therefore suitable as an LSP layer for CFRP composites.

Figure 6.39 illustrates the lightning strike test setup, recorded by a high-speed camera to capture instantaneous arc dynamics. The fixture grounds the specimen on all four side edges and the upper surface. When the protective layer possesses high surface conductivity, the lightning current is rapidly diverted along the laminate surface to the ground, thereby validating the LSP performance of the protective film.

Figure 6.40 compares the damage morphologies under a 100 kA A-component lightning strike for: (a) an unprotected CFRP panel, (b) a panel protected by copper mesh, and (c) a panel protected by a graphene film. The bare CFRP panel exhibits extensive fiber protrusion and fracture, primarily along the fiber direction. With copper mesh protection, damage is confined to minor fiber breakage in the central region, along with localized mesh ablation. By contrast, the graphene-protected specimen shows no observable fiber damage, outperforming the copper mesh. However,

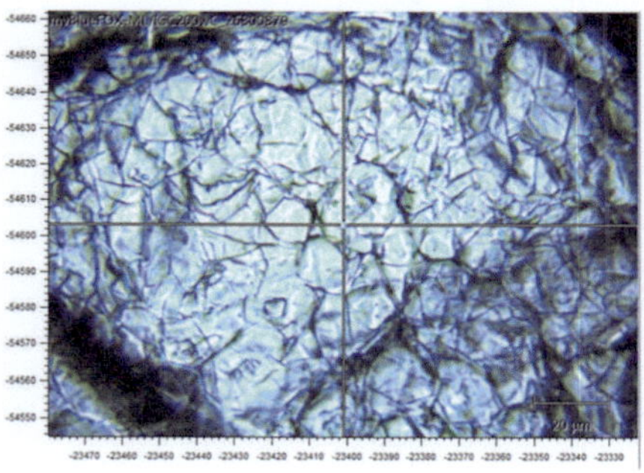

(a) Measurement location

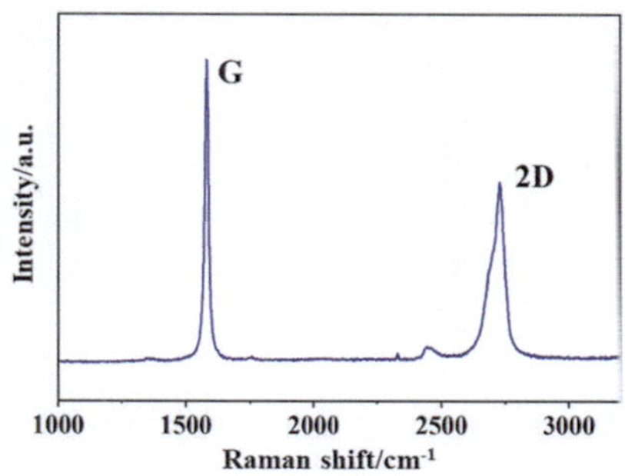

(b) Raman spectrum post-Joule heating

Fig. 6.37 Raman analysis of ultra-high-temperature-reduced graphene films

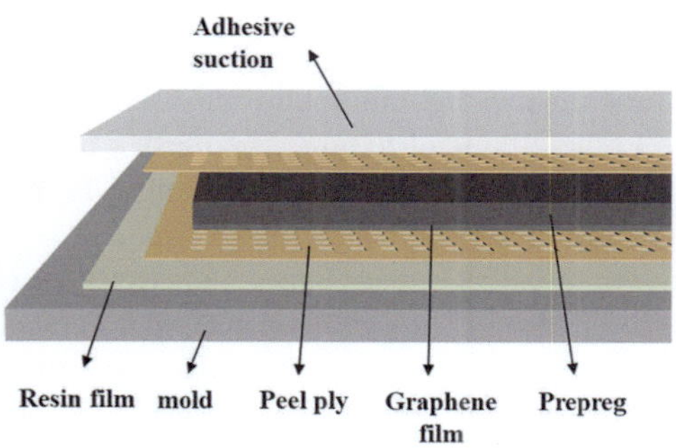

(a) Schematic of layup and cure

(b) Photograph of cured panel

Fig. 6.38 Graphene film cured with CFRP prepreg

(a) Overall layout (b) Specimen clamping and grounding scheme

Fig. 6.39 Lightning damage test platform

(a) Bare plate (b) Copper mesh-protected (c) Graphene film-protected

Fig. 6.40 Lightning damage under 100 kA A-component current

the graphene film itself displays large-area structural disruption post-strike, indicating that while current channeling is highly effective, the film's intrinsic resilience under extreme heating requires further optimization.

During the recorded lightning strike sequences captured by high-speed imaging (Fig. 6.41), the following observations were made:

(i) On copper mesh-protected specimens, the lightning current was rapidly conducted along the mesh surface to the grounded fixture. The mesh's porous architecture facilitated venting of resin vaporization recoil, confining failure modes to localized melting or vaporization at the arc attachment site.
(ii) On graphene film-protected specimens, the excellent in-plane conductivity similarly enabled swift current dispersion. However, the film's dense, hierarchical structure inhibited recoil gas escape, which led to more extensive structural damage of the protective layer and ejection of debris from the strike center. Thus, the graphene film's protective action relies on sacrificial damage of the film to absorb and dissipate lightning energy.

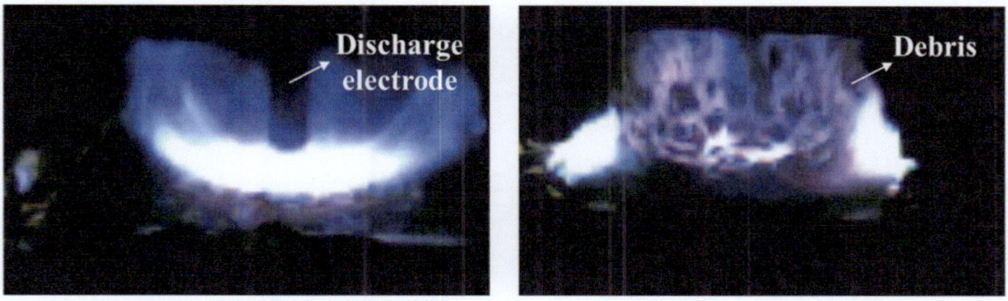

(a) Copper mesh protection (b) Graphene film protection

Fig. 6.41 High-speed camera capture of lightning strike sequences

Table 6.6 presents the ASTM D3359 adhesion rating method used to assess bonding strength via peel testing. For aeronautical airworthiness, the protective coating must achieve a 5B rating (no detachment).

Figure 6.42 illustrates the post-peel condition of the graphene film. Extensive film detachment is evident in the grid regions, failing to meet the 5B criterion.

Despite superior lateral conductivity, graphene films present two critical challenges as LSP layers on CFRP:

(i) Inadequate resin infiltration into the dense hierarchical network leads to poor interfacial bonding, increasing the risk of film delamination under lightning-induced mechanical recoil.
(ii) Impeded venting of resin vaporization gases causes recoil pressure buildup, which compromises the film's structural integrity and halts continuous current diversion.

Enhancing both resin-graphene interpenetration and venting pathways will be essential for achieving durable, sustainable lightning protection with graphene films.

6.4.3 Lightning Damage to Carbon Fiber/Graphene Hybrid Membrane Protective Structures

A three-dimensional (3D) porous conductive network, comprised of one-dimensional carbon fibers and two-dimensional graphene, not only enhances resin infiltration during co-curing, thereby strengthening interfacial bonding, but also retains graphene's superior conductivity, promoting rapid lightning-current evacuation. Microstructural design of this hybrid network is therefore critical for macroscopic performance optimization and improved lightning-strike resistance in CFRP composites.

Table 6.6 ASTM D3359 adhesion rating methods

Adhesion rating	Degree of spalling	Dample drawing
5B	0%—no flaking	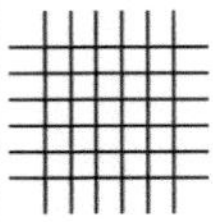
4B	Less than 5%	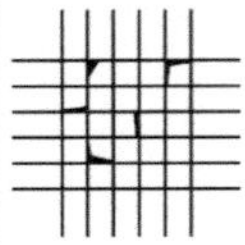
3B	5—15%	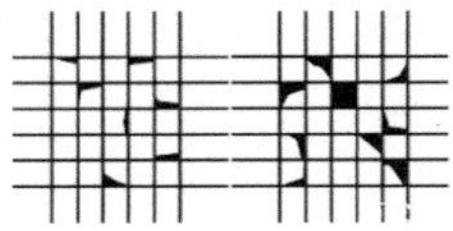
2B	15—35%	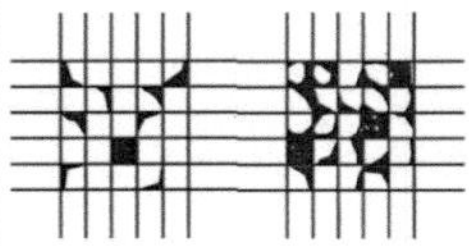
1B	35—65%	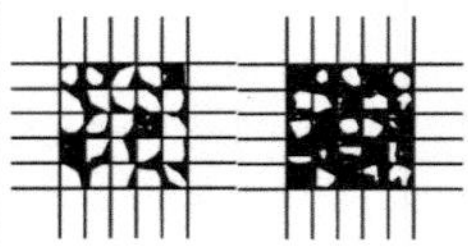
0B	More than 65%	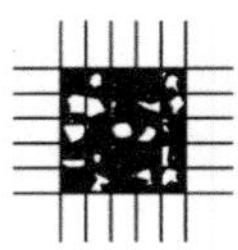

Carbon Fiber/Graphene Hybrid Film Preparation and High-Temperature Reduction

The solution-casting method is commonly employed to fabricate the 3D porous preform. In this process, short-cut carbon fibers are dispersed in a graphene oxide (GO) solution; upon solvent evaporation, a crosslinked porous network forms via van der Waals and intermolecular interactions between the fibers and GO. This network provides structural support for subsequent thermal reduction of GO. However, short-cut carbon fibers are inherently hydrophobic and tend to agglomerate, as

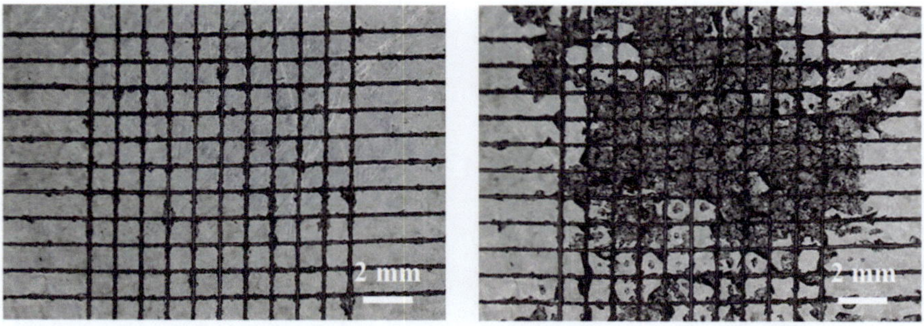

(a) Before peeling (b) After peeling

Fig. 6.42 Graphene film peel-off images

shown in Fig. 6.43. Agglomeration compromises material homogeneity and structural integrity, weakening the protective layer's ability to conduct lightning current. Moreover, inhomogeneous conductive paths can lead to localized overheating and premature failure, undermining overall performance and safety.

Carbon fiber felt (CFF), a 3D porous carbon material typically produced by high-temperature annealing of PAN-based fiber mats, which offers a low-density scaffold ($\approx$0.12 g/cm^3), yielding a surface mass density of ~120 g/m^2 at 1 mm thickness (versus 195 g/m^2 for commercial copper mesh). The porous CFF serves as a skeleton, into which GO sheets are infiltrated via vacuum filtration or vacuum infusion. Thereafter, the electrothermal effect of the CFF activated by electrical heating reduces GO in situ and promotes covalent crosslinking between graphene and carbon fibers. The specific preparation flowchart is shown in Fig. 6.44.

Fig. 6.43 Agglomeration of short-cut carbon fibers in GO solution

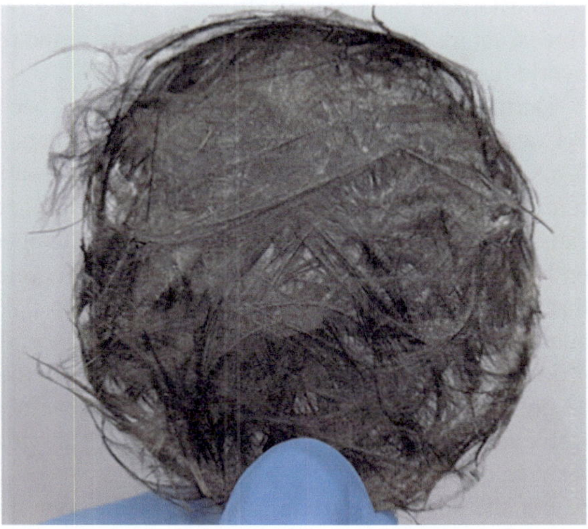

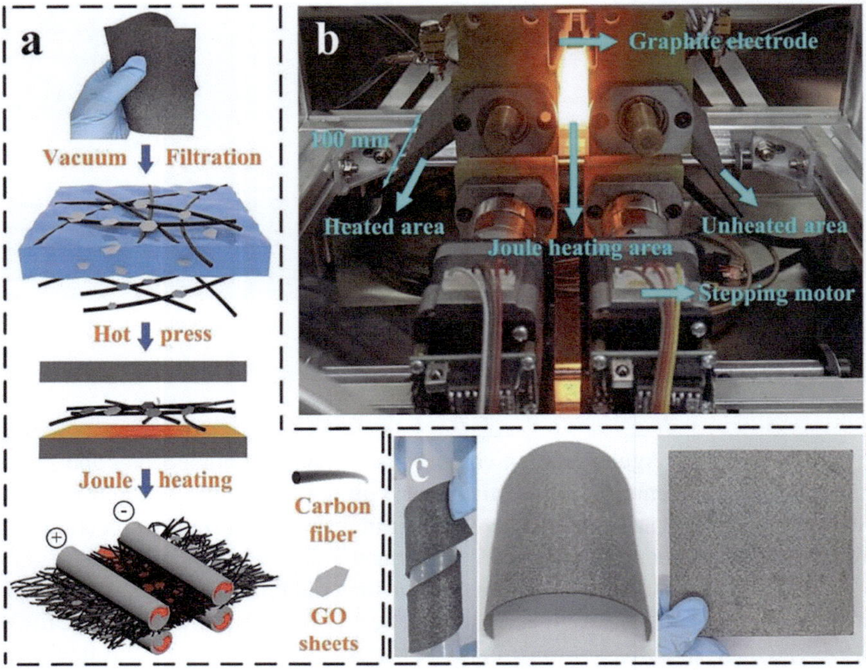

Fig. 6.44 Preparation flowchart of carbon fiber/graphene hybrid membrane: **a** Process schematic; **b** Physical setup; **c** Fabricated hybrid samples

The pristine carbon fiber felt (CFF) exhibits a relatively loose interior, with inter-fiber pores forming flow channels for the graphene oxide (GO) solution. The surface of the short-cut carbon fibers displays numerous grooves and a characteristic skin–core structure. The microscopic morphology of the resulting carbon fiber/graphene hybrid film is shown in Fig. 6.45.

During vacuum infiltration, some GO flakes become adhered to fiber surfaces via van der Waals and π-π interactions, while others lodge between adjacent fibers, forming an initial 3D cross-linked preform. In the subsequent electrothermal reduction, oxygen-containing functional groups on GO sheets are progressively removed, causing surface wrinkles to flatten. At elevated temperatures, non-sp^2 carbon atoms on both GO and the carbon fibers become highly reactive, forming covalent bonds that create additional conductive pathways. GO flakes that were initially loosely attached become tightly welded to the fiber surface, thereby smoothing the original grooves and enhancing the intrinsic conductivity of the carbon fibers. Figure 6.46a1–a3 presents TEM images illustrating the lattice-structure transformation of carbon fibers and GO before and after Joule heating, providing a microscopic view of the welding behavior.

For the short-cut carbon fibers, the skin–core structure exhibits an outermost layer of carbon atoms in a disordered arrangement. When GO sheets, held by van der Waals forces, adhere to these fibers, the resulting lattice remains poorly aligned

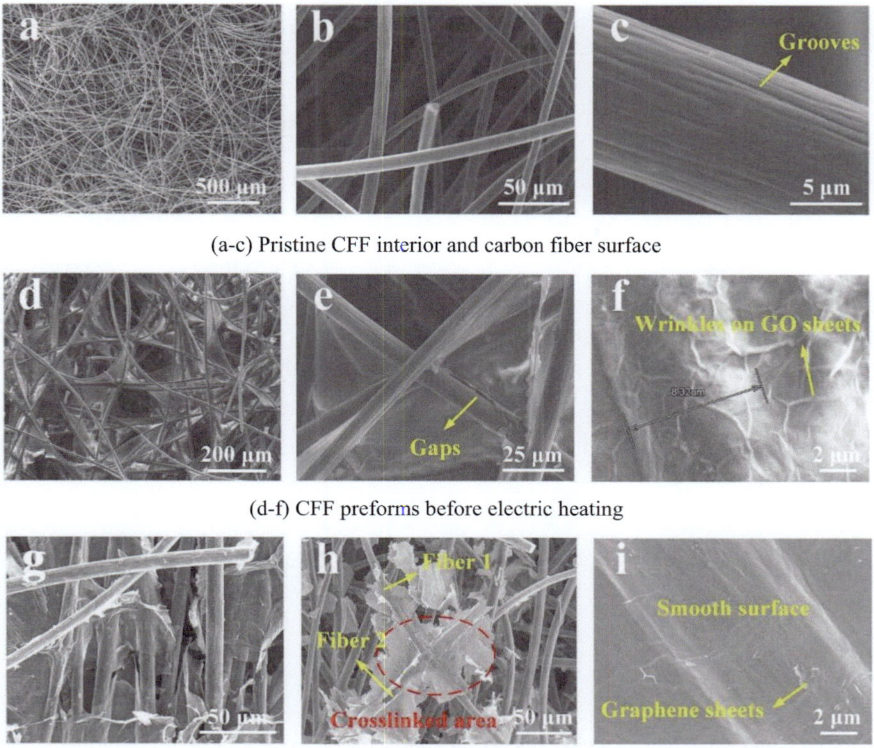

(a-c) Pristine CFF interior and carbon fiber surface

(d-f) CFF preforms before electric heating

(g-i) CFF hybrids after Joule heating reduction

Fig. 6.45 Microscopic morphology of carbon fiber/graphene hybrid film

and the adhesion to the CFRP substrate is weak. However, after Joule heating, the originally disordered lattice transforms into a highly aligned crystalline structure, and the interlayer spacing decreases from 0.75 to 0.35 nm due to the removal of oxygen-containing functional groups (Fig. 6.46b). As illustrated in Fig. 6.46c, this lattice realignment is driven by the interaction between the graphene sheets and the surface-conducted current. The surface-dominant current flow, induced by Joule heating in the skin layer of the carbon fibers, generates a torque that rotates the graphene flakes around the fiber axis. This results in an ordered stacking, often referred to as a moiré pattern, which represents a thermodynamically stable graphene assembly.

Furthermore, regions of high contact resistance between carbon materials create localized high-temperature "hot spots" before covalent bonding occurs. These conditions facilitate cross-linking reactions (Fig. 6.46d): heteroatom defects (e.g., nitrogen) and amorphous carbon are removed, generating highly reactive dangling bonds. Bergmann cyclization between adjacent dangling bonds forms new carbon–carbon linkages, integrating additional carbon atoms into the graphite lattice. The versatility of these dangling bonds enables the formation of diverse microcrystalline

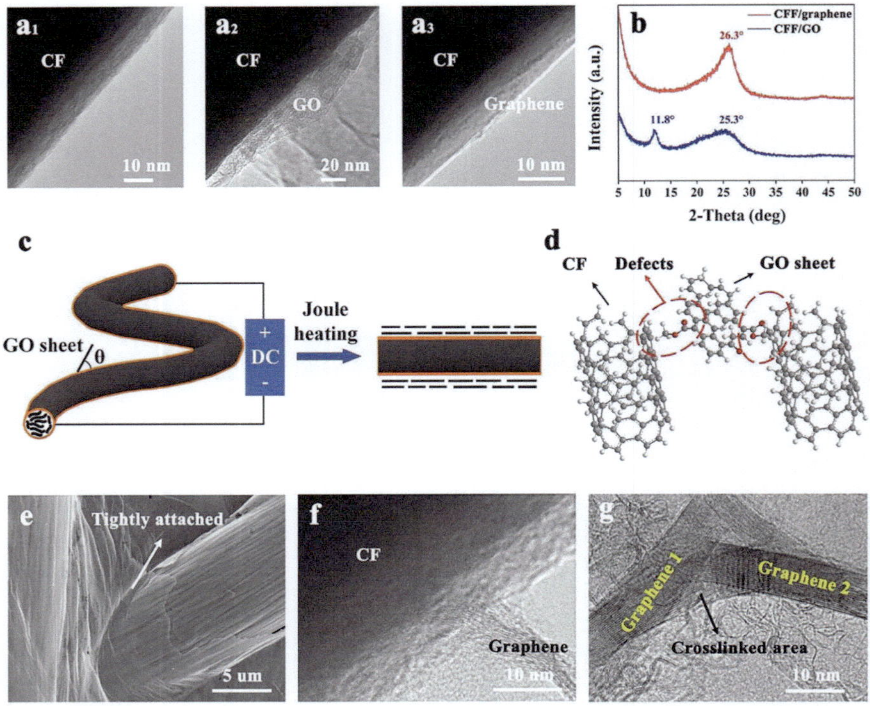

Fig. 6.46 Microstructural evolution and bonding mechanisms: **a1–a3** TEM of carbon fibers and graphene oxide before/after Joule heating; **b** XRD patterns pre- and post-heating; **c** Schematic of graphene arrangement transformation; **d** Bonding mechanism between short-cut carbon fibers and graphene; **e** SEM of graphene-crosslinked fibers; **f** TEM of graphene-crosslinked carbon fibers; **g** TEM of crosslinked graphene sheets

structures, including curling, crossing, and bifurcation, enhancing the contact interface and significantly reducing resistance. SEM images (Fig. 6.46e) reveal large-area graphene sheets tightly bridging adjacent fibers, while TEM micrographs (Fig. 6.46f–g) confirm that graphene flakes are covalently crosslinked with fibers and with each other. This cross-microcrystalline network facilitates electron and phonon transport, thereby enhancing both electrical and thermal conduction throughout the hybrid protective structure.

Since CFF serves as the backbone and graphene functions as the crosslinking medium, the graphene loading critically influences the conductivity of the hybrid film. Insufficient graphene leads to incomplete covalent crosslinking, while excessive graphene adds mass without further conductivity gains and can impede resin infiltration during co-curing. Therefore, identifying the optimal graphene content is essential. Figure 6.47a presents the conductivity of hybrid films as a function of graphene content. When the graphene loading reaches 19wt%, conductivity improves from 488 (pristine CFF) to 2552 S m^{-1}. Beyond this threshold, a continuous graphene overlayer forms on the CFF surface, and further conductivity gains plateau, while the dense layer hinders resin penetration.

The temperature dependence of conductivity during Joule heating is shown in Fig. 6.47b, demonstrating a positive correlation. Higher film temperatures facilitate rapid evacuation of lightning-current energy. Both electric-heating temperature and dwell time affect film properties. Figure 6.47c, d show Raman spectra and conductivity at different heating temperatures. The I_D/I_G ratio decreases with increasing temperature, indicating defect repair in graphene, and the 2D peak intensity at 2800 °C confirms an enhanced crystallinity and optimal conductivity. Temperatures above this were not explored due to the carbon-fiber decomposition limit (3316 °C).

Similarly, Fig. 6.47e, f compare Raman spectra and conductivity at varying heating durations (30 s vs. 300 s) at 2800 °C. The spectra and conductivities are nearly identical, indicating that dwell time beyond 30 s has negligible impact on electrical performance. Figure 6.47g compares the electrical and thermal conductivities of pristine CFF and the modified hybrid film after resin curing. Thanks to covalent crosslinking, which creates additional conduction pathways for electrons and phonons, the hybrid film's electrical conductivity increases by 5.2 × (2690 vs. 518 S $m^{-}1$), and thermal conductivity by 1.8 × (0.49 vs. 0.27 W m^{-1} K^{-1}), despite the presence of an insulating resin. Moreover, the covalent network not only enhances conductivity but also improves ablation resistance, as evidenced by the thermogravimetric analysis in Fig. 6.47h.

The covalent cross-linking network also enhances the mechanical stability of the hybrid films. Figure 6.48a presents the tensile properties of modified hybrid membranes with varying graphene contents. As graphene loading increases, more graphene sheets act as cross-linkers between adjacent carbon fibers, thereby restricting fiber mobility under load and improving mechanical performance. For instance, at 19 wt% graphene, the tensile strength increases from 0.27 MPa (pristine CFF) to 0.68 MPa, and the modulus rises from 5 to 110 MPa. Concurrently, the elongation at break decreases significantly, and the necking phenomenon is eliminated due to the stiffer composite network. Microscopically, fibers in the pristine CFF are loosely entangled, allowing some fiber straightening under tension (Fig. 6.48b, c). In contrast, the graphene-crosslinked fibers exhibit brittle fracture surfaces (Fig. 6.48d, e), reflecting limited fiber pull-out. In situ tensile microscopy (Fig. 6.48f_1, f_2) shows graphene sheets bridging adjacent fibers before loading; upon fiber separation, graphene residues remain attached to both fibers, demonstrating that covalent bonds are strong enough to fracture the graphene flakes themselves (Fig. 6.48g). This enhanced interfacial strength ensures the durability and reliability of the modified CFF as a protective layer on CFRP surfaces.

Lightning Protection Performance Test of Carbon Fiber/Graphene Hybrid Film

Under a 100 kA A-component lightning strike, the damage morphologies of pristine CFF (no graphene) and CFF with 19 wt% graphene are compared in Fig. 6.49.

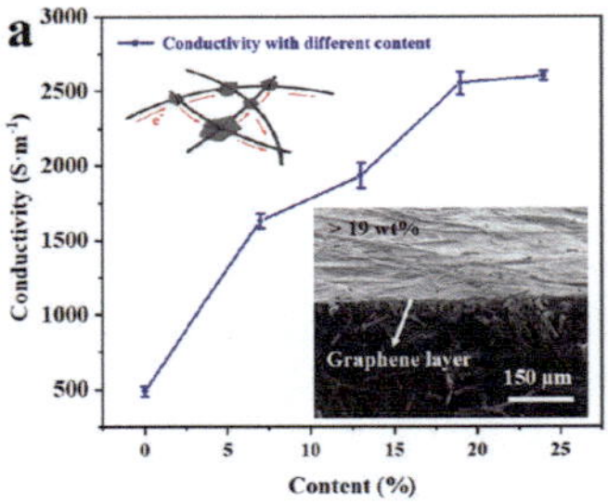

(a) Conductivity vs. graphene content

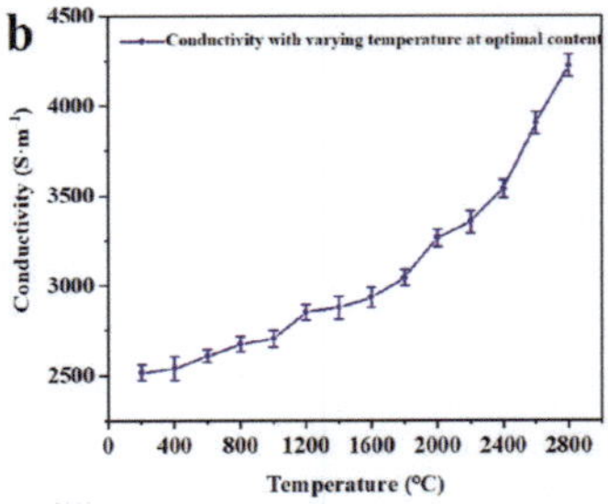

(b) Conductivity vs. temperature during heating

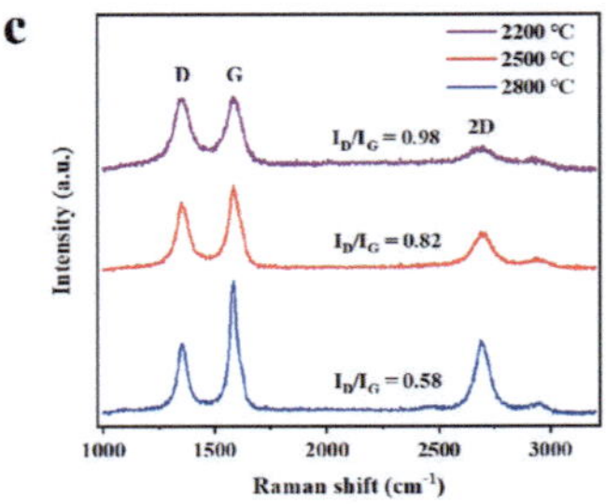

(c) Raman spectra at different heating temperatures

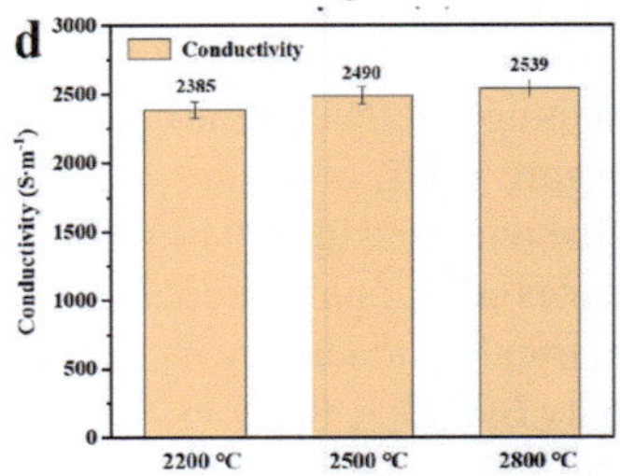

(d) Conductivity at different heating temperatures

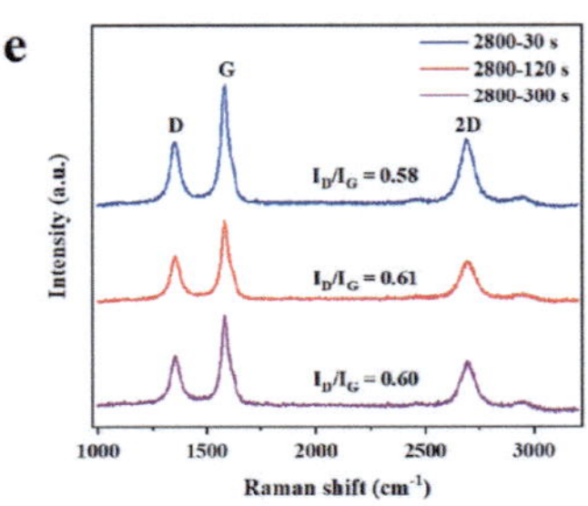

(e) Raman spectra at different heating durations

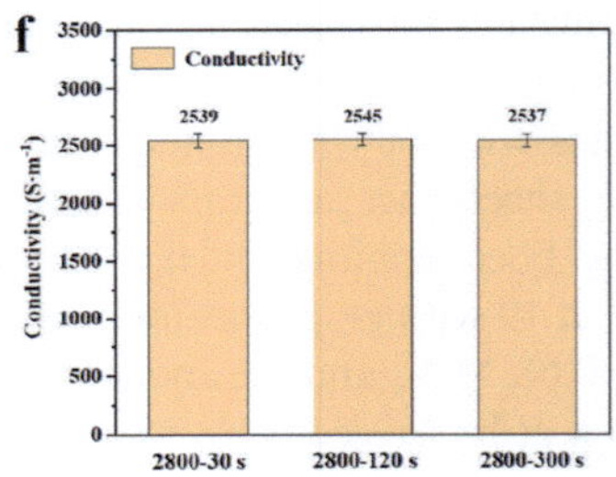

(f) Conductivity at different heating durations

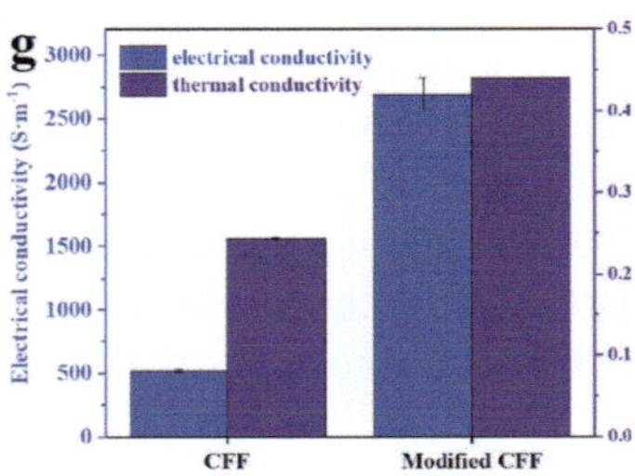

(g) Electrical and thermal conductivities comparison

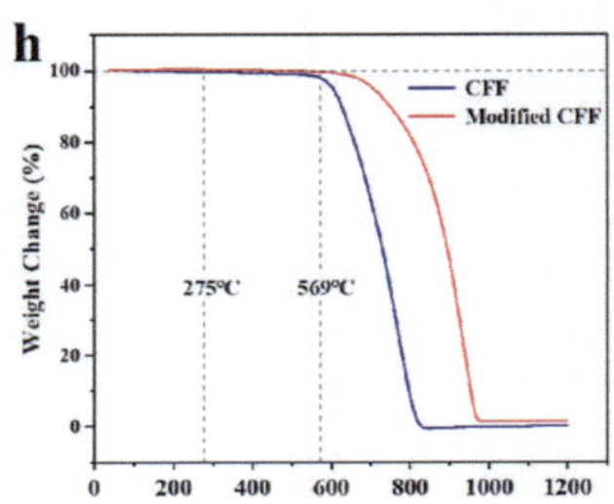

(h) Thermogravimetric curves

Fig. 6.47 Effects of graphene content and Joule heating on pristine CFF and modified hybrid films

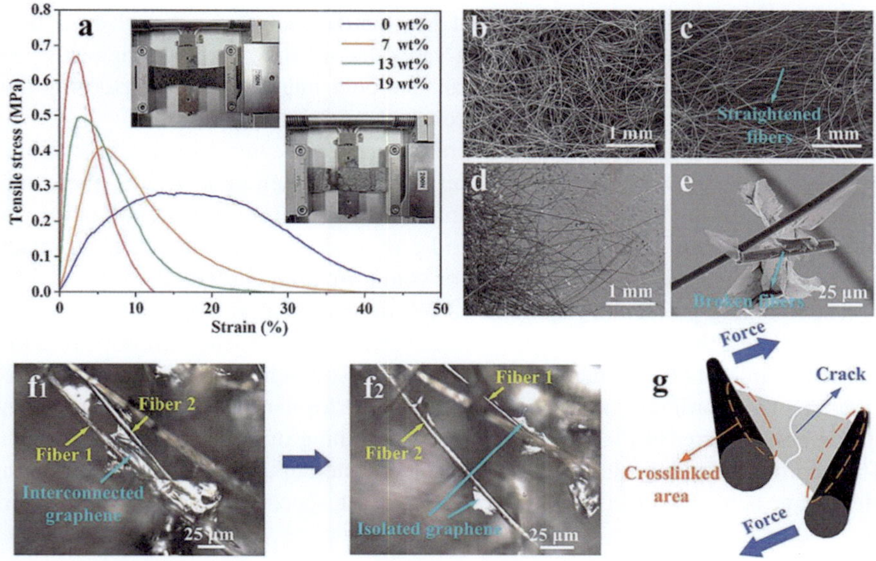

Fig. 6.48 Mechanical reinforcement by graphene cross-linking: **a** Stress–strain curves and failure modes for hybrid films with increasing graphene content; **b–c** Fiber morphology in pristine CFF under no load and tensile load; **d–e** Fiber fracture morphology in graphene-enhanced CFF; **f1–f2** Optical micrographs of graphene bridges between fibers before and after in situ tensile loading; **g** Schematic of graphene-flake fracture mechanism under tensile stress

The unmodified CFF displays fiber bulging and fracture on the surface, indicating current penetration into the CFRP substrate due to discontinuous conductive pathways. In contrast, the graphene-enhanced hybrid film, with its improved electrical conductivity and ablation resistance, effectively diverts the lightning current and blocks thermal penetration through the composite thickness. Nevertheless, localized structural damage still occurs at the strike center.

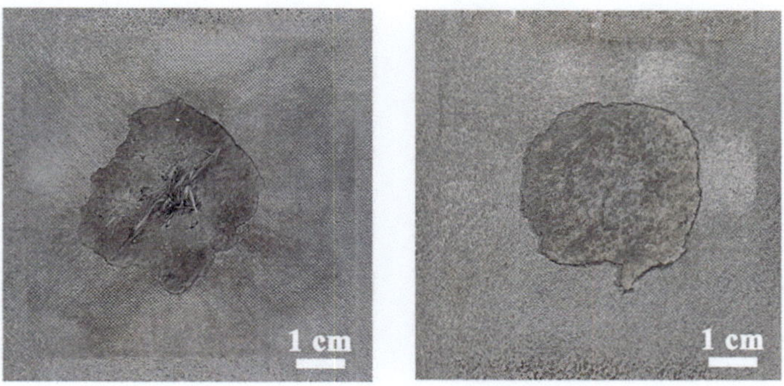

(a) Pristine CFF (b) Carbon fiber/graphene hybrid film

Fig. 6.49 Lightning damage morphologies under 100 kA A-component strike

High-speed imaging of the strike event (Fig. 6.50) reveals that, although the hybrid film suffers fragmentation due to resin-vapor recoil, the resulting fragments are substantially larger than those from graphene-only membranes, underscoring the mechanical reinforcement provided by covalent cross-linking.

Figure 6.51 presents CT scans of internal damage. The unmodified CFF exhibits both in-plane and through-thickness damage, with penetration to 2–3 plies, and numerous voids at the CFF-composite interface, likely from resin vaporization. In contrast, the carbon fiber/graphene hybrid-protected specimen maintains an intact internal structure, with no delamination or void formation observed.

The lightning protection requirements dictate that the lightning current must be dispersed along the protective layer to prevent penetration into the composite interior and retain at least 80% of the residual mechanical strength, ensuring material reuse after repair. Accordingly, a residual compression performance test was performed on specimens subjected to a 100 kA A-component lightning strike, as shown in Fig. 6.52. The unprotected composites exhibited extensive internal delamination, resulting in yield-stage behavior under compression. In contrast, specimens protected by copper

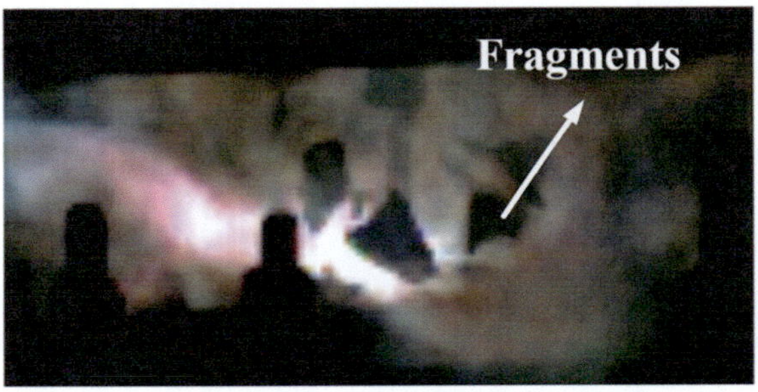

Fig. 6.50 Structural fragmentation of hybrid films during lightning strikes

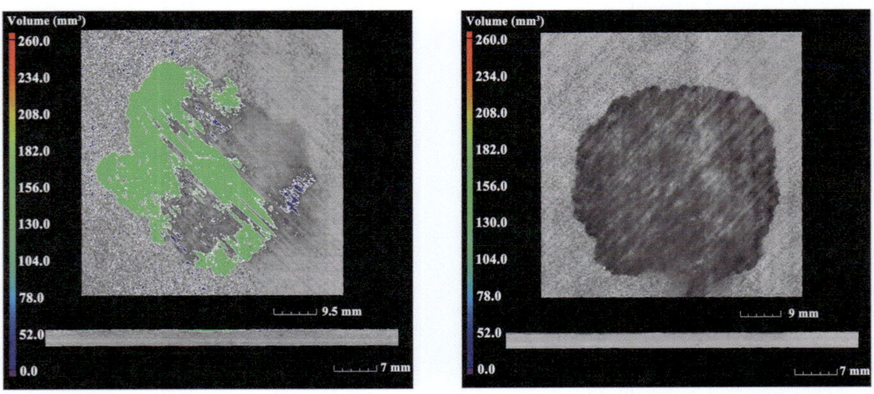

(a) Pristine CFF (b) Carbon fiber/graphene hybrid film

Fig. 6.51 CT scans of lightning damage (100 kA A-component)

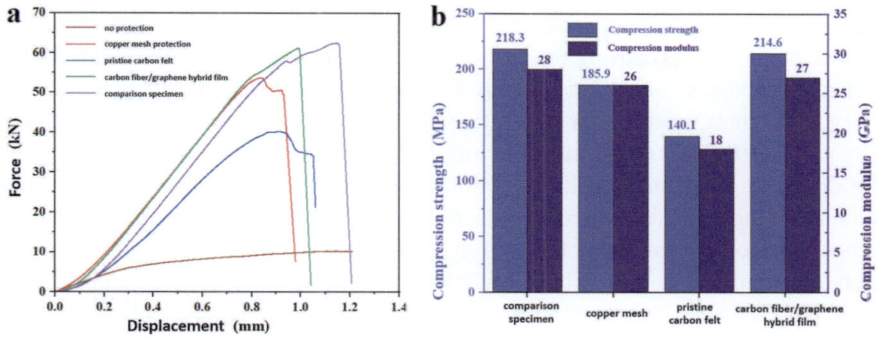

Fig. 6.52 Residual compressive properties after 100 kA A-component lightning strike

mesh, pristine CFF, and the carbon fiber/graphene hybrid film retained 85%, 65%, and 98% of their original compressive strength, respectively, confirming that the carbon fiber/graphene hybrid film fully satisfies the lightning strike protection criterion.

Lightning Protection Mechanism of Carbon Fiber/Graphene Hybrid Film

Figure 6.53 illustrates the damage pathways in CFRP composites during lightning strikes and the corresponding protection mechanisms of three protective schemes.

During a lightning event, two primary damage processes occur: (i) Current Conduction through the composite generates intense Joule heating internally. (ii) Surface Arcs form expanding plasma channels, ionizing air and releasing extreme thermal energy directly on the laminate surface. Thus, in addition to high electrical conductivity, thermal stability and ablation resistance of the protective layer

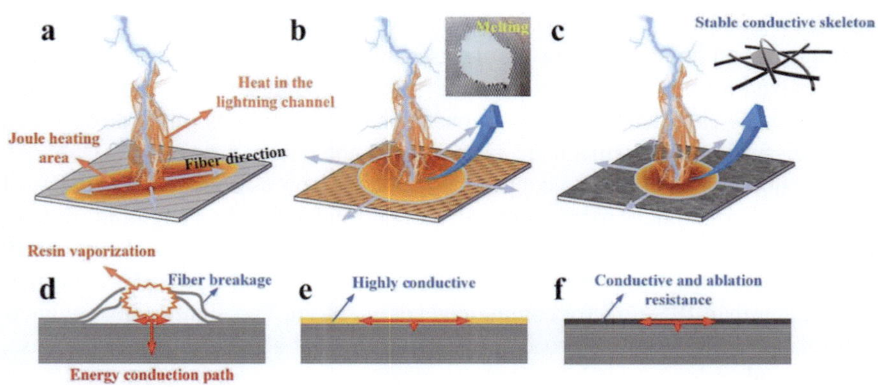

Fig. 6.53 Lightning protection mechanisms for different layers: **a, d** Bare CFRP panel; **b, e** Copper mesh-protected panel; **c, f** Carbon fiber/graphene hybrid-protected panel

are equally critical for mitigating lightning damage. Unprotected Composites lack surface conductivity, causing the entire arc heat to impinge on the resin and fibers. Damage propagates along the fiber direction, where thermal and electrical conductivity are highest, resulting in severe fiber fracture and delamination. Copper Mesh Protection effectively shunts the lightning current, reducing Joule heating. However, the discharge heat flux locally melts or vaporizes the copper mesh, re-exposing the underlying composite to residual current and heat. Moreover, copper's negative temperature coefficient leads to electron scattering and diminished conductivity at high temperatures. Carbon Fiber/Graphene Hybrid Film remains structurally stable under extreme heat, as neither carbon fibers nor graphene melt or combust. The 3D covalent network formed by fibers and graphene shields the composite by: blocking ablation, which is the hybrid layer resists plasma erosion and deflects arc heat, and maintaining conductivity, which is the network rapidly evacuates current even at elevated temperatures. Therefore, the hybrid film provides the most comprehensive protection, combining robust electrical pathways with superior thermal resistance, ensuring minimal damage to the underlying CFRP composite.

In addition to direct structural damage, lightning strikes induce rapid fluctuations in the surrounding magnetic field, which can disrupt onboard navigation systems and precision instruments. Therefore, beyond assessing the direct LSP performance of the carbon fiber/graphene hybrid film, its ability to mitigate indirect electromagnetic effects must also be evaluated. Figure 6.54 presents the electromagnetic shielding performance of unprotected, copper mesh-protected, and hybrid film-protected CFRP specimens, which are measured in the X-band (8.2–12.4 GHz) waveguide setup, reported as total shielding effectiveness (SE_T), reflection effectiveness (SE_R), and absorption effectiveness (SE_A).

The bare CFRP panel exhibits the lowest SE_T, averaging 41.05 dB, due to its lack of conductive pathways. Copper mesh-protected specimens show a modest increase in SE_R, thanks to enhanced surface conductivity, but SE_A remains unchanged, as the mesh's open structure permits electromagnetic waves to penetrate and interact with the underlying composite. In contrast, hybrid film-protected specimens demonstrate significant gains in both SE_R and SE_A. The covalent cross-linking network between short-cut carbon fibers and graphene sheets establishes continuous conductive pathways, extending the internal propagation path of incident waves and increasing conductive losses. Meanwhile, the graphene sheets function as internal reflectors, further dissipating electromagnetic energy within the porous network.

Consequently, compared to copper mesh, the 3D porous conductive network of the carbon fiber/graphene hybrid film provides both enhanced direct lightning protection and superior electromagnetic shielding against indirect lightning effects. At 19 wt% graphene, the hybrid film's surface mass density is only 141 g/m^2, substantially lighter than 195 g/m^2 for commercial copper mesh. This lightweight yet high-performance solution offers broad application potential across aerospace, defense, and other industries.

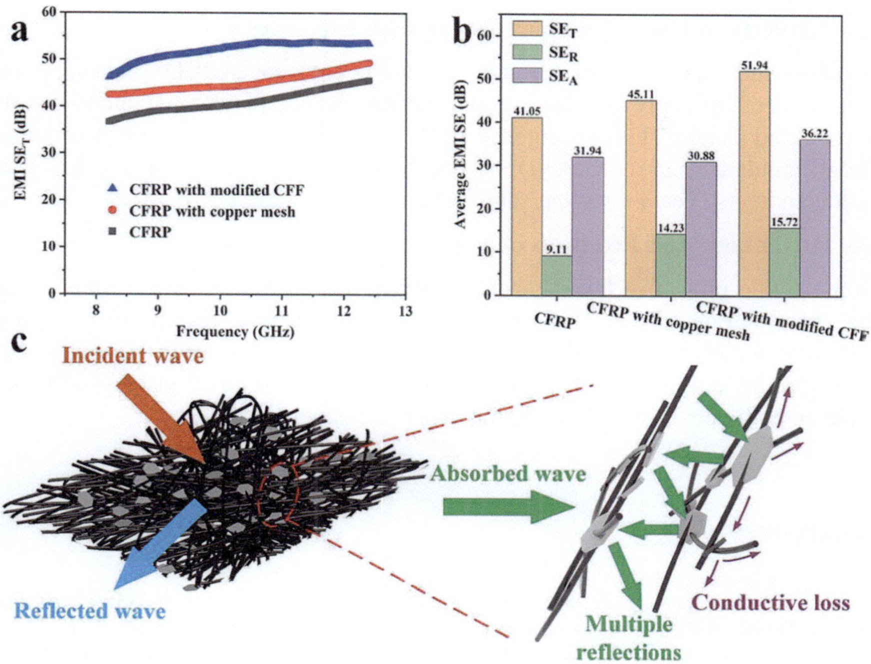

Fig. 6.54 Electromagnetic shielding effectiveness: **a** Total shielding effectiveness (SE_T); **b** Reflection (SE_R) versus absorption (SE_A) components; **c** Shielding mechanism schematic for hybrid film

6.5 Summary

In this chapter, multiple continuous, uninterrupted, time-sequence lightning direct-effect experiments were conducted on CFRP composites. The relationships among damage morphology, damage characteristics, and the parameters of sequential lightning current components were analyzed using visual inspection, infrared thermography, and ultrasonic C-scans. Based on these investigations, a multifactor assessment of the direct effects of both single-component (A) and multi-component lightning currents was performed. The main findings are:

(1) **Effectiveness of Copper Mesh Protection**

Under low to moderate energy inputs (e.g., A-only or A + B components), the 0.25 mm copper mesh conducts the majority of the lightning current, thereby substantially reducing the damage area on the composite surface. With the inclusion of higher-energy or long-duration components (notably component C), the damage area increases markedly, and the protective effect of the mesh diminishes. For damage depth, the copper mesh layer reduces penetration by approximately 30% under A-only or A + B current sequences, demonstrating a significant protective benefit at these energy levels.

(2) **Limitations Under Complex Component Sequences**

In three-component (DBC or ABC) or full four-component (ABCD) sequences, the damage depth of copper mesh-protected specimens becomes comparable to that of unprotected panels. This convergence arises because elevated surface temperatures (from cumulative current energy) melt and ablate the copper mesh, causing it to lose conductivity and cease diverting subsequent current pulses.

(3) **Advantages and Prospects of Non-metallic Protection**

Non-metallic protective layers, such as carbon fiber/graphene hybrid films, offer low mass, strong adhesion, and excellent thermal stability, making them highly promising for advanced composite LSP applications. Future research will emphasize zone-specific non-metallic structures and self-healing protective systems to further enhance both direct and indirect lightning strike resilience.

References

1. Welch J M. Repair design, test, and process considerations for lightning strikes[C]. Joint FAABoeing-Airbus damage tolerance workshop, Amsterdam, The Netherlands, 2007.
2. Kawakami H. Lightning Strike Induced Damage Mechanisms of Carbon Fiber Composites[J]. Dissertations & Theses - Gradworks, 2011.
3. Liu Z, Yue Z, Wang F, et al.Damage zoning characteristics of composite laminates with different protections subjected to lightning strike [J]. Acta Materiae Compositae Sinica, 2015, 32(1): 284–294.
4. Kawakami H, Feraboli P. Lightning strike damage resistance and tolerance of scarf-repaired mesh-protected carbon fiber composites[J]. Composites Part A: Applied Science and Manufaturing, 2011, 42(9): 1247–1262.
5. Hirano Y, Yokozeki T, Ishida Y, et al. Lightning damage suppression in a carbon fiber-reinforced polymer with a polyaniline-based conductive thermoset matrix[J]. Copmosites Science and Technology, 2016, 127:1–7.
6. Kumar V, Yokozeki T, Okada T, et al. Effect of through-thickness electrical conductivity of CFRPs on lightning strike damages[J]. Composites Part A: Applied Science and Manufacturing. 2018, 114:429–438
7. Kumar V, Sharma S, Pathak A, et al. Interleaved NWCNT buckypaper brtween CFRP laminates to improve through-thickness electrical conductivity and reducing lightning strike damage[J]. Composite Structures, 2019, 210:581–589.
8. Han J, Zhang H, Chen M, et al. The combination of carbo nanotube buckypaper and insulating adhesive for lightning strike protection of the carbon fiber/epoxy laminates[J]. Carbo, 2015, 94:101–113.
9. Rajesh P S M, Sirois F, Therriault D. Damage response of composites coated with conducting materials subjected to emulated lightning strikes[J]. Material & Design, 2018, 139:456–55.
10. Guo Y, Xu Y, Wang Q, et al. Eliminating lightning strike damage to carbo fiber composite structures in Zone 2 of aircraft by Ni-coated carbon fiber nonwoven veils[J]. Composites Science and Technology, 2019, 169:95–102.
11. Zhang J, Zhang X, Cheng X, et al. Lightning strike damage on the composite laminates with carbon nanotube films: Protection effect and damage mechanism[J]. Composites Part B: Engineering, 2019, 168:342–352.
12. Chakravarthi D K, Khabashesku V N, Vaidyanathan R, et al. Carbon Fiber–Bismaleimide Composites Filled with Nickel-Coated Single-Walled Carbon Nanotubes for Lightning-Strike Protection[J]. Advanced Functional Materials, 2011, 21(13): 2527–2533.

13. Tang L-C, Wan Y-J, Yan D, et al. The effect of graphene dispersion on the mechanical properties of graphene/epoxy composites[J]. Carbon, 2013, 60: 16–27.
14. Abliz D, Ziegmann G, Duan Y, et al. Effect of CNT Concentration on Mechnical Properties of Composites Manufactured by Compression Resin Transfer Molding (CRTM)[C]. 16th EUROPEAN CONFERENCE ON COMPOSITE MATERIALS, 2014.
15. Yokozeki T, Goto T, Takahashi T, et al. Development and characterization of CFRP using a polyaniline-based conductive thermoset matrix[J]. Composites Science and Technology, 2015, 117: 277– 281.
16. Sun J. Damage mechanism of aeronautic carbon fiber reinforced polmer composite subjected to lightning strike[D]. Xi'an. Xi'an jiaotong university, 2019.